IEE CONTROL ENGINEERING SERIES 51

Series Editors: Professor D. P. Atherton
Professor G. I. Irwin

ADVANCED ROBOTICS
&
INTELLIGENT
MACHINES

Other volumes in this series:

ADVANCED ROBOTICS
&
INTELLIGENT MACHINES

Edited
by

J. O. Gray
&
D. G. Caldwell

The Institution of Electrical Engineers

Published by: The Institution of Electrical Engineers, London,
United Kingdom

© 1996: The Institution of Electrical Engineers

The Institution of Electrical Engineers,
Michael Faraday House,
Six Hills Way, Stevenage,
Herts. SG1 2AY, United Kingdom

British Library Cataloguing in Publication Data

A CIP catalogue record for this book
is available from the British Library

ISBN 0 85296 853 1

Printed in England by Bookcraft, Bath

Contents

SECTION 2 Applications

8   Robotics in the Nuclear Industry   133
P.E. Mort and A.W. Webster

SECTION 3 Advanced Concepts and Procedures

16   The concept of robot society and its utilisation in future robotics   255
        A. Halme, P Jakubik, T Schonberg and M Vainio

22 Parallel Processing, Neural Networks and Genetic Algorithms for Real-Time Robot Control 355

A.M.S. Zalzala

Preface

The term 'advanced robotics' came into general use during the 1980s to describe the emerging developments in sensor based robotic devices which exploit relatively low cost computing power to achieve levels of functionality which often appear to mimic intelligent human behaviour. Such devices are often semi-autonomous in nature with quite sophisticated human computer interfaces. They clearly represent a significant technical advance on the familiar pick and place industrial robot and have potentially a wide range of applications in the manufacturing, nuclear, construction, space, underwater and health care industries. Developments in the field utilise results from the domains of cognitive science and artificial intelligence, as well as employing aspects of mechanical and electronic engineering science, real time computing, control and sensor techniques. It is thought that the successful integration of the required wide range of enabling technologies to produce viable marketable devices provides one of the most interesting challenges in current engineering.

The object of this book is to provide a brief overview of developments in the basic technologies, survey some recent applications and highlight a number of advanced concepts and procedures which may influence future directions in the field. Within the text there is some emphasis on developments within the United Kingdom, although the activities described are being studied in many other laboratories within the United States, Japan and the European Union. Currently, the topic is characterised by intense research interest, funded by significant programmes at national and international level with results being presented at frequent specialist international conferences. The material of this book thus cannot possibly aspire to be comprehensive or, indeed, contain the latest results of a very dynamic and exciting technical field. However, it does aspire to address many of the key elements of the technology and provide a wide-ranging introduction for postgraduate students and practising engineers attracted to the challenges of the domain.

Some of the earlier concepts considered in the development of semi-autonomous robots such as automatic collision avoidance, world modelling, navigation, trajectory planning, sensory feedback and enhanced human computer interfaces have now been incorporated into practical machines in a number of industrial and service sectors where important issues relating to safety and reliability must be appropriately addressed. More recent concepts drawn from the field of cognitive science have yet to be fully explored in robotic studies and advances in micro engineering provide opportunities for innovative robotic development and exploitation.

The book is divided into three sections each of which is provided with editorial comment to place the contents in an appropriate context. In the first section Chapter 1 provides a brief overview of the subject while the rest of the chapters cover aspects of the basic generic technologies involved in the evolution of advanced robotic techniques. Section 2 is concerned with a range of practical applications and contains a number of surveys of uses in specific domains. Section 3 focuses on emerging technologies such as microrobotics, learning systems and co-operant robotic behaviour much of which will no doubt be developed and used to enhance existing and proved technology, increase the functionality of present systems and add value to marketable machines and processes.

Each chapter is furnished with an extensive bibliography to facilitate further study.

SECTION 1 Technologies for advanced robotics

Editorial comment

Chapter 1 provides an introduction to the topic and a brief overview of the various generic technologies, application domains and emergent techniques. As such it attempts as far as possible to provide a template within which all the following chapters in the text can be placed in an appropriate context. Chapter 2 reviews some of the fundamental concepts in machine intelligence and early studies in rule-based control and machine learning. Suitable architectures for intelligent control are briefly surveyed and the reader can find additional material in Chapters 18 and 19. Further developments in learning systems are covered in Chapters 21 and 22. Chapter 3 deals with advanced control techniques for robotic arms and includes procedures for reactive collision avoidance and the control of arm systems with inherent redundancy. This technique allows quite complex and dextrous devices to be successfully deployed in relatively unstructured environments. Chapter 4 focuses on gripping procedures for robot end-effectors and provides a systematic approach to the design of grippers for the safe acquisition and manipulation of objects with bounded parameters but of arbitrary shape. Such a facility is vital for the successful deployment of robots capable of coping with a range of target objects which may or may not be predictable in shape.

A key factor of modern robots is the incorporation of sensors in direct feedback loops for enhanced performance and this is a common theme in the remaining three chapters of this section. Force feedback control which is considered in Chapter 5 is an essential requirement for the proper deployment of robot operated tools and for the safe operation of the robot arm in unstructured environments where interaction is necessary but impact energies are to be minimised. The procedure of active compliance control is considered in some detail as it is regarded as being the most easily implemented technique for practical robot devices. The application of additional sensors, including audio, video and multifunctional tactile transducers offers the possibility of wide-ranging feedback configurations in teleoperation including the creation of so-called telepresence effects. This latter objective, as discussed in Chapter 6, provides an enhanced human/machine interface, which will, by exploiting natural human capabilities, optimise operator performance when controlling complex devices which are remotely located. The need for such procedures will clearly increase with the complexity of the devices under control and the concomitant required to reduce cognitive loading and enhance operator efficiency.

The final chapter in this section considers the general aspects of sensing and sensor management with emphasis on applications to mobile automata. Such devices will normally have a range of sensors each with its own performance envelope and it is important to optimise overall system performance and minimise the effects of noise and uncertainty. The fusion of the outputs of a number of sensors is now a common feature of advanced robotic devices and there is a requirement to optimise signal processing techniques to ensure that the robot controllers are presented with the best hypothesis of the state of the robots environment in the presence of significant sensor uncertainties. The concepts presented in this chapter are central to the development of complex automata which are capable of flexible operation in unstructured and possibly hazardous environments.

SECTION 2 Applications

Editorial comment

Given the potentially hazardous nature of its materials and its processes it is not surprising that the nuclear industry was an early user of advanced teleoperated robotic procedures and a review of the installed systems in one large nuclear processing company is outlined in Chapter 8. Extensive use is now made of stereo vision systems, computer based three dimensional modelling techniques and direct force feedback to provide operator 'feel' during joystick manipulation. Once a three dimensional model of the robot's environment has been created it is a relatively easy step to provide computer generated stereo images to create a virtual reality environment for the operator and this, combined with stereo video overlays, trajectory control, automatic collision avoidance and direct force feedback provides a very powerful teleoperated system for safe operation in complex environmental topologies. It has been shown that a not dissimilar overall system design philosophy can be applied to robotic surgery which has similar overriding requirements for safety. Accurate three dimensional models of target environments can be created by computer tomography and the resulting data-base used for the precision guidance of robot operated instruments which can be equipped with sensors for the generation of direct force feedback effects. More recent developments include the use of virtual reality operator environments for key hole surgery procedures. The overall aspiration is to increase the accuracy and quality of surgical procedures, minimise trauma and speed patient recovery. An overview of recent trends in this evolving field is given in Chapter 9.

The next two chapters in this section also deal with specialist applications of robotic techniques but have a common theme of transport and mobility. Chapter 10 focuses on the development of the intelligent autonomous car which is essentially a fully instrumented mobile robot with a capability of automatic collision avoidance and having advanced control and navigation capabilities. The final objective here is to enhance both the safety and utilisation of the existing motorway networks. Both of these factors could have a very significant economic impact on transport policy and costs. Although wheeled and tracked vehicles provide the basis for most transportation systems there are some topologies where legged devices may provide practical solutions. Such devices present complex problems in gait control and stability but have the advantage of imposing a relatively small 'footprint' on any environment and the ability to deal with relatively large discrete obstacles in a relatively elegant manner. Some work has been undertaken to develop anthropomorphic bipedal robots but the hexapodal configuration described in Chapter 11 offers advantages in load carrying and stability and practical hexapodal robots have been developed for possible use in forestry environments and planetary exploration.

The remaining chapters of this section are concerned with aspects of automation in manufacturing with some emphasis on the use of robotic devices in industries such as garment manufacturing, food production and agriculture which hitherto have not been regarded as suitable application domains for robotisation.

Traditionally robots have been used for the operation on and manipulation of rigid devices in automated manufacture thus precluding their use in, for example, the large clothing manufacturing sector which deals essentially with the handling and processing of sheet like limp materials. The techniques presented in Chapter 12 outline the advances made in the automated handling of flexible materials by using innovative gripper systems and vision sensors. Combining these techniques with input procedures such as laser scanning and 3-D modelling could provide a comprehensive CAD/CAM system for garment manufacture which would have a very significant impact on the industry.

Meat, fish and poultry products also lack rigid structures and the generally well defined dimensionalities usually associated with robot manipulated objects. Advances in robotic sensors and feedback procedures now allow applications in the food processing sector with advantages including enhanced cleanliness due to the absence of human operators, improved yield due to accurate and consistent cutting procedures and generally improved quality of the final product. Another advantage is that the processing operation can take place in environmental conditions such as low temperature which enhance product shelf-life but which would be quite inimical to human operators. The success of any application depends on the development of innovative grippers, the development of mechanisms which are optimised for speed and capacity rather than accuracy and the availably of high speed image processing techniques to deal with product variability during fast production runs. Somewhat in contrast the robotic milking procedure outlined in Chapter 14 considers a relatively simply but elegant engineering solution to a process that has existed since the dawn of agriculture. Simple but cleverly arranged sensors locate the target zone for the application of a safe compliant end- effector. This application could perhaps not be further from the traditional use of robotics in high speed automation but in common with robotic surgery represents the application of automation procedures in a process which is generally regarded as being difficult and requiring unique human skills and sensitivity.

The final chapter in this section returns to more traditional aspects of automation and a describes a practical application of modern image processing techniques in the automated products of semi-conductor wafers. The chapter includes an extensive review of existing procedures in the important field of robot sensors.

Section 3 Advanced concepts and procedures

Editorial comment

The aim of this section is to present some of the more recent research themes in robotic systems which may well mature into practical techniques in robotics and intelligent automation systems. One of the more interesting concepts now discussed widely in the research literature is that of using multiple robotic devices in some co-ordinated (but not necessarily closely synchronised) way to accomplish a complex task. The objective may be to replace a single complex robot with a group of simpler and, perhaps, more robust devices to provide a degree of redundancy in any process; alternatively, the object may be to create a group of co-operating semi-autonomous machines each with a different functionality operating within an unstructured environment to undertake a task such as surface mining or construction. Some of the basic concepts and possible application domains are discussed in Chapter 16, where allusions are made to biological paradigms. This theme is further developed in Chapter 17 where the focus is on the exploitation of miniature and microrobotic devices. Silicon technology is normally associated with the development of computing devices but the material can be machined and fabricated to produce electric motors, mechanisms and drives which have the physical dimensions of the order of a few micrometers. These components can be assembled into microrobots having minute end- effectors and sensors with the computing elements being integrated with the robot's silicon infrastructure. Possible applications lie in process and machine scavenging, environmental surveillance and invasive medicine. Whether the set of multiple robots consists of a swarm of microrobots or groups of very large mobile machines, a sound methodology must be evolved for co-ordinating their activities to achieve the specified task. Traditional, hierarchical, deterministic procedures do not appear to address the problem of complex interactions amongst multiple robots in an efficient way and some recent emphasis has been placed on the study of behavioural or reactive control architectures. These are reviewed briefly in Chapter 18 and the theme is continued in Chapter 19 which describes a so-called behavioural synthesis architecture suitable for the co-operant operation of a set of mobile robots. While demonstrating the efficacy of the behavioural approach, it is clear that there is much to be gained in evolving hybrid system architectures which combine behavioural characteristics with top level planning and trajectory control. Previous chapters in this section have concentrated on mobile robots and Chapter 20, in contrast, focuses on the control problem associated with the development of co-operative behaviour in coupled manipulators. The object here is to produce a safe, practical working prototype using existing hardware and software elements. The techniques employed include global sensing based on laser rangefinding, predictive modelling and fast parallel processing using transputer devices.

The remaining chapters of this section are concerned with aspects of learning systems based on the ubiquitous neural network. There are clearly many situations where it is advantageous for a robot to extrapolate from *a priori* knowledge and learn about variations in its environment. The study of such learning procedures is probably one of the most intensely researched aspects of robotics at this time. Chapter 21 presents a broad introduction to neural networks and their application to automation. Various practical implementations are reviewed. Chapter 22 returns to the problem of real-time robot control which includes the implementation of neural networks and genetic algorithms. The advantages of parallel processing are emphasised and this is now a common feature of many advanced real-time control systems. Further successful developments will almost certainly incorporate powerful, distributed computing elements embedded within an architecture that facilitates both learning procedures and planning operations and permits efficient remote and safe human interventions as and when necessary.

J O Gray, November 1995

Acknowledgements

There are many people and organisations to be acknowledged for their contributions and assistance in preparing and publishing this book. Our fear is that we may overlook some whose contributions were significant enough to merit inclusion. To those we may have overlooked, if there are any, we apologise in advance.

We would like to thank all the authors for their time and patience in preparation of draft and final manuscripts. We are particularly indebted to Sandra Gorse and Maureen Simhon for their efforts in getting the book into a coherent structure and their patience through the many revisions.

Finally on behalf of the authors, we could like to thank the IEE for their assistance in the production of this book and in particular Fiona MacDonald, Robin Mellors-Bourne and John St.Aubyn.

J O Gray
D G Caldwell

November 1995

Contributors

D.P Barnes
Department of Electronic & Electrical
Engineering
University of Salford
Salford   M5 4WT

J.M. Bishop
Department of Cybernetics
University of Reading
Whiteknights
Reading   RG6 2AY

C.L Boddy
Intelligent Systems Solutions Ltd
University Road
Salford   M5 4PP

D.G. Caldwell
Department of Electronic &
Electrical Engineering
University of Salford
Salford   M5 4WT

R.W. Daniel
Department of Engineering Science
Oxford University
Parks Road
Oxford   OX1 3PJ

J.S. Dai
Department of Aeronautical &
Mechanical Engineering
University of Salford
Salford   M5 4WT

P. Dario
Scuola Superiore S. Anna
ARTS Lab
Via Carducci 40
56127 Pisa
Italy

B.L. Davies
Imperial College of Science,
Technology & Medicine
Department of Mechanical Engineering
Exhibition Road
London   SO 2BX

G. Dodds
Department of Electronic &
Electrical Engineering
Queen's University of Belfast
Ashby Building
Stranmillis Road
Belfast   BT9 5AH

E. Grant
Department of Computer Science
University of Strathclyde
Glasgow

J.O. Gray
Department of Electronic &
Electrical Engineering
University of Salford
Salford   M5 4WT

R.C. Hall
Silso Research Institute
Wrest Park
Silso
Bedford   MK45 4HS

A. Halme
Automation Technology Laboratory
Helsinki University of Technology
Otakaari 5A
SF-02150 ESPOO
Finland

K. Khodabandehloo
AMARC
University of Bristol
1st Floor, Fanum House
23-32 Park Row
Bristol   BS1 5LY

A. Lush
Department of Computer Science
University of Wales
Aberystwyth
Dyfed   SY23 3DB

P.E. Mort
British Nuclear Fuels
STD.B148
Sellafield
Seascale
Cumbria   CA20 1PG

P.J. Probert
Department of Engineering Science
Oxford University
Oxford   OX1 3PJ

P.M. Taylor
Department of Electronic &
Electrical Engineering
University of Hull
Hull   HE 7RX

R.H. Tribe
Lucas Advanced Engineering Centre
Dog Kennel Lane
Shirley
Solihull   B90 4JJ

J.E. Vaughan
Electrical Engineering Department
Cockcroft Building
University of Brighton
Brighton   BN2 4GJ

A.M.S. Zalzala
Department of Automatic Control
& Systems Engineering
University of Sheffield
Mappin Street
Sheffield   S1 4DU

Chapter 1

Recent developments in advanced robotics and intelligent systems

J.O.Gray

1.1   Introduction

During the 1980s a series of initiatives was launched within the European Community to foster the development of studies in and exploitation of advanced robotic activities. These initiatives were broadly in response to technical challenges presented by our Japanese colleagues but were also influenced by the increasing availability of low cost computing power which would clearly enhance the functionality of existing mechanisms and extend the application domain into new and potentially gainful market sectors.

For the purpose of the initiative an advanced robot was defined as a machine or system capable of accepting high level mission oriented commands, navigation to a work place and performing complex tasks in a semistructured environment with a minimum of human intervention. Such a semi-autonomous device would exhibit various attributes of intelligent operation the totality and integration of which would certainly challenge existing capabilities. The application areas envisaged usually encompassed the nuclear industry, space, underwater, construction, health care and general service functions such as surveillance and cleaning.

In the United Kingdom a National Centre was established and an ambitious core research programme was defined [1] and within a wider European context various advanced robotic projects were sponsored within the framework of the ESPRIT CIME programme with some emphasis being placed on industrially led application projects [2]. Parallel work was undertaken in the United States at Carnegie Mellon University [3] and elsewhere while work in Japan continued apace. It is perhaps timely to review the progress of the field, compare actual technical achievements with earlier aspirations and to discover what type of market is actually emerging.

1.2   The semi-autonomous robot

Advanced robotics is essentially concerned with the development of sensor based control of mechanisms and automata and the evolution of suitable system architectures to generate appropriate levels of functionality to implement the tasks defined in the original broad specification. It was fashionable to describe the performance of such devices as exhibiting aspects of intelligent behaviour although it was far from clear initially (as it is still now) what was meant by the term intelligence. Certainly there was a requirement to work in an unstructured environment, avoid unplanned collisions, undertake task oriented operations with, ideally, only a high level interaction with the human operator. This represented a significant advance on existing robotic devices and, in the performance of simple tasks such as path finding and conflict resolution, modern automata do exhibit to the casual observer aspects of cognitive behaviour which may be interpreted as machine intelligence although it can be argued that few robots developed so far are particularly intelligent [4]. Semi-autonomous operation implies that a human operator remains in the control loop. This appears to be a sensible engineering and legal requisite when robotic devices are operating in an environment in which personnel may be in close proximity. The human operator will generally act in an essentially supervisory role with most of the lower level tasks being implemented by sensor based controllers embedded within an overall systems architecture which integrates real-time operation and aspects of artificial intelligence with a degree of complexity commensurate with the specified performance criteria. The development of such architectures and the integration of the required enabling techniques of control, actuators, sensors and artificial intelligence provide some of the most interesting intellectual challenges in current engineering research.

The efficient interaction of a human operator with such potentially complex machines should ideally involve procedures which maximise the use of human skills and minimise cognitive loading particularly in hazardous situations. Various types of feedback are commonly employed to achieve these goals including visual and force feedback components. If three-dimensional models of the robot and its environment are available then computer generated displays can create a virtual world in which a suitably instrumented operator can be immersed to explore the limits of both the machines and his own capabilities. Changing scenarios can be developed by overlaying direct stereo video feedback and additional sensor/actuator combinations can be used to generate a range of tactile feedback signals (contact pressure, force, texture, slip, hardness and temperature) to enhance the sense of reality and provide additional information to the operator concerning the state and performance of the remote robotic device. The objective is not just to control remote mechanisms by slaving their operation to the natural movement of human joints (fingers, wrists, arms and head position) but to create the sensation of telepresence to allow the operator to master complexity by utilising inherent human skills and natural reflex capabilities. Clearly the challenge here is not just to develop appropriate technology but also to comprehend and optimise human psychological performance in such a situation.

1.3   Systems architectures

1.3.1   Concepts

Modern intelligent automata have specifications which include the ability to accept high-level commands, undertake situation analyses and planning functions, navigate to a required operational point and perform quite complex mechanical manipulation with the minimum of human interaction. Practical engineering constraints usually require the device to be fault tolerant and degrade gracefully in the presence of major system failures. Increasingly there is also a requirement for devices to co-operate with other automata in some organised way so that complex tasks can be performed by co-ordinating the activities of a group of devices. Such co-ordination is reasonably straightforward in a well structured and predictable environment, but less so in open and unstructured terrain.

The engineering specifications usually demand that the device be equipped with a range of sensors, the signals from which must be interpreted, integrated and utilised for control functions. Such an interpretation cannot generally be direct but must be influenced by other factors such as an overall situation analysis based on a generated or embedded world model, the fundamental mission directives of the vehicle and inputs, if any, from the human operator. There is thus a clear need for an overarching structure or schema in which data flows and control signals can be organised in some logical way and prioritised in an appropriate functional or temporal framework, which can be designed to known standards, verified, commissioned and maintained as a useful industrial entity.

Various approaches have been adapted to provide a comprehensive structure which will encompass aspects of artificial intelligence (as exemplified by task planning navigation/collision avoidance and situation analysis strategies), sensors and signal processing and human computer interface methodologies.

Traditionally hierarchical deterministic procedures have found acceptance within the industrial environment because of the ready utilisation of standard computing procedures and interfaces, accepted signal synchronisation protocols and the basic transparency of the overall processing system provided by the use of finite state machines. A detailed description of suitable hierarchical architecture for real-time intelligent control is given by Albus [5] and a typical implementation of a three-layer architecture is shown in Figure 1.1.

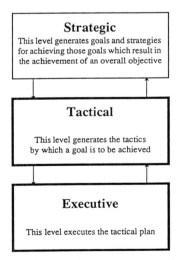

Figure 1.1 *Three layer systems architecture*

The lowest or executive level is concerned with real time control functions and it receives macro commands, such as course trajectories from the central tactical layer which generates these commands by interpreting instructions from the highest or strategic layer. In terms of development in advanced robotic devices, it is this highest layer which is the most interesting as it must combine aspects of situation analysis planning and an interpretation of operator requirements. Such functions generally require a more, or less, complex model of the real world (built up from either *a priori* knowledge or sensor data), task planning agents, appropriate advisory agents for the human operator and a communication link with the human computer interface. A simplified functional representation of a typical hierarchical architecture is shown in Figure 1.2.

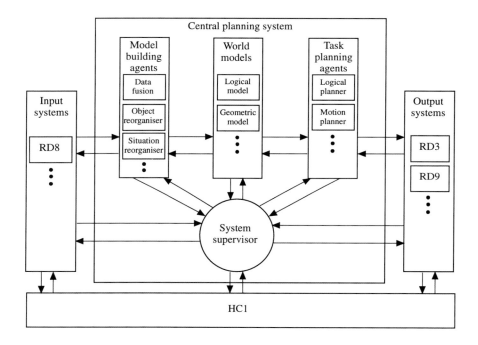

Figure 1.2 *Generic robot architecture*

Although widely utilised by workers in the field, the computational overheads and communication requirements of such hierarchical control architectures can be large. The maintenance of world model consistency and accuracy and the general representational problems of unstructured and potentially unpredictable environments are now important research areas for this type of control architecture.

An alternative architecture for real-time control, which has been gaining in impetus over recent years, is that of the behavioural or reactive control architecture [6]. Behavioural control architectures typically break down the overall task into a set of behaviours or 'competences' each of which selectively assists or assumes the control functions of other behaviours. The behaviours are usually organised into a layered architecture, with the lowest level being assigned the responsibility of, for example, obstacle avoidance. Once such a level is built and tested, others may be incrementally added, e.g. to wander, explore and to build maps, so that the overall competence of the mobile robot is improved upon.

Priority of operation is pre-specified and accordingly a layer with a high level of priority can inhibit the operation of a layer with a low level of priority and the overall architecture consists of an assembly of such layers, as shown in Figure 1.3 with the overall behaviour of the robot being determined by how the various outputs are combined. There is no overall world model to be shared between the various layers, intercommunication is of a simple protocol, narrow bandwidth type, each layer has its own relatively simple computing engine, parallel operation is possible and the overall computing overheads are generally low.

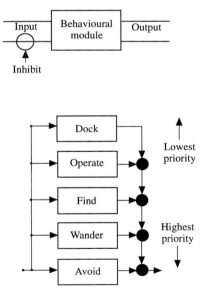

Figure 1.3 *Subsumption architecture*

The basic modularity of the design allows for an extensible systems approach and the ready ability to perform in unstructured and potentially unpredictable environments is a major strength of the behavioural approach.

Although quite complex behaviours can be generated from a combination of simple behaviour modules [7], as the number of modules or layers increases, the number of inhibiting and superseding connections also increases, leading to complexity and difficulties in conflict resolution. Overall behaviour patterns are generally handcrafted and as complexity increases, there is the possibility of unwanted emergent behaviour which is difficult to predict and often difficult to analyse.

The hierarchical approach, with its emphasis on a logical functional decomposition and its distribution of data flows and control functions amongst the individual components, and behavioural control in which operation is determined by active interaction with the physical world and the integration of a set of simple behaviours to achieve complex capabilities represent contrasting design paradigms with concomitant advantages and disadvantages. The literature is replete with examples of different types of system architectures for advanced robotic devices [8], many of which seek to combine the advantages of both the hierarchical and behavioural approaches [9][31]. To cope with a wide variety of tasks, a flexible architecture will ideally be required which can be changed or programmed for the desired set of functionalities and which could be based on the well known concepts of object oriented programming now commonly used by the computer science community [32]. The field is far from being mature and is the focus for much research activity, particularly in the artificial intelligence and cognitive science communities.

1.3.2   Engineering aspects

System architectures provide a general schema which defines structural composition and the flow of control and data signals between the various components of the robot and it can encompass aspects of sensing, modelling planning and the real-time operation of practical controllers. The system is usually complex and implemented on a distributed set of computing elements linked by networks with suitable bandwidth capabilities. System design and development is usually nontrivial and until recently hindered by a scarcity of suitable software tools capable of integrating AI modules with a real-time control environment. There are, as yet, no performance benchmarks, no universally accepted standards for structures or communication protocols and, commonly, development is undertaken using a number of diverse computing platforms, ranging from powerful work stations for AI system development to sets of distributed microcomputers for data processing and control. The software support tools used for development have generally been equally diverse and individualistic in nature, and while such a situation is acceptable for research and laboratory prototyping, it has implications for the verification, installation and maintenance of marketable robotic systems.

Advanced robots should have the ability to operate in unstructured environments where hazards may be encountered. It is thus important that their design is inherently robust in every aspect, that the device will cope with a span of difficult scenarios and that in extremis, performance will degrade in a predictable and safe manner. In application areas such as health care, transportation, construction, surveillance, agriculture and the service industries, advanced robotic devices will come into close proximity to humans without the usual safety barriers normally associated with manufacturing robots. Safety and predictability are thus prime engineering concerns and a conservative approach still favours open, transparent architectures and the use of verifiable finite state machines.

Current engineering practice argues for the use of a modular structure which can be readily extended or reconfigured without a major redesign, thus allowing, for example, the use of a range of actuator or sensor modules, as appropriate. The functionality of each module can be precisely specified and, ideally, standard communication protocols should be utilised for all outputs and inputs. System verification, installation and maintenance procedures are thus greatly simplified and the reliability of operation is enhanced.

The development and imposition of standards do, inevitably, constrain innovation but they also represent the maturing of the subject domain and the evolution of marketable devices which are fundamentally safe, reliable and suitable for their purpose.

1.4 Applications

Over the last five years or so there has been a gradual expansion in the number of engineering applications of advanced robotic concepts and although the field is not yet mature, these applications are continuing evidence of a growing confidence in the utility of such devices.

The nuclear industry has always been an obvious application domain for robotics with an emphasis being placed on remote handling systems. Recent developments include the use of stereo vision and direct force feedback employing six axis force sensitive joysticks based on configurations such as a Bilateral Stewart Platform. Trajectory control, real-time collision avoidance and the extensive use of geometric modelling procedures for risk assessment and training are technical features now being introduced into material handling systems. A number of designs for relatively agile mobile robots is now being considered, such as that shown in Figure 1.4 for intervention within unstructured environments. Such devices must be mechanically robust, have predictable characteristics and operate in a totally stable and safe manner. There is much scope for the development of innovative human computer interfaces for the remote control of such devices and to ensure their certain extrication in any credible operational situation.

The medical profession tends to be entrepreneurial in its use of technology and has rapidly adapted some of the generic technologies associated with advanced robotics. Applications include robot controlled prostate surgery [10], the precision robotic machining of bones for hip surgery utilising geometrical models derived from precisely computed tomography scans [11], use of autonomous guided vehicles to aid the disabled [12], and aspects of microsurgery that employ micromanipulators, virtual reality and force feedback concepts [13]. The safety of such systems and concomitant legal aspects are key issues here and it is refreshing to see that these have been addressed in some detail by key workers in the field [14].

The study of mobile automata has been the focus of much robotic research over many years and this work is now being utilised in practical applications. The ESPRIT Project PANORAMA [2] has resulted in the development of practical perception and navigation systems suitable for the control of vehicles in general outdoor use with a particular application being the development of intelligent autonomous systems for cars which could have significant implications for modern high density motorway utilisation, both by enhancing safety and the overall throughput of the existing roadway systems.

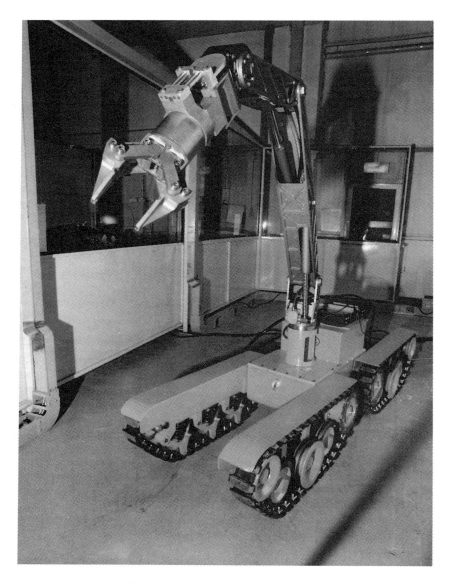

Figure 1.4 *Robot for the nuclear industry*

Fully instrumented vehicles have already been tested in realistic situations and there is commercial interest in the exploitation of the existing technology. In the manufacturing arena, similar generic technologies have resulted within the ESPRIT Project MARIE [2] in the development of sophisticated AGVs for use in a CIM environment or in the handling of hazardous materials. Automatic inspection is now another common application domain and includes practical examples such as the robot crawler used for monitoring corrosion inside large storage tanks [15] or wall climbing robots for use in nuclear power plants [16]. In the United States, free ranging fully instrumented mobile robots are now employed in mail delivery services within large industrial complexes and in Italy [17] for the delivery of pharmaceutical and other medical products within hospital environments, while autonomous cleaning robots with comprehensive model based navigation capabilities for use in industrial and commercial areas are now under development within the ESPRIT Project ACRO [2].

Applications in aerospace combine advanced teleoperated devices and mobility for satellite maintenance, constructional purposes and planetary exploration [18], with the latter encouraging the evolution of innovative mobility mechanisms such as legged robots to cope with difficult terrains [19]. The latter are quite complex in both design construction and operation but they are generally more agile than the conventional multiple articulated tracked vehicle of the type shown in Figure 1.4 and have applications in the nuclear industry where there will, inevitably, be buildings with stairs, and in agriculture where a machine with a relative low foot print area has distinct environmental advantages [20]. Most of the devices produced so far appear to have a relatively limited functionality but the field is currently being well researched and the work is attracting increasing commercial interest. A typical test bed for the development of suitable control algorithms is shown in Figure 1.5.

Economic forecasts produced by the Japanese Robot Society suggest a substantial market for robotic applications in the manufacturing area, particularly with respect to construction and their technical literature is replete with specific examples of practical machines for use within construction sites for the performance of such tasks as assembly, transportation and surface finishing procedures. Such sites are particularly difficult for machine operation and are generally characterised by very rough unstructured topographies. Mining is another fruiful area for the application of robotics as it combines a hazardous environment within a relatively structured topography and is certainly suitable for remote teleoperation[33]. In Europe, a project is underway to develop generic system architectures within the ESPRIT ROAD ROBOT [2] project for the control of a set of heavy duty mobile platforms engaged in such activities as road construction and paving. The objective is to create an outdoor assembly line operation using conventional CIM principles but using components such as large mobile diggers and pavers which are normally manually operated and manually co-ordinated in operation. In a contiguous technical theme the objective of ESPRIT Project ATHENA [2] is to operate a fully instrumented 30-tonne compactor vehicle via a remote, sophisticated computer graphical interface which overlays video and computer generated geometrical data of landfill sites for waste disposal within which the vehicle moves.

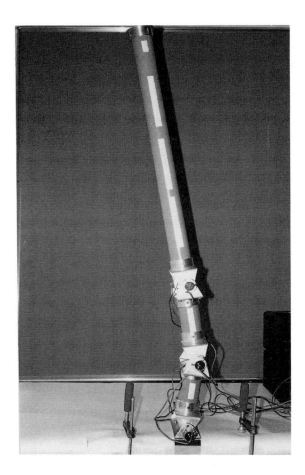

Figure 1.5 *Triple inverted pendulum (University of Salford)*

The control architecture incorporates both hierarchical and behavioural aspects and uses model based navigation and sensor fusion techniques. The transfer of advanced robotic concepts into these unconventional domains is supported in each case by strong financial and environmental arguments and there is clearly much scope for future practical implementations of the technology in these and related fields.

Almost alone amongst the manufacturing nations, Japan appears convinced of the economic viability of the totally automated factory and has embedded this concept within the technical spectrum of its recent Intelligent Manufacturing Systems initiative which has now been co-sponsored by a group of nations as an international study programme. The technical spectrum originally defined encompasses almost every aspect of manufacturing from simple metal cutting to advanced anthropomorphic mobile automata which will operate between individual production cells to ensure the smooth flow of products, even in the event of minor systems failure and uncertainty. Japan is backing this initiative with very substantial research funding and policy makers there are convinced that a future market exists for such complex automata.

Current international practice tends to focus on the use of sensor based robots for enhancing conventional production procedures, such as welding and assembly, by providing enhanced operational flexibility and a degree of fault tolerance which can be generated by sensor feedback. The improved technology has, however, allowed the application of automation techniques to production sectors such as food and agriculture, which previously mainly utilised manual procedures. The ESPRIT Project ROBOFISH [2] for example is concerned with integrating a range of technologies, including computer vision, image processing and distributed control systems for the automatic processing of freshly caught fish. Other applications include the use of sensor based robots in agriculture for crop gathering and milk production and the use of CAD/CAM procedures in the clothes industry where some quite innovative gripper techniques have been developed for the precision manipulation of the very thin, light, flexible materials used in manufacturing [21].

The brief range of applications cited above is not meant to be exhaustive and some significant application domains such as underwater surveying and maintenance activities have not been mentioned. However, the examples given do indicate the extent to which early concepts on advanced robotics technology are now being implemented as practical engineering solutions to significant technical problems. Currently, these engineering implementations are quite diverse in nature but generic features are emerging which may eventually provide cohesion for the subject area and hasten the evolution of acceptable international engineering standards which, of course, will further promote the acceptance of the technology within a wider range of market sectors.

1.5   Advanced concepts and procedures

1.5.1   Decentralised autonomous robotic systems

An alternative approach to using a single complex robot to perform a task is to use a group of simpler robots controlled in some distributed manner to achieve the same objective. Each unit in the group could be an essentially autonomous entity and as inter-relationships need not always be precisely co-ordinated in space and time, the overall behaviour is often referred to as being co-operative in nature. There are many potential applications for such multiple co-operant automata as, for example, in the industrial environment, where a group of independent devices could be used to move objects which because of their size, shape or weight, could not be transported by a single device. Other applications could include the use of multiple co-operating autonomous underwater vehicles for ocean bed surveying, product disassembly (which is quite difficult to achieve with conventional robotic procedures), area cleaning and surveillance.

A key argument used for the development of such distributed systems is the replacement of a single complex (and hence relatively fragile) unit by a set of simpler units, thus providing robustness and graceful system degradation as the performance of any one unit is not necessarily affected by the dysfunction of a companion unit, as long as sufficient units are available to provide for redundancy. Each unit in the group need not necessarily be identical, thus allowing a richness in the functionality of any particular combination.

The simplicity of individual robot functionality is not necessarily reflected in the required controlling paradigm and the field is currently one of particular interest to the research community [22]. Some early work at Salford [23] concentrated on the development of an enhanced subsumption architecture called a Behavioural Synthesis Architecture, in which a behavioural vector is constructed from elements determined by the state of the robot (self), its relationship with other robots (species), the nature of the environment (environment) and an overall global directive (world). The structure of this architecture is illustrated in Figure 1.6 and it has been implemented on a pair of instrumented mobile robots with the objective of achieving a benchmark task of moving a pallet to a 'specific target point for docking over a planar object strewn environment' as shown in Figure 1.7.

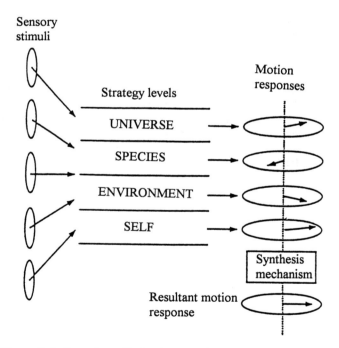

Figure 1.6 *Overview of behaviour synthesis control architecture*

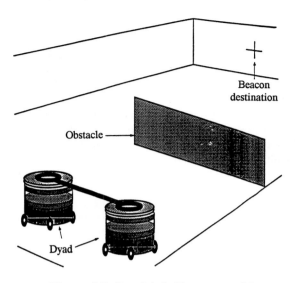

Figure 1.7 *Bench hybrid system architecture*

Recent work [24] has focused on the development of a hybrid architecture, as outlined in Figure 1.8 which makes sensible use of *a priori* knowledge and provides a mechanism for the behaviour scripts which control the robots. The concept of a robot society, the elements of which co-operate to achieve a complex task is of particular interest to the emerging field of microrobotics and will provide many challenges for researchers in this field.

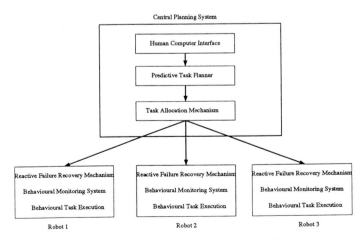

Figure 1.8 *Proposed hybrid system architecture*

1.5.2   Miniature and microrobotics

Advances in micromachining and microfabrication techniques has led to the development of miniature electric motors, gear trains and actuators with overall dimensions of a few millimetres or less. Combining this technology with silicon processing power and miniature silicon based sensors results in the development of extremely small robotic devices and the generation of a new field of study in robotic engineering. There appears, as yet, to be no accepted standard of nomenclature and the term microrobotics is loosely used in the literature to encompass devices with a wide span of dimensions. For the purposes of this brief review it is appropriate to define miniature robots as having dimensions of a few cubic centimetres, micro robots to have dimensions ranging from millimetres to micrometres and nanorobots to have a size commensurate with a biological cell with overall dimensions of a few hundred nanometres.

Miniature robotics clearly have application in precision machining and in the performance of delicate surgical operations and mobile autonomous devices can be appropriately instrumented to monitor temperature, pressure and gas composition in environmental surveys. Winged insect like robots have been proposed [25] with a wing span of a few centimetres which could be used for environmental monitoring or area surveillance. Whatever the engineering feasibility of this concept it is clear that a range of suitable actuators and sensors in the micrometre range is being actively developed [26] utilising existing technology from industrial VLSI processing procedures.

The work space for microrobots is clearly only visible to human operators by means of microscopes and such devices can manipulate objects the size of a single biological cell. This facility has suggested uses in the bioengineering field for cell separation and selection. Groups or swarms of such devices could be used for industrial cleaning purposes but despite the formidable computing power available in each element overall communications and control paradigms pose significant technical problems. Applications have also been suggested in the medical field where such devices could perform surgical or other functions internally without invasive surgical operations. Unfortunately at such small diameters viscous forces become very significant and normal electromechanical devices operate with some difficulty.

Development in nanorobotics are currently limited by the existing technology but any eventual successful design will probably employ new forms of fabrication technology.

1.5.3   Learning systems

Few engineering research areas have received more attention recently than machine learning systems where attention has focused mainly on the exploitation of neural networks [27]. Such networks have generally proved to be excellent classifiers of scenes or situations where sensor data are sparse and they have a capability of learning the characteristics of nonlinear functions in cases where analytical derivation may prove to be untractable. Research workers quickly applied the evolving technology to robotic systems [28] where the useful nonlinear function classification capabilities were initially applied to such problems as inverse kinematics and dynamic modelling for control.

Applications have been extended to address such aspects as collision avoidance, trajectory planning and reactive navigation [29]. To allow such networks to adapt to situations which lie outside original training domains, genetic algorithms, which are essentially adaptive search routines, have recently been employed to dynamically optimise neural interconnections to deal with changes either in the external environment or the robots' own characteristics. The object of this work is to develop a robot which requires little, if any, *a priori* information about its environment and which will use sensor information to learn about the external world and aspects of its own functionalities.

Alternative innovative learning systems have been presented [30] and no doubt others will evolve in due course. Any successful practical and robust learning procedure will clearly enhance the functionality of a future generation of robotic systems allowing a greater degree of autonomous operation and providing a wider scope for industrial and other applications.

The field is clearly a culturally rich one for cognitive scientists whose key objective is the study of the learning process itself and who perceive robotics as an amenable applications domain which gives a practical dimension to their computer simulation studies. The benefit to engineers will lie in the ability to extract and apply those features which will enhance existing and proven technology, increase the functionality of present architectures and add value to marketable machines and systems.

1.6   Future developments

The ubiquitous availability of inexpensive computing power and increasing availability of hardware and software development tools have rendered feasible the design and implementation of complex automata with a high degree of functionality and possessing architectures incorporating the integration of sensor based real-time control and innovative concepts based on artificial intelligence (as defined by the computer science community). As such, the devices will be used mainly to complement human skills in applications where it is either undesirable or impossible to use human operators. They will be, for the immediate future, semi-autonomous in nature and thus involve human computer interfaces which will be optimised to cope with the complex functionality of the devices. Such interfaces will involve aspects of teleoperation, telepresence and virtual reality in the immediate future and the technology will be embedded to enhance the performance of machine systems.

In order to enhance the capability of the operator, the devices will have an ability to accept and implement high level commands, execute task and trajectory planning operations and effect simple manoeuvres such as local collision avoidance and beacon guided automatic docking. Such complexity has implications for traditional engineering concepts of reliability, verification, safety and maintainability. Many such issues are already well addressed within the aerospace industry which has a long established culture in all safety related matters. This culture and its associated rigour for standards and quality assurance must eventually be adopted by the robotics industry as it moves into the market for advanced automata, particularly for safety critical applications in the nuclear industry and medicine. Inevitably there will be a future trend towards enhanced decision making capabilities within the machine, for example, in the extended use of innovative and flexible architectures and opaque computing procedures based on neural networks or equivalent methodologies. Such a trend will raise interesting engineering, safety and legal problems which have yet to be addressed but will be relevant to all types of intelligent systems in whatever way we define the term intelligence.

Clearly these are significant issues, posing exciting challenges to engineers as they exploit the ever increasing power of silicon technology in the production of devices and processes in the coming decade.

1.7   References

1.   GRAY, J.O. Advanced Robotics and Integrated Approach. Mechatronics and Robotics I ed. P.A. McConall, P.Drews and K.H.Robrock. IOS Press. Amsterdam. ISBN.90 5199 057X pp 43-50. 1991.

2.   ESPRIT Computer Integrated Manufacturing Initiative. Annual Synopsis. Published by the Commission of The European Communities Brussels. ISBN.9-826 4816-8. 1991.

3.   WHITTAKER, W.L. Field Robotics for the Next Century. Mechatronics and Robotics I ed. P.A.McConall, P.Drews and K.H.Robrock. IOS Press. Amsterdam. ISBN.90 5199-057X pp 3-9.

4.   MILLER, D.P. Intelligent Mobile Robots. Perception and Performance. Proceedings 6th International Conference on Advanced Robotics. Tokyo. Published by Japanese Industrial Robot Association pp 419-422, November 1993.

5.   ALBUS, J. and QUINTERO, R.   Towards a Reference Model. Architecture for Real Time Intelligent Control Systems (ARTICS.) Robotics and Manufacturing, eds. M Jamshidi and M Saif ASME Press Series, N.Y. pp 243-250. 1990.

6.   BROOKS, R.A. A Robust Layered Control System for a Mobile Robot. IEEE Journal of Automation and Robotics, Vol.R.A2 No.1. pp14-23, March 1986.

7.   BROOKS, R.A. A Robot that Walks. Emergent behaviour from a carefully evolved network. MIT. Artificial Intelligence Laboratory Report 1091, July 1989.

8.   CORFIELD, S.J. et al Architectures for Real Time Intelligent Control. Control of Autonomous Vehicles. IEE Computing and Control Engineering Journal. Vol.2 No.6. London pp 254-262. 1991.

9.   LYONS, D.M. and HENDRICKS, A.J. Planning for Reactive Robot Behaviour, IEEE Int. Conference on Robotics and Automation. Nice, May 1992.

10.   NG.W.S. et al Experience in Trans-Urethral Resection of the Prostate IEEE. EMB.S.JI. pp 120-125, March 1993.

11.   MITTLESTADT H. et al Robotic Surgery achieving predictable results in an uncertain   environment.   Proceedings   6th   International   Conference   on   Advanced Robotics. Tokyo. Published by Japanese Industrial Robot Association pp 367-372, November 1993.

12.   DARIO, P.et al An Autonomous Mobile Robot System for the Assistance to the Disabled. IBID. pp 341-346, 1993.

13.   HILL, J.W. et al Telepresence Surgery Demonstration System. Proc. IEEE Conference on Robotics and Automation. Vol.3. San Diego. USA. pp 2302-2307, May 1994.

14.   DAVIES, B. Safety of Medical Robots. Proceedings 6th of the International Conference on Advanced Robotics. Published by the Japanese Industrial Robot Association. Tokyo. pp311-317, November 1993.

15.   SCHEMPF, F. Above Ground Storage Tank Inspection Robot. Proceedings IEEE Conference on Robotics and Automation Vol.2. IEEE Computer Society Press. San Diego. pp 1403-1408, May 1994.

16.   BRIONES, D. et al Wall Climbing Robot for Inspection in a Nuclear Power Plant IBID. pp 1409-1414, 1994.

17.   COLLINS, G. et al   A Friendly Interactive Transport System for a Hospital Environment. Proceedings 24th Symposium on Industrial Robots. Published by the Japanese Industrial Robot Association. Tokyo. pp 487-502, November 1993.

18.   LAVERY, D.   Perspectives on Future Space Telerobotics Technology. Proceedings 6th International Conference on Advanced Robotics. Pub. by JIRA. Tokyo. pp 443-448, November 1993.

19.   SIMMONS, R. and KRUTOV, E. IBID. pp 429-434, 1993.

20.   HALME, A. et al   Proceedings 1st IFAC. International Workshop on Intelligent Autonomous Vehicles. Southampton. pp 1-7, April 1993.

21.   TAYLOR, P. et al   Electrostatic Grippers for Fabric Handling. Proceedings IEEE Conference on Robotics and Automation. Philadelphia, pp 431-433.

22.   FUKUDA, T. et al   Optimisation of Group Behaviour on Cellular Robotic Systems in a Dynamic Environment. Proceedings IEEE Conference on Robotics and Automation. San Diego. pp 1027-1032, May 1993.

23.   BARNES, D.P. and GRAY, J.O.   Behavioural Synthesis for Co-operant Mobile Robot Control. Proceedings IEE International Conference Control 91. Vol.2. Edinburgh. pp 1135-1140, March 1991.

24.   EUSTACE, D., AYLETT, R and GRAY, J.O.   Combining Predictive and Reactive Control in Multiagent Systems. Proceedings IEE International Conference Control 94. IEE publication 389. University of Warwick. Vol.2. pp 989-995, March 1994.

25.   YAYOI KUBA et al   Study of wings of flying microrobots. Proceedings IEEE Conference on Robotic and Automation, Vol. 1, San Diego, pp 834-839, May 1994.

26.   BOHRINGER, K.F. et al   A theory of manipulation and control for microfabricated arrays. IEEE Workshop on Micro-electro mechanical Systems, IEEE Press, Oiso, Japan, Jan. 1994 pp 102-107.

27.   LAU, C. et al   Neural networks theoretical foundation and analysis IEEE Press. ISBN 087942-280-7, 1992

28.   HORNE, B. et al   Neural networks in robotics - a survey. Journal of intelligent and robotic systems .3.3-66, 1990

29.   DUBROWSKI, A. et al   Self supervised neural systems for reactive navigation. Proceedings IEEE Conference on Robotics and Automation, San Diego, pp 2076, May 1994.

30.   MYSTEL, A.   Multiscale models and controllers, Proceedings IEEE/IFAC Joint Symposium on CACSD, edited by S.U. Matteson, J.O. Gray and F.E. Cellier, IEEE Press, Tuscon, Vol. 3, pp 13-26, March 1993.

31.   ARKIN, R.C. Integrating Behavioural, perceptual and world knowledge in Reactive Navigation Robotics and Autonomous Systems 6. pp105-122., 1990.

32.   THORPE, C.E., and HERBERT, M.   Mobile Robotics.   Perspectives and Reality Proceedings ICAR, Barcelona, pp497-506. ISBN84.7653-576-7. Sept. 1995

33.   SHAFFER, G. and STENTZ, A.   A Robotic System for underground coal mining.   Proceedings IEE International Conference on Robotics and Automation, pp633-638, Vol.1.   May 1992.

Chapter 2

Machine intelligence: architectures, controllers, and applications

E. Grant

2.1 Introduction

Rather than enter into a long discussion regarding a definition for machine intelligence, or even the word intelligence itself, this chapter will focus on mechanisms and technologies for implementing machine intelligence. Nevertheless the ability to learn must be one criterion for describing intelligent behaviour. Another one closely related to robotics is a machine that is able to act autonomously in the presence of uncertainty. The ability of a robot to adjust its actions based on sensed information [1,2,3] could be seen as another prerequisite for intelligence. The actions taken by the machine would be considered to be intelligent if the same action would be taken by a human given the same conditions.

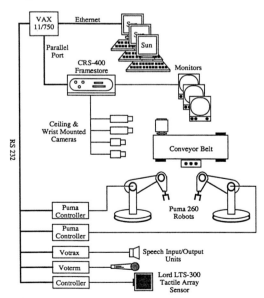

Figure 2.1 *The Freddy 3 advanced robotic research test-bed*

In advanced robotics systems [2, 4], robots equipped with sensors (vision, tactile, proximity, speech recognition and voice synthesis), robot controllers, conveyors, vision processing equipment, and computers, all networked together through a variety of high bandwidth (Ethernet) and low bandwidth (RS 232, IEE488) communication channels are an ideal domain for researching into machine intelligence, Figure 2.1 Advanced robotics systems commonly incorporate the three major subsystems for machine intelligence: sensors, actuators and control, Figure 2.2. However, the interconnection of physical systems, or the task undertaken by the system, does not make a machine intelligent. Intelligence comes from the manner in which the system is controlled or from the reasoning and decision making that the machine performs. Therefore 'intelligent-control' is closely associated with machine intelligence [5].

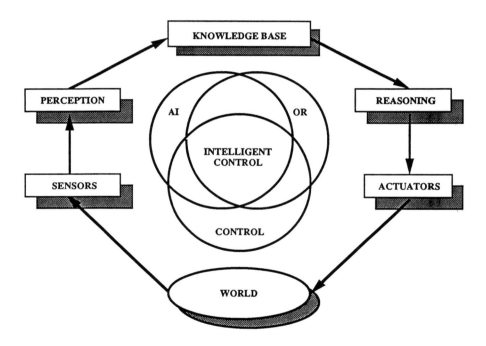

Figure 2.2 *Architecture for intelligent control*

Intelligent control systems integrate several kinds of data, including task specification, and the task state from sensory data integration. An intelligent control system must handle information about its own state and the state of the environment too, and be capable of reasoning under uncertainty. This can involve the use of both heuristic and algorithmic programming methods.

We begin by reviewing three hierarchically ordered control architectures used for implementing intelligent control [5,6,7], Figures 2.3, 2.4 and 2.5. After this we concentrate on controlling an unstable dynamic system with a variety of rule-based and machine learned programs. The final section deals with 'human-in-the-loop' control. In particular, we investigate how hypermedia-based interactive interfaces can be used in robot task planning.

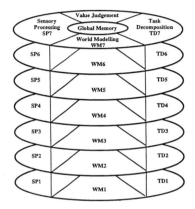

Figure 2.3 *The NIST architecture*

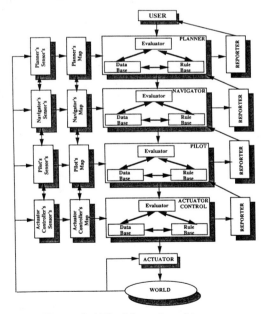

Figure 2.4 *The Meystel architecture*

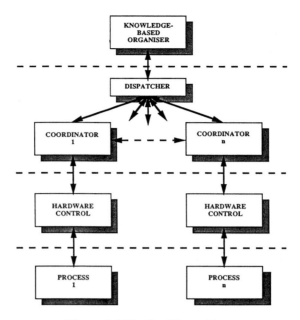

Figure 2.5 *The Saridis architecture*

2.2   Architectures for intelligent control

Saridis [5] states that intelligent machines require the use of 'generalised' control strategies to perform intelligent functions such as the simultaneous utilisation of memory, learning or multi-level decision-making in response to 'fuzzy' or qualitative commands. He proposed that these intelligent functions should be implemented using 'intelligent control'.

Intelligent control is a combination of high-level decision making: using computers, and advanced mathematical modelling and synthesis techniques of systems theory. These, together with linguistic methods, attempt to deal with imprecise or incomplete information from which appropriate control actions evolve. The control functions in an intelligent machine have been implemented as a hierarchy of processes [5,6,7]. The upper layers concentrate on abstractions, decision-making and planning, while the lower levels concentrate as time-dependent sub-tasks, like processing data from sensors, or operating an actuator. Hierarchical decomposition is applied to complex control problems to reduce them to smaller sub-problems.

In the hierarchical control architectures of both Albus [6] and Meystel [7] each layer possesses the same processing nodes, see Figures 2.3 and 2.4. Although these processing nodes have different names in these two architectures they include a knowledge base, sensory processing, task decomposition and communication. Alternatively, Saridis [5] recognises that each layer in a hierarchy need not perform the same activity over time. Both he and Albus [6] recognise that middle layers are frequently hierarchies of linguistic, or heuristic decision structures that handle imprecise or 'fuzzy' information.

In his work Saridis proposes a hierarchical structure of three basic levels of control, this hierarchy is distributed according to his 'Principle of Precision with Decreasing Intelligence'. Each level consists of several layers of processing modules, see Figures 2.5 and 2.6, that communicate with each other with strings of symbols.

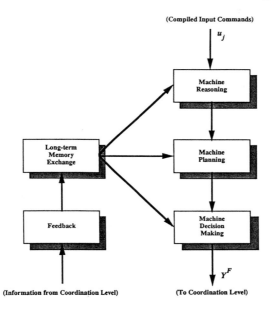

Figure 2.6 *Saridis Processing Modules*

The organisation level

Performs task planning, learning reasoning and information processing from long-term memory.

The co-ordination level

Performs the same functions as above but for lower level information processing and short-term memory.

The control level

Executes tasks through actuation and feedback control.

At the National Institute of Standards and Technology (NIST) there have been several implementations of the architecture described in [6]. These include being applied to manufacturing control in the NIST Automated Manufacturing Research facility (AMRF) [8], and in the NIST/DARPA Multiple Autonomous Undersea Vehicle (MAUV) [9].

- Automated Manufacturing Research Facility (AMRF)
    - Horizontal and Vertical Machining
    - Small Batch Workcell
    - Deburing and Cleaning
    - Surface Measurement
    - Optical Scattering

- 7-Dof Independent Robot Joint Control

- MAUV

- TMAP and TEAM

- NASREM

Meystel [7] has also used an autonomous undersea vehicle as a demonstrator whereas the architecture of Saridis was applied widely on tasks associated with space station control including:

- The mathematical theory of intelligent control

- Multisensor Fusion

- Task planning and integration

- Multi-arm manipulation

- Adaptive and learning control

- Reliability and safety

- Parallel computation and information management

2.3   Machine learning

Since the mid-1970s, artificial intelligence (AI) methods have been continuously developed and applied by industry, business, and commerce. Expert systems are the most successful implementation of AI. Applications include: medical diagnosis, mineral analysis, control of complex processes, and fault diagnosis and monitoring;

systems for fault diagnosis and monitoring account for half their applications [10, 11, 13, 18]. However, the difficulties surrounding the development of the production rules for expert systems, going from the general to the specific, including knowledge acquisition, descriptive knowledge representation and context handling, have led to the development of a sub-division of expert system technology known as 'machine learning'. Compared to expert systems, machine learning can be described as going from the specific to the general. Here, we will look at how the production rules for 'rule-based' control can be produced manually, and automatically, and we will discuss two approaches for achieving machine learned control (MLC). First we will describe a controller based on the machine learning algorithm BOXES [15], an algorithm that uses a reinforcement learning approach. Second we will discuss the implementation of neural networks for control; both reinforcement learning and competitive learning are considered [16, 17].

2.3.1 *Rule-based control*

In many expert systems the rule interpreter, or inference engine as it is sometimes called, uses an exhaustive backward-chaining strategy in which the rules are expressed in a highly focused 'IF-condition-THEN-result' form, the result is known or concluded, and, knowing the answer, the system searches backwards for the rule whose condition produces the result [10, 11, 14, 18].

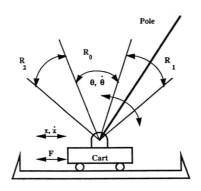

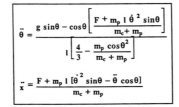

Figure 2.7 *Pole and cart*

Such an expert system is termed goal directed. Goal-directed systems are therefore the result of going from the general to the particular since the search is, in fact, directed toward a goal (the known result). An inference mechanism based on a backward chaining strategy is commonly equipped with an extensive explanation facility to reduce the amount of conflict resolution needed. To be user friendly, an expert system must also possess a user interface that interrogates the human expert in a manner that is unambiguous. The information obtained from the user interface must then be stored as facts in a database and the relationship between individual facts must be stated. The rule structure determines which rule should be examined next, in a forward-chaining expert system, or it further questions the user in a backward-chaining system. Finally, the representational structure must also include an uncertainty handling mechanism that is based on measurements of belief.

A human can derive a set of control rules through the process of interpretation, Makarovic [18] derived a rule by examining a system's differential equations of motion, see Figure 2.7. It can be seen that the Makarovic rule:

if theta_dot > THRESHOLD theta_dot then push RIGHT

if theta_dot<-THRESHOLD theta_dot then push LEFT

if theta > THRESHOLD theta then push RIGHT

if theta < -THRESHOLD theta then push LEFT

if x_dot >THRESHOLD x_dot then push RIGHT

if x_dot <-THRESHOLD x_dot then push LEFT

if x >THRESHOLD x then push RIGHT

if x < -THRESHOLD x then push LEFT

worked well where the system's specifications remain constant. However, when system parameters change, the Makarovic rule cannot guarantee success. This was because of the arbitrary choice of threshold values by Makarovic. One set of threshold values is not ideal for a system configuration that alters. In contrast, a rule derived from observing the systems performance [18],

if (theta(k) >THRESHOLD)
    then
    if ((theta(k)<theta(k-1))
   and (|theta(k)-theta(k-1)| > |theta(k-1)-theta(k-2)|)
    then
     apply a RIGHT force

```
        else
          apply a LEFT force

if (theta(k) <-THRESHOLD)
        then
        if ((theta(k)>theta(k-1))
          and (ltheta(k)-theta(k-1)l > ltheta(k-1)-theta(k-2)l)
        then
          apply a RIGHT force
        else
          apply a LEFT force

if (ltheta(k)l <=THRESHOLD)
        then
        if (x(k)>=0)
          then
          if ((x(k)<x(k-1) and (lx(k)-x(k-1)-x(k-2)l))
            then
              apply a RIGHT force
            else
              apply a LEFT force
if (x(k) <0)
        then
        if ((x(k)>x(k-1) and (lx(k)-x(k-1)l>lx(k-1) -x(k-2)l))
        then
          apply a RIGHT force
        else
          apply a LEFT force
```

was written without any threshold values being placed on observation. Here the condition part of the rules only deals with the sign of errors, and with the sign of variations of observed system state variables. It reflects the human control heuristics, and can adapt itself to varying system configurations and perform consistently well in all circumstances.

2.3.2 *Machine learned control*

Machine learning, as presented here, is classified into two areas, artificial-intelligence type learning on symbolic computation, and neural nets. These are chosen because we have first-hand experience of applying them. This is not an exhaustive list since genetic algorithms is another commonly used learning approach. An effective machine learning system must use sampled data to generate internal updates, and also be capable of explaining its findings in an understandable way, for example symbolically, and communicating its explanation to a human expert in a manner that improves the human's understanding. Expert systems based on artificial-intelligence learning have surpassed the performances of human experts and, they have been able to communicate what they have learned to human experts for the purposes of verification. Artificial-intelligence type learning originated from an investigation into the possibility of using decision trees or production rules for concept representation. Subsequently the work was extended to use decision trees and production rules in order to handle the most conventional data types, including those with noisy data, and as a knowledge acquisition tool.

2.3.3   Reinforcement learning

Reinforcement learning, Michalski [13], is similar to supervised learning in that it uses feedback for adaptation. However, unlike supervised learning, the feedback reinforcement learning gets is only an indication of the value of the systems action. Often this is delayed. Reinforcement is feedback on the correctness of an action, it is not information on what the correct action is. Reinforcement learning is useful in cases where supervisory information is not available.

Also, reinforcement learning falls into two categories, (i) non-associative type, which only receives a reinforcement signal from the environment, and (ii) associative reinforcement learning, where the system receives both a reinforcement signal, and sensory information, on the state of the environment. Sensors are used to discriminate between different situations. This we considered more suited to our particular needs with the pole-and-cart application.

In this section, we will discuss the application of rule-based, MLC, and neural network controllers to control a pole-cart system by using AI techniques.

2.3.4   'BOXES'

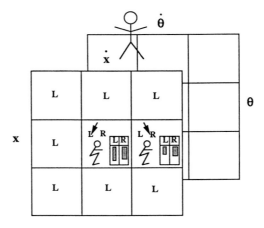

Figure 2.8 *'BOXES'*

In Michie and Chamber's learning algorithm 'BOXES' [15, 18], the physical state space is partitioned into boxes. The algorithm learns to set correct decisions for each box through trial-and-error, see Figure 2.8.

- Local Demon Totals (consider S decision 'left')

$$LL = (LL' \times DK) + \sum_{i=1}^{N} (T_F - T_i)$$

$LU = (LU' \times DK) + N$
$RL = RL' \times DK$
$RU = RU' \times DK$

- Global Demon Totals

$GL = (GL' \times DK) + T_F$
$GU = (GU' \times DK) + 1$

Merit = GL/GU

Target = $(C0 + C1) \times$ Merit $(C0 >= 0; C1 >= 1)$

- Control Action (s = 'left' or 'right')

Value(L) = $\dfrac{(LL+K) \times Target}{(LU + K)}$

Similarly , Value (R) is claculated. Control action S is decided depending on whether

Value(L)<>Value(R)

Unfortunately, state space partitioning prior to experimentation is arbitrary because it is reliant on human knowledge. If the original partitioning is wrong, the algorithm cannot learn to correct it. In the following, we will show how our rule can be used to partition the state space in pole-cart application.

2.3.5 *Basic concepts*

Control surface

A control surface can be defined as the problem of generating an optimal response as a function of each class to which the input situation is assigned. The function, or mapping, which associates each possible input situation with the optimal control actions is termed a control surface. The control task is therefore one of implementing control surfaces so we can formulate the design problem in a form suitable for pattern recognition interpretation [19]. Where control systems operate with partially specified control surfaces, appropriate actions taken in some situations can be used to generate actions for unknown situations through learning. In pole-and-cart the control surface is a non-linear function of four state variables.

Control situations

A control situation is a region in plant parameter space, state space, or augmented state space, for which a unique choice of control actions, or a single set of feedback gains exist. It leads to a satisfactory performance for all points contained in that region.

In learning control there are two separate sub-sections of control situations, (i) learning the optimal control action for each control situation, and (ii) learning to partition the problem space. Here, after consideration, we deal with the former. The latter is reported on by Zhang and Grant [20, 21]. Control situations may result from pre-selected threshold quantisation of the state variables, as in 'BOXES' [15, 18] or learned on-line, as in Andersons' work on strategic learning [20, 21, 22]. The introduction of control situations helps in the solution of an optimisation problem, because it reduces the scale of the problem to the optimisation of a small region.

Memory

In learning systems a separate memory is associated with each control situation. There are two forms of memory; short-term memory (STM), which remembers information for as long as the system is in the same control situation, and long-term memory (LTM), which remembers information arising from a particular control situation. A good example of the use of STM and LTM is the 'BOXES' algorithm [15, 18].

Sub-goals

It is often the case that the main goal of a control system cannot be formulated precisely, or it is compiled over a number of sampling periods. But, it is often difficult to single out which control choice was responsible for performance improvement. Samuel [23] in his checker playing machine, partially solved this problem through establishing sub-goals, which are related to the main goal. Of importance in the context of our learning control work is that the sub-goal can provide the necessary reward/punishment signal for reinforced learning.

Computational methods

The control surface can be specified by mathematical equations based on a number of example mappings learned, and these equations can then be evaluated by a computer.

The input data for the calculation of the equation come from coded information about control situations.

All the parameters must be included in the set of equations specifying the control surface. If the parameters change, the control surface can be re-mapped. If the number of parameters changes the algorithm must change; re-mapping here involves adapting the control surface specification, which is structural rather than parametric. The control situation consists of pattern vectors that are components and arguments, it does not use much memory but it is time inefficient.

<u>Look-up tables</u>

Assume that the controller contains a 'state-space memory', essentially a look-up table, which provides a value for a control output when indexed by a value of input control situation. After the control surface is stored in memory, coded information about a control situation acts as a pointer to a piece of memory specifying the required control action. Because memory space is limited if a control situation arises that is not specifically contained in memory, the interpolation methods are used to obtain the action. This is commonly the fastest form of obtaining a control decision and although it uses considerable memory, it is very fast, see Figure 2.9.

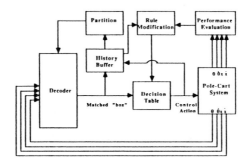

Figure 2.9 *The 'Boxes' controller*

2.3.6 *Neural network-based reinforcement learning*

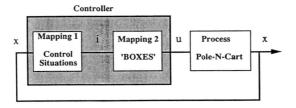

Figure 2.10 *The RLLN controller*

The learning controller consists of a two-layered neural network for implementing the input-output transfer function and an evaluation network, a look-up table, which provides the necessary reinforcement signal for evaluative feedback via a goal oriented performance index. The high-level architecture for the teaching controller is shown in Figures 2.10 and 2.11.

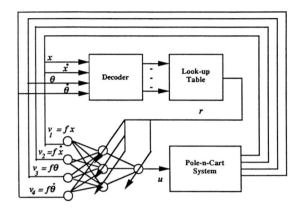

Figure 2.11 *The RLNN high-level architecture*

Neural networks

Neural networks implement information storage with synaptic weights storing information and distributed patterns acting as keys, they combine the benefits of both the computational method and look-up tables. With neural networks, information about control situations is coded in terms of distributed patterns, hence they can support distributed representation and reduce the storage requirements associated with control surface dimension.

A neural network can specify control actions for a given situation not visited during learning, according to similarity of conditions. This associated structure automatically generalises according to degree of similarity. The trade-off between computation time and storage space is better resolved by neural networks.

Establishing the look-up table

In reinforcement learning at every time step during learning, control actions are evaluated with respect to a sub-goal. The action which maximises the sub-goal is regarded as the optimal control action and is rewarded; all other actions are punished. In the pole-and-cart, learning is difficult because the effects of choosing different actions cannot be tested. So, here we evaluate alternative control actions with respect to a small region of the state space. We also assume that they have the same reinforcement value.

Decoding the state variables

The term 'decoder' describes the process of accepting an input situation and transforming it into one activity from a choice of a large number. Hence evaluation signals are stored as a look-up table where an input situation appears as an activity on a single pathway to a storage location. The storage location contains the appropriate evaluation specification. This approach was motivated by 'BOXES' [15, 18] in which the four-dimensional state-space is divided into disjoint regions ('boxes') by quantising the state variables. However, unlike [20, 21, 22] we attempt to estimate the evaluation of different control actions with respect to a sub-goal. This estimate then provides the necessary reinforcement signal for our neural network based reinforcement learning control (RLNN), see control surface Figure 2.12.

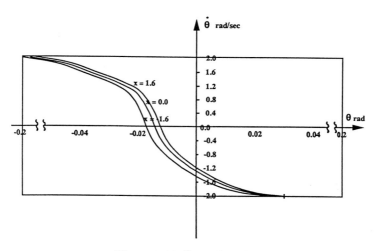

Figure 2.12 *Control surface*

2.4   Competitive learning

In applying machine learning to control problems, one of the difficulties is that there are an infinite number of points in the control systems state space. A general approach to reducing the control problem size is to make an approximation that neighbouring points within a small region in the state space are assigned to the same or similar control choices. In this way, the task of learning control can be managed as if consisting of two pattern recognition tasks:

- Learning the threshold locations per partition
- Learning the number of partitions

with the a goal to:

- develop an algorithm to threshold and partition.

Any partitioning methodology should maintain a fine balance between generality and learning speed regardless of the proposed learning system.

Commonly, the former type of learning receives more attention from machine learning practitioners. Many of the machine learning control algorithms partition the system state space according to human experience, prior to learning. Little effort was made to expand these algorithms to include the learning of the state space partitioning along with the learning of appropriate control action for each partition in the state space. In the following, we present our research into competitive learning of state space partitioning based on neural networks. The particular system we had in mind was the learning control system based on the 'BOXES' machine learning algorithm [15, 18]. Competitive learning is a learning paradigm which performs unsupervised learning. Competitive learning paradigm uses a layer of processing units that compete with one another, resulting in a network that can be used for pattern classification applications. It divides a set of input vectors into a number of disjoint clusters in such a way that the input vectors within each cluster are all similar to one another.

Competitive learning models developed in the early 1970s through contributions of Christoph von der Malsburg [24] and Stephen Grossberg [25]. Authors such as Shun-ichi Amari [26], Kohonen [27, 28] and Rumelhart [29], have further developed many variations of competitive learning. Recently, a number of neural networks paradigms have adopted competitive learning and inhibitory connections as their key elements. For example, the Kohonen feature map employs a competitive learning scheme along with other structures. Networks such as ART also contain a substructure that is similar to the basic competitive learning scheme.

For reasons of study in this report, we will in the following two sections give a brief description of Kohonen's work on self-organising maps and a basic idea of Grossberg's ART network. Of particular interest for this chapter is the usefulness of these networks in processing sensory data in learning control systems.

2.4.1   Kohonen feature maps

An important phenomenon that occurs in biological systems is that the neurons are placed in an orderly fashion on the surface of the brain. This placement often reflects some physical characteristics of the external stimulus being sensed [30]. It is very likely that organisations of some of the biological maps are created through a process of self-organisation. Kohonen [27, 28] developed an algorithm which produced what he called self-organising feature maps and showed how a topological mapping that occurs in the brain can be modelled by a neural network. This not only performed pattern recognition tasks using a parallel architecture, it also gave a dimensionality reduction, see Figure 2.14.

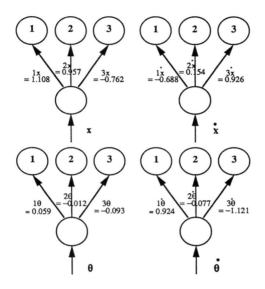

Figure 2.14 *Partitioning based on self-organising (Kohonen) mapping*

Kohonen's self-organising feature map is a two-layered network that uses an input layer along with a competitive layer of processing units. The processing units are laid out in a spatial structure (usually two-dimensional), and they are trained by unsupervised learning. Continuous-valued input vectors are presented sequentially in time without specifying the desired output. Kohonen's algorithm not only updates the weight vector of the unit that receives the greatest total input, it also updates the weight vectors of adjacent units. This encourages adjacent units to respond to similar input vectors, and this can be viewed as a way of performing gradient descent on a cost function that has two terms. The first term measures how accurately the weight vector of the most active unit represents the input vector. The second term measures the dissimilarity between the input vectors that are represented by adjacent units.

After enough input vectors have been presented, weights will specify cluster, or vector, centres that sample the input space. In this way, the point density function of the vector centres tends to approximate to the probability density function of the input vectors [27, 28]. In addition, weights will be organised such that topologically close nodes are sensitive to inputs that are physically similar.

2.4.2   Adaptive resonance theory (ART)

Human beings are able to continuously remember new things without forgetting or modifying the old things that are already stored in the brain.

Many existing artificial neural network models have failed to solve this problem. Too often, learning new patterns erases or modifies previous training. When a fully trained network must learn a new training pattern that is not within the previous training set, the connection weights may be disrupted so badly that the learning of previous training patterns is washed away. In this case, complete retraining is required. This retraining has to be taken under the previous training set with the new pattern included in that set.

In a real-world case, the network will be exposed to a constantly changing environment. It may never see the same training vector twice. Under such circumstances, a backpropagation network will often learn nothing; it will continuously modify its weights to no avail, never arriving at satisfactory settings.

The ART network and algorithm based on the Adaptive Resonance Theory was developed by Carpenter and Grossberg [31] to solve the stability problem. The ART network is useful for pattern classification applications. Incoming patterns are classified by the units that they activate in the recognition layer of the ART network. However, if the activated unit suggests a classification that does not match the current input vector within a specified tolerance, a new category is created by adding a unit that represents the classification of that input vector, otherwise, the weights of the activated unit are adjusted to make it more sensitive to the input vector. This means that new patterns from the environment can create additional classification categories, but a new input pattern cannot cause an existing memory to change unless the two match closely. In this way, new patterns can be learned without modifying old patterns that have been learned previously.

As shown in Figure 2.15, the basic ART network consists of two levels: F_1 is the input signal level, F_2 is the classification level where the competitive learning paradigm is employed.

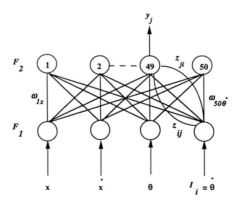

Figure 2.15 *The ART neural network*

2.5   Hypermedia front-ends

In the drive toward the development of intelligent machines we are constantly reminded that it is not always possible to completely automate, represent mathematically or transfer adequately, the inherent skills of the human into the computer such that these can be replicated by a robot performing some desired task when dealing with uncertainties. The reasons for this are well documented, principal among them is the requirement that robots are still programmed off-line at too primitive a level to allow them to integrate the sensory data that would allow reasoning under uncertainty. Knowing that little could be done to overcome the physical characteristics of the robot arms to make a dramatic improvement in performance, robot practitioners have concentrated on developing advanced robotics through concentrating on intelligence.

To minimise the uncertainty factor, and make a move towards increased intelligence, in any given robotics task, robot practitioners found it necessary to provide the robot with transducers to provide more sensory data. These sensors commonly replicate the senses of the human, thus over the years much attention has been given to the senses such as vision, taction, sound, and olfactory sensing. Although in the past we contributed to the sensory integration debate, by integrating vision and taction data for an object recognition task, this was done within an advanced robotics research testbed. A significant technical element of this testbed was the development of expert systems' *rule induction* techniques. Rule induction is one approach that has been used in the evolution of machine learning algorithms; briefly, it is a method of generalising from specific examples.

Being associated with such a comprehensive robotics research environment there was one remaining major technical element that we were so reliant upon, the control architecture. The testbed architecture had been developed as part of a laboratory-wide networked system, i.e. part of an open system architecture. As a direct consequence of using an open system architecture there was a move to develop object-oriented software, re-usable and reconfigurable software, and in particular a move was made to hypermedia technology.

Hypermedia arose from the electronic transformation of all forms of media, (multimedia) in digital form, to allow every type of message, such as that from the variety of robot transducers described above (vision, taction, sound and olfactory sensing) and including text into the form of another. Much of this has been achieved through the development and use of optical storage media, which allow the efficient use of memory for storing images and sound. As stated, an equally important element in the development of this technology is the treatment of texts as hypertexts and this has contributed significantly to the overall evolution of hypermedia.

Here we have adopted hypermedia as a platform upon which we conduct robot task planning for object grasp. Such tasks are achieved not through on-line joint-level, or manipulator-level programming, via a teach pendant, or through an iterative process of presenting the robot with numerical values that represent the 6 degrees of freedom, three Cartesian and three orientation vectors (X, Y, Z, O, A, T), within its working space. This programming of the robot is commonly done from a computer terminal with successful completion of the desired task being verified through visual feedback and an assessment by the human operator. We believed that the hypermedia domain presented us with a more comprehensive means of interacting with the robot, and facilitated a more effective means of off-line robot programming. Many tasks that are undertaken by robots today are not of the independent, intelligent and autonomous type that is the ultimate goal of robotics, and most have some degree of teleoperation and human-in-the-loop control, see Figure 2.15.

Robot World - Images & Simulations

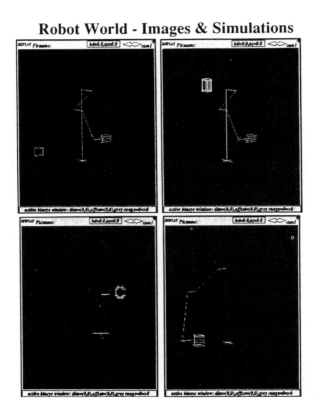

Figure 2.15 *Human-in-the-loop task planning*

We believed that a contribution could be made to the advancement of robotics through developing more flexible, user-friendly, highly interactive, hypermedia based user interfaces. Initially, the hypermedia based interface was developed for use with the vision control system of the advanced robotics research testbed, Freddy3 [2], which used to reside at the Turing Institute Ltd. Further, the interface was based on the hypermedia user interface building tool that was produced at the Institute. This software product possessed all the associative and expressive powers of hypermedia, allowing the work to be compared with other types of interfaces, and hypermedia type interfaces for similar robotics workcells.

2.6   References

1.   BAIRD, S. and LURIE, M., Precise robot assembly using vision in the hand, In: Robot Sensor, Ed: A. Pugh, Vol.1, pp 85-94, Pub: IFS, Bedford, England, 1984.
2.   Freddy3 User Manual, The Turing Institute Ltd., Glasgow, UK, 1986.
3.   SHEPHERD, B. A., High level programming of vision guided robot assembly tasks, Ph.D. Thesis, University of Strathclyde, Glasgow, UK, 1992.
4.   BRADY, J. M., Artificial intelligence and robotics, In: Artificial Intelligence, 26(1), 1985, pp 79-121.
5.   SARIDIS, G. N., On the revised theory of intelligent machines, CIRSSE Report 58, ECSE Department, Rensselaer Polytechnic Institute, Troy, NY, 12180-3590, USA, 1990.
6.   ALBUS, J. S., A theory of intelligent systems, Proceedings of the 5th IEEE International Symposium on Intelligent Control, Eds: A. Meystel, J. Herath, and S. Gray, 5-7 September, Philadelphia, PA, USA, pp 866-875, 1990.
7.   ISIK, C. and MEYSTEL, A., Pilot level of a hierarchical controller for an unmanned mobile robot, IEEE Journal of Robotics and Automation, Vol. 4, June 1988, pp 241-255.
8.   SIMPSON, J. A., HOCKEN, R. J., and ALBUS, J. S., The Automated Manufacturing Research Facility at the National Bureau of Standards, Journal of Manufacturing Systems, Vol.1, No.1, 1983.
9.   ALBUS, J. S., System description and design architecture for multiple autonomous undersea vehicles, NIST Report 1251, Gaithersburg, MD, September, 1988.
10.   ALTY, J. L. and COOMBS, M. J., Expert Systems: Concepts and Examples, NCC Publications, Manchester, UK, 1984.
11.   BARR, A. and FEIGENBAUM, E. A. (Eds), The Handbook of Artificial Intelligence, Volume 1 - 4, William Kaufmann, Los Altos, CA, 1981.
12.   FIKES, R. and KEHLER, The Role of Frame-Based Representation in Reasoning", Principles of Expert Systems, IEEE Press, New York, pp. 94-110, 1988.
13.   MICHALSKI, R. S., CARBONELL, J. G. and MITCHELL, T. M., Machine Learning: An Artificial Intelligence Approach, Tioga Publishing Company, Palo Alto, CA, 1983.
14.   QUINLAN, J. R. (Ed.), Applications of Expert Systems, Vols 1 and 2, Turing Institute Press/Addison-Wesley Publishing Company, Sydney, Australia.
15.   MICHIE , D.M. and CHAMBERS, R.A. BOXES: an experiment in adaptive control, Machine Intelligence 2, ed: E. Dale and D. M. Michie, Pub: Oliver and Boyd: Edinburgh University Press, Edinburgh, Scotland, UK, 1968.
16.   GRANT, E. and BING ZHANg, Neural network based rienforcement learning, Proceedings of the 31st IEEE Conference on Decision and Control, 16-18 December, Tucson, AZ, USA, 1992, pp 856-861.
17.   ZHANG, B. and GRANT, E. Neural network based competitive learning for control, 4th IEEE International Conference on Tools with Artificial Intelligence, Arlington, VA, 10-13 November, 1992
18.   GRANT, E. Learning in control, Introduction to Intelligent and Autonomous Control, Eds: K. Passino and P. Antsaklis, Kluwer Academic Publishers, Norwell, MA, 02061, USA, 1992, 31 pp.

19. MENDEL, J.M. and FU, K.S. eds: Adaption Learning and Pattern Recognition Systems: Theory and Applications, Academic Press, 1970.

20. ANDERSON, C.W., Strategy learning with multilayer connectionist representations, Proceedings of the 4th International Workshop on Machine Learning, Irvine, CA, pp 103-114, June 1987.

22. ANDERSON, C.W., Learning to control an inverted pendulum using neural networks, IEEE Control Magazine, vol. 9, no.3, pp 31-37, April 1989.

23. SAMUEL, A.L. Some studies in machine learning using the game of checkers, IBM Journal of Research and Development, vol. 3, pp 210-229, 1959.

24. MALSBURG C., Self-organisation of orientation sensitive cells in the straite cortex, Kybernetik, Vol. 14, pp 85-100, 1973.

25. GROSSBERG S., Competitive learning: From interactive activation to adaptive resonance, Cognitive Science, Vol. 11, pp 23-63, 1987.

26. AMARI S. and TAKEUCHI A., Mathematical theory on the formation of category of detecting nerve cells, Biological Cybernetics, Vol. 29, pp 127-136, 1978.

27. KOHONEN T., Self-organisation and associative memory, Series in Information Sciences, Vol. 8, Springer Verlag, Berlin, 1984.

28. KOHONEN T., Self-organised formation of topologically correct feature maps, Biological Cybernetics, Vol. 43, pp 59-69, 1982.

29. RUMELHART E. and ZIPSER D., Feature discovery by competition learning, Parallel Distributed Processing: Experiments in the Microstructure of Cognition, Vol, 1, MIT Press, Cambridge, Mass, USA, 1986.

30. BULLOCK H. et.al., Introduction to nerveous systems, W.H. Freeman and Company, San Francisco, USA, 1977.

31. CARPENTER G.A. and GROSSBERG S., The ART of adaptive pattern recognition by a self-organising neural network, IEEE Computer, March, pp 77-88, 1988.

Chapter 3

Advanced control systems for robotic arms

C.L.Boddy, D.J.F.Hopper and J.D.Taylor

3.1   Introduction

This chapter describes some of the research work that has been undertaken at the National Advanced Robotics Research Centre (NARRC), situated on the campus of the University of Salford, in the last few years. It outlines some of the broad problems in controlling robotic arms before proposing solutions for some particular examples. The development of these solutions has been performed to provide a generic framework for addressing such problems.

The robot control problem can be divided into two main areas: kinematic control, the co-ordination of the links of the kinematic chain to produce desired motions of the robot, and dynamic (or kinetic) control, driving the actuators of the mechanism to follow the commanded positions/velocities.

For advanced robotic arms to operate with some increased degree of autonomy within non-structured, partially unknown, or hostile environments, systems will be required to enable the robots to identify potential problems in their environments and implement limited responses in real time. This requires that the kinematic control of the robot arm is oriented towards multiple task performance in real time in order to incorporate sensor information. In particular this has involved work in two areas, 'reactive' collision avoidance for robot arms, and the control of redundant robot arms.

The modern industrial robot is typically controlled, at the lowest level, by decoupled Proportional, Integral and Derivative (PID) controllers: a standard industrial control technique. The application of such simple linear controllers is made possible by using high gear ratios between the drive motors and links, decoupling the non-linear payload from the actuator, and a relatively slow motion range. The current trend in industrial applications is towards lighter and faster robots where the applicability of these techniques is limited, and there is therefore a requirement for more sophisticated control algorithms.

3.2   Kinematic and dynamic control

Inverse kinematics comprise the computation needed to find joint angles from a given Cartesian position and orientation of the end-effector. This computation is fundamental in the control of robot arms. It is, in general, a non-linear algebraic computation which has been shown for the general case of a 6 degree-of-freedom (dof) arm to require the solution of a 16th order polynomial equation. Most industrial robot arms are designed so that the principal of wrist partitioning can be used [1], that is where the three wrist joint axes intersect at a point, which reduces the problem to that of a second order solution. It should be noted that even for many 6 dof robot arms, multiple solutions exist relating to different configurations (e.g. elbow up/down for a PUMA) but these are of a finite number (e.g. 8 for a PUMA) and in general the robot arm cannot move between configurations unless passing through a singularity.

Trajectory planning, including inverse kinematics, is performed within the tactical layer of the ARRL advanced robot architecture. The ARRL robot control hierarchy, shown in Figure 3.1, consists of three distinct levels:

a) The Strategic level - which fulfils the role of the 'decision maker', by intelligently formulating the best approach for achieving the requested task and then subdividing this requirement into a number of elementary operations that best utilise the available resources;

b) The Tactical level - which develops the tactics by which individual elementary operations issued by the strategic level are precisely translated into detailed instructions to the lower control level;

c) The Executive level - which is expected to accurately and precisely execute the task delegated to this level by the tactical level.

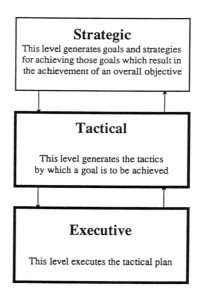

Figure 3.1 *ARRL control hierarchy*

The lower two layers of the ARRL control hierarchy are characterised by the real time nature of their operation and inherently have to deal with the physics of the robotic system. The function of the tactical layer is to take as input the gross trajectory specifications from the strategic layer and generate in real time intermediate joint positions that will satisfy the position and dynamic requirements. The joint positions generated must take into account the physical capabilities of the robot arm so that they are presented to the executive layer as set points which are achievable within each control cycle.

The ARRL robot arm tactical layer was initially developed for the RD3 end-effector collision avoidance demonstrator shown in Figure 3.2 [2]. Subsequent to the completion of this project further work has been carried out to increase the functionality and robustness of operation of the tactical layer.

Figure 3.2 *ARRL's reactive collision avoidance demonstrator*

At the dynamic level of control, represented in the ARRL control hierarchy by the executive layer, ARRL has also been pursuing a work programme on the control of a SCARA robot arm. The aim of this work is to implement a high performance non-linear multivariable controller and assess its performance against the single-loop controllers currently used on robot arms. Such controllers should lead to all-round performance improvements from any industrial arm; for example faster, more efficient motions coupled with better tracking accuracy. In addition, system identification techniques could lead to payload mass and inertia identification allowing further performance improvements.

3.3   Kinematic control design

In machine design a mechanism is synthesised that functions with the minimum complexity of the structure. Most industrial robots have been designed based on this common sense rule. It is generally claimed that a six dof non-redundant robot arm is a general purpose device since it can 'freely' position and orient an object in the Cartesian workspace. Welding robots, which usually do not need rotation about the welding torch, have five actuators. Kinematic redundancy occurs when a robot arm possesses more degrees of freedom than the minimum number required to execute a given task, and it is fundamental to understand why it becomes necessary to introduce kinematic redundancy into a robot with its commensurate increase in mechanical and control complexity. Although nominally a six degree-of-freedom non-redundant robot arm can 'freely' position and orient an object in the Cartesian workspace, its geometry, in fact, has a number of kinematic flaws such as limited joint ranges, singularities, and workspace obstacles which prevent the arm from doing so. It is therefore desirable for a true general purpose robot arm which is not constrained to structured, well designed environments, to dispose of additional degrees of freedom to overcome the above limitations.

Dexterity in a robotic arm implies the mechanical ability to carry out a variety of non-trivial movement tasks. If a robot arm is to have high dexterity it should have seven or more degrees of freedom rather than six or less. The seven degree-of-freedom human arm constitutes an excellent model of a dextrous redundant arm.

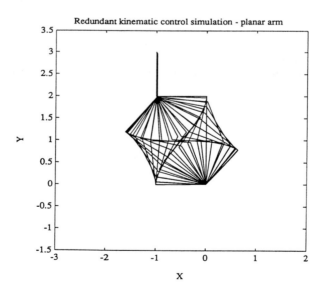

Figure 3.3 *Self-motion of a 4-link planar arm*

The fundamental problem of controlling kinematically redundant robot arms is that the inverse kinematic problem admits infinite solutions. In other words the problem is under-constrained since there are more joint angles to assign than there are Cartesian end-effector co-ordinates. This implies that, for a given constant location of the arm end-effector, it is possible to induce a self-motion of the structure i.e. a motion of the joints without effecting the end-effector position.

This is shown in Figure 3.3 for a 4-link planar arm where the end-effector maintains constant position and orientation whilst the arm links sweep out the range of available internal motion. It is thus clear that if a motion task trajectory is commanded to the end-effector, it is possible in principle to continuously modify the joint motion in such a manner that not only the end-effector task is executed correctly, but also an additional constraint task is accomplished e.g. collision avoidance.

The advantages of such kinematic configurations are clearly many-fold. The available workspace of the robot arm is significantly increased, since on conventional arms much of the internal workspace is physically unreachable due to mechanical joint limits. Thus the available dexterity at any position in the workspace is also increased, i.e. the range of continuous movement available at the end-effector. Over a work volume this may imply, for example, that one dextrous robot arm may be capable of replacing two or three conventional robot arms in a work-cell. A property related to this is that a redundant robot arm can reach spaces with tricky access yet retain full 6 dof motion at the end-effector. Lastly, such robot arms have the ability to perform sub-tasks such as resting the robot arm against solid surfaces to reduce end-effector compliance during force operations, collision avoidance whilst maintaining desired end-effector motions, joint torque optimisation, or avoidance of singularities (a major problem with conventional robot arms).

It is therefore, perhaps, surprising that more such robot arms are not commercially available given the broad range of tasks that they can perform. The fundamental problem with controlling the motions of such robot arms lies in the infinite number of solutions available for the configuration of the arm for any given desired position and orientation of the end-effector. In other words, the crucial problem of translating desired 'world' trajectories into 'joint' trajectories is under-constrained. But, of course, it is this very lack of constraints that gives kinematically redundant robot arms their desirable properties.

The position and orientation of the end-effector with respect to a fixed reference frame is given by the vector variable x and is related to the joint angles by the forward kinematic model

$$x = f(\vartheta) \tag{3.1}$$

and the differential kinematic model

$$\dot{x} = J\dot{\vartheta} \tag{3.2}$$

where $\vartheta \in R^n$ is the vector of joint variables, $f(\vartheta)$ represents the forward kinematic vector function and $J \in R^{m \times n}$ is the configuration dependent Jacobian matrix of the end-effector, formally given by $\delta f / \delta \vartheta$. The dot denotes time derivative. An important point of note is the occurrence of kinematic singularities when Rank(J)<m when the

robot arm loses its ability to move along or rotate about some direction of the task space. The onset of this condition is often detected by the manipulability measure of Yoshikawa [3].

The desired motion of the end-effector is denoted by the reference trajectory vector $x_d(t)$ and constitutes the task to be accomplished by the robot arm. The kinematic control problem can be formulated as to find a joint space trajectory $\vartheta(t)$ such that $f(\vartheta(t)) = x_d(t)$ is satisfied.

It is clear that in the case of a redundant robot arm $(m<n)$, with respect to a given task, the inverse kinematic problem admits infinite solutions. This suggests that redundancy can be conveniently exploited to meet additional constraints.

In a conventional robot arm control system the kinematic control problem is satisfied by finding the analytic function

$$\vartheta = f^{-1}(x) \tag{3.3}$$

and solving for ϑ at each cycle of the kinematic controller. However, except for kinematically simple robot arms (e.g. wrist partitioned robots) there is no guarantee that such a function exists, or indeed is a practicable solution. For example, the general solution for a 6 dof arm requires the solution of a sixteenth order polynomial. In the control of kinematically redundant robot arms with $n_{dof}>6$ and also the use of non-simple kinematic design to improve the available workspace (e.g. non-spherical wrist) the employment of analytic solutions can be seen to be not practicable. Therefore, the use of the linear velocity relationship is indicated as an alternative.

Probably the most well known solution is that of 'pseudo-inverse' control based on the equation

$$\underline{\dot{\vartheta}} = J^+ \underline{\dot{x}} \tag{3.4}$$

where $\dot{\vartheta}$ is the vector of required joint velocities, $\dot{x}$ is the commanded Cartesian velocity vector, and $J^+ = J^T (JJ^T)^{-1}$ is the Moore-Penrose pseudo-inverse. However, this solution has associated problems including the fact that it does not avoid, or control, singularities. A 'damped least squares' inverse based on a modified solution has been proposed which circumvents this problem. This solution can also be readily generalised to

$$\underline{\dot{\vartheta}} = J^+ \underline{\dot{x}} + (I - J^+ J)\underline{\dot{w}} \tag{3.5}$$

where $\underline{w}$ represents a desired joint sub-task which is projected by (I-J+J) onto the null space of the robot arm Jacobian, J. This formulation is very revealing since it clearly separates the two available forms of motion, the end-effector task motion, and the mechanism self-motion available via the null space. The null space of the Jacobian represents the space of instantaneous joint motions which will result in zero motion of the end-effector and is a crucial concept in the control of redundant structures.

3.3.1 Whole arm collision avoidance control system

By incorporating proximity sensors over large areas on such robot arms, the methods used for ARRL's end-effector collision avoidance control system can be used to generate elemental repulsions [2] and thus suitable $\underline{w}$'s may be constructed to provide robot arm collision avoidance. An example of this is shown in Figure 3.4, based on the above equations, for a planar robot arm following a desired trajectory with collision avoidance as a sub-task.

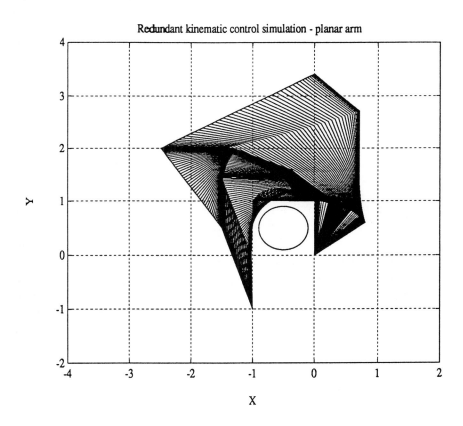

Figure 3.4 *Pseudo-inverse collision avoidance of a 4-link planar arm*

Further work in collision avoidance work has enabled the NARRC to develop a whole-arm collision avoidance system for a 7 dof Robotics Research Corporation (RRC) arm shown in Figure 3.5. The control system described below [4] is currently being implemented on the actual arm - but is based on extensive simulation work in the animation package IGRIP.

Figure 3.5 *RRC K-1607HP robotic arm at NARRC robotics laboratory*

The control system differs from the pseudo-inverse formulation given above and is based on the configuration control technique [5], with the improvements described in Seraji and Colbaugh [6] for the handling of kinematic and algorithmic singularities. The configuration control approach allows more flexibility in the handling and weighting of sub-tasks, but in fact is closely associated with the pseudo-inverse approach.

A solution to equation (3.2) is required. The configuration control approach requires the selection of *r=n-m* (*r* being the degree of redundancy) explicit additional task variables *z*, where

$$z = g(\vartheta) \tag{3.6}$$

and $g(\vartheta)$ is an additional kinematic function(s) defined in joint or task space. In configuration control, the user also specifies the desired variations of the additional kinematic functions denoted by the reference vector z_d. The end-effector position and orientation, and the kinematic function *z* are combined to form the configuration vector

$$X(t) = \begin{bmatrix} x(t) \\ z(t) \end{bmatrix} = \begin{bmatrix} f(\vartheta) \\ g(\vartheta) \end{bmatrix} = h(\vartheta) \tag{3.7}$$

The configuration variables $X_1 .. X_n$ define a set of task related generalised co-ordinates for the arm and (3.7) represents the augmented forward kinematic model for the arm. The associated differential relationship is given by

$$\dot{X} = \begin{bmatrix} J_e(\vartheta) \\ J_z(\vartheta) \end{bmatrix} \dot{\vartheta}(t) = J_h(\vartheta)\dot{\vartheta}(t) \tag{3.8}$$

where J_e is the configuration-dependent Jacobian matrix of the end-effector and J_z is the Jacobian matrix associated with the additional kinematic function. The solution to (3.8) may then be found by matrix inversion. This solution can be made robust to both kinematic and algorithmic singularities by the use of a damped inverse [6], and also introducing a relative weighting matrix W between tasks,

$$\dot{\vartheta}9t) = \left[J_h^T W J_h + \lambda I \right]^{-1} J_h W \dot{X}_d \tag{3.9}$$

The damping factor, λ, can be scheduled to zero away from singularities, and in this system is varied with respect to a measure of closeness to singularity, the manipulability index [3].

A typical complete kinematic control equation with one active collision avoidance task and one active joint limit avoidance task would be:

$$\dot{\theta} = (J_e^T w_e J_e + J_c^T w_c J_c + J_l^T w_l J_l + \lambda I)^{-1} (J_e^T w_e (\dot{x}_d + k_e \dot{e}) + J_c^T w_c C + J_l^T w_l L) \tag{3.10}$$

Where the *e*, *c* and *l* subscripts refer to end-effector, collision avoidance, and joint limit task Jacobians and weights respectively. The joint limit avoidance task is described in Seraji and Colbaugh [6]. The generation of the collision avoidance task Jacobian and demand is described in the next section. The positional error term, *e*, is scaled and added to account for linearisation errors.

3.3.1.1 Collision avoidance task

Each sensor on the arm is considered to have two states, active and inactive, and is active when the sensed distance is lower than a threshold maxdist. When the sensor is inactive its output is essentially ignored by all ensuing procedures. When a sensor is active a repulsive action is generated unique to that sensor which is of a magnitude proportional to the reading (which is assumed to be filtered to reduce random noise) and in an opposite sense to the direction of the object. These repulsive actions, $\underline{a}_k^i$, representing the repulsion of the *i*th sensor on the *k*th link in the kinematic chain, are generated from the proximity readings $\underline{r}_k^i$ as

$$\underline{a}_k^i = \left(\begin{array}{l} (\max dist - \underline{r}_k^i \ if(\underline{r}_k^i < \max dist)) \\ 0 \ if \ (\underline{r}_k^i > \max dist) \end{array} \right) \tag{3.11}$$

This concept is illustrated in Figure 3.6.

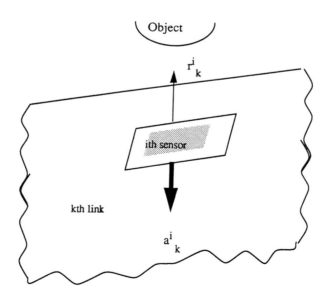

Figure 3.6 *Proximity sensor, associated reading, and repulsive action.*

A repulsion screw may then be generated, as proposed by Espiau and Boulic [7], associated with each link of the kinematic chain

$$\underline{R}_k = \begin{pmatrix} \sum \underline{a}_k^i \\ \sum \underline{x}_k^i \times \underline{a}_k^i \end{pmatrix} \tag{3.12}$$

where $\underline{x}_k^i$ is the position of sensor i in the link k co-ordinate frame, and x denotes the vector cross product. The repulsion screw consists of two parts, the first 3-vector being the summed magnitude of the repulsive actions associated with each individual sensor (effectively the component used in [2]), whilst the second 3-vector comprises the total moment of the repulsive actions about the link co-ordinate frame. The effective 'action' of the repulsion on a typical link on a robot arm is shown in Figure 3.7. In this way, a repulsion screw associated to each link can be generated independently of sensor type or number. Using some of the ideas developed in [2] some intelligence can be built in to the individual repulsive actions to account for velocity of approach. For example, there is no repulsive action if the velocity is negative and small, and a repulsive action if the velocity is small and positive.

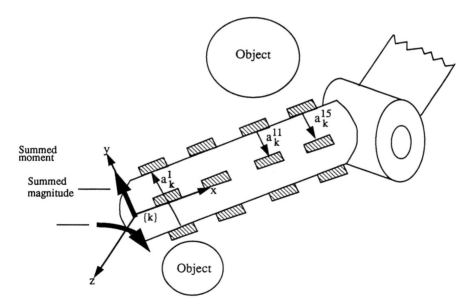

Figure 3.7 *Repulsion screw components*

The repulsion screws are used to calculate the secondary task, i.e. collision avoidance rows of the Jacobian

$$J_{c_k} = \frac{R_k^T \cdot J_k}{\|\underline{R}_k\|} \tag{3.13}$$

where $\underline{R}_k$ and J_k are the repulsion screw and link Jacobian of the kth active link.

The collision demand signal, for each active link, is calculated from a scaled norm, (3.14), of the link repulsion screw. The scaled norm is used to handle the mixed units present within the screw. The rate of change of the repulsion screw norm is calculated via a simple difference and both the norm magnitude and "velocity" are used to calculate the demand signal. The full expression for the link collision demand is given in (3.15)

$$\|\underline{R}_k\|_s = k_r \cdot \left\| \sum \underline{a}_k^i \right\| + k_m \cdot \left\| \sum \underline{x}_k^i \times \underline{a}_k^i \right\| \tag{3.14}$$

where the ks are the vector/moment scale factors

$$C_k = k_p \cdot \|\underline{R}_k\|_s + k_v \cdot (\|\underline{R}_k\|_s)_v \tag{3.15}$$

where C_k is the collision demand of the kth active link, the ks are gains and $(\|\underline{R}_k\|_s)_v$ is the rate of change of the link repulsion screw norm, which is used to vary the magnitude of the repulsion depending on the approach velocity.

3.3.1.2 Simulation results
For the purposes of evaluating the control system an IGRIP (a robotic animation package) model of an RRC K-1607 arm was generated, and has been used in various environments. The Ray_trace facility in IGRIP has allowed 'idealised' sensors to be modelled, which are only capable of simplistic sensing along a straight line, normally set to be perpendicular to the surface of the robot. The number of simulated sensors is 60, particularly concentrated around the elbow and end-effector regions. The sensor distribution around the elbow is shown in Figure 3.8, each 'tag point' or frame represents one sensor. The sensor distribution is such that there are five links which register repulsive actions.

The animation of a simulated move is shown in Figure 3.9. The end-effector trajectory has been specified such that it passes through one of the columns in the workspace, so that collision avoidance is required both for the trailing mechanism and the end-effector. The trajectory is successfully completed excluding the segment which is impassable. All of the links of the arm successfully avoid the other workspace obstacles. In Figure 3.10a the norms of the repulsions are shown for the elbow (dotted line) and last link. The peak shown for the last link can be seen to cause the end-effector to avoid the column on its path. The variation of the elbow roll angle as the trailing links avoid the other columns is shown in Figure 3.10b.

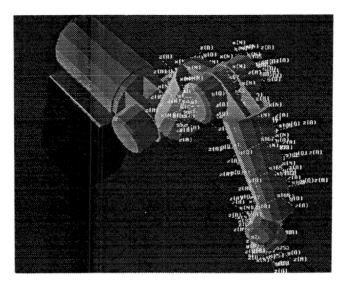

Figure 3.8 *Sensor simulation 'tag' points.*

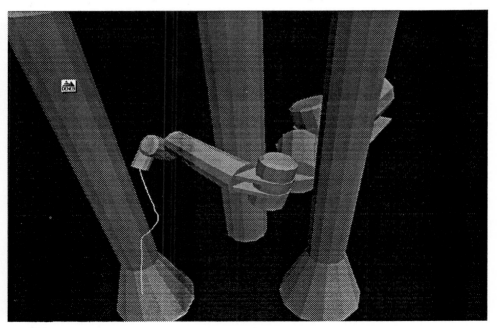

Figure 3.9 *Mode 2 collision avoidance in 3-column environment.*

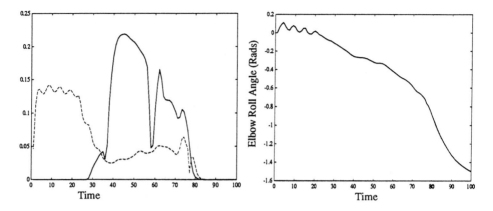

Figure 3.10a *Norms of elbow repulsions*   **Figure 3.10b** *Elbow roll angle*

3.4   Servo control design

The modern industrial robot is typically controlled by decoupled Proportional, Integral, and Derivative (PID) controllers: a standard industrial control technique [8]. The application of such simple linear controllers is made possible by using high gear ratios between the drive motors and links (effectively decoupling the time-invariant inertia of the motor from the complex, time-varying, non-linear inertia of the arm) and a relatively slow motion range (reducing the interactive coriolis and centripetal forces). The current trend in industrial applications is towards lighter and faster robots where the applicability of these techniques is limited due to the increasing dominance of the payload of unknown mass and inertia and the higher proportion of non-linear interactive forces. The use of direct drive mechanisms [9] where the effective gear ratio is 1:1 is an extreme example (e.g. ADEPT, Hirata) reflecting the full robot arm dynamics to the motor, but has advantages in its inertial configuration and avoidance of the requirement for a gear-train.

The NARRC currently has a COMAU SAR4.10 SCARA-type robot, shown in Figure 3.11, which is being developed to evaluate improved control techniques. This arm was selected for the study because, being a fast pick and place arm, it can achieve high speeds. This ensures that some non-linear effects are amplified making comparisons of controller performance easier. Furthermore, if the controllers being studied give improved high speed performance then the benefits of this may be passed on to industry; in this application 'speed equals money'.

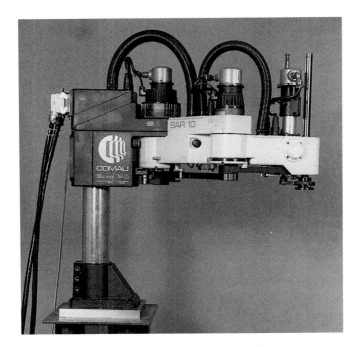

Figure 3.11 *COMAU SAR 4.10 in NARRC robotics laboratory*

Simulation also plays a strong role in this research area. The schematic diagram in Figure 3.12 shows how a generic single link module may be expanded and made to represent any of the four links of the COMAU arm. Even the third axis, which is a prismatic joint, may be represented by this single link module because a revolute to linear conversion can be employed. This single axis module is broken down into a drive module and a link module and the drive module is broken down further into a sensor module (e.g. a resolver and tacho, including noise, quantisation and compliant coupling), an actuator module (e.g. a brushed DC motor) and a speed reducer (e.g. a harmonic drive gearbox, including non-linear gearbox compliance). A stiction, viscous and coulomb friction module is also shown as part of the single link module [10]. The simulation environment includes the packages SD_Fast (a symbolic manipulation package for generating dynamic equations of motion), ACSL (a non-linear simulation package), Pro-Matlab (a linear analysis package), and custom written 'C' software for trajectory planning. Code developed in this environment will then be ported onto the real-time control system (currently under development) for implementation and evaluation.

Conventionally, an error signal is generated, from the set point trajectory and the sampled encoder or resolver readings, and simple discrete differentiation and integration are used to generate the error integral and derivative signals (some controllers still use an analogue version). The motor drive signal is then calculated as:

$$\dot{e}^p = \frac{(e_n^p - e_{n-1}^p)}{t_s} \qquad (3.16)$$

$$\int e_n^p = \int e_{n-1}^p + e_n^p \cdot t_s \qquad (3.17)$$

$$u_i = G_p \cdot e + G_i \cdot \int e + G_d \cdot \dot{e} \qquad (3.18)$$

where e is the error, the G s are the control gains, and u is the control signal to the amplifier.

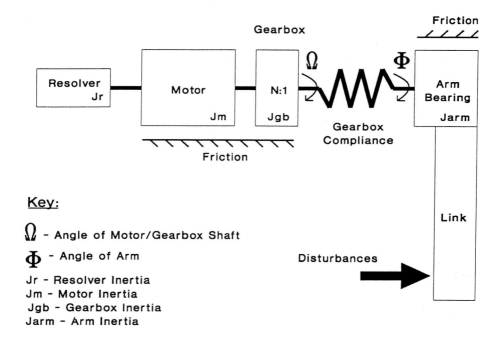

Figure 3.12 *Single link dynamic module for the Comau arm*

In order to improve the performance of the robot arm a method of either compensating for or rejecting the non-linear disturbance torques generated by the robot arm dynamics must be found. Using high gain linear controllers is one alternative, but the regions of stability can be very limiting. Due to the complex interactions of many robots the scheduling of gains could also prove very costly in terms of tuning and memory.

The dynamics of a robot arm (under the assumption of rigid links) are given by [11]:

$$\tau = M(\theta) \cdot \ddot{\theta} + C(\theta, \dot{\theta}) + G(\theta) \qquad (3.19)$$

where τ is the vector of torques at the joints, M is the inertia matrix, C is the vector of coriolis and centripetal forces, and G is the vector of gravity forces. Each actuator is a linear time-invariant second order system (third-order if electrical dynamics are included).

Since the robot arm dynamics are not prohibitively expensive to compute on modern hardware a number of schemes based around a 'model inverse' have been proposed [12]. The feed-forward compensation technique was originally proposed in the context of off-line compensation. It utilises the trajectory planner to generate the joint accelerations, velocities and positions required to execute the desired Cartesian path. These are used to calculate the required torques, allowing for link interaction, to drive the robot arm links with the required accelerations etc. Normally the sampling rate of the feedback-loop will be considerably higher than the rate of points generated by the trajectory planner, so effectively the compensation term will remain fixed for a number of samples.

The computed-torque technique is similar in that a non-linear compensation term and a linear feedback term are calculated. In this case, the non-linear term is calculated at the same rate as the feedback term and the measured encoder/resolver positions and velocities, rather than the planned values, are used along with the planned accelerations (although obviously a measured signal could be used if available).

The total torque is composed of inertia, coriolis/centripetal, and gravity components

$$\tau_{nl} = M(\theta_{act}) \cdot \ddot{\theta}_d + C(\theta_{act}, \dot{\theta}_{act}) + G(\theta_{act}) \qquad (3.20)$$

The required link torque is then transformed back to the motor side of the gearbox and compensation terms for the actuator inertia and friction are added. A linear feedback torque is also generated from the position and velocity errors.

The linear feedback torque may be decoupled by pre-multiplying by the actual link inertia matrix

$$\tau_{pv} = M \cdot (G_p \cdot (\theta_d - \theta_{act}) + G_v \cdot (\dot{\theta}_d - \dot{\theta}_{act})) \qquad (3.21)$$

where τ_{nl} is the non-linear computed torque term, τ_{pv} is the linear feedback term, M_{li} is the link inertia matrix, G_p is the proportional (or positional) gain, G_v is the velocity gain, θ_d is the desired position/velocity and θ_{act} is the measured position/velocity. Note in this case a linear feedback controller is used which uses both position and velocity reference trajectories generated by the trajectory planner and error signals for both these quantities are generated.

The performance of the conventional PID control scheme and the computed torque control schemes, controlling the simulation model described above, are shown

in Figures 3.13 and 3.14, respectively. The plots show tracking errors in Cartesian co-
ordinates. The behaviour of the linear controller is as expected, with maximum
deviations occurring mid-motion when the dynamic interactions are at a peak. The
tracking performance of the computed torque controller shows a marked order of
magnitude improvement.

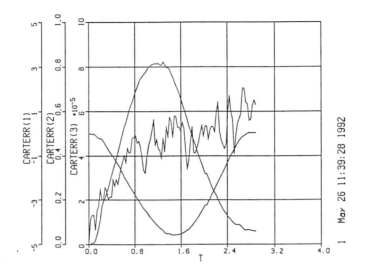

Figure 3.13 *PID controller tracking error*

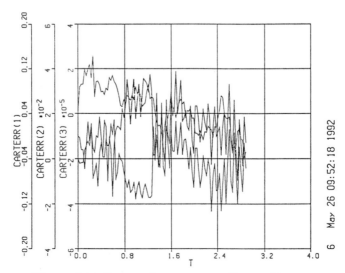

Figure 3.14 *Computed torque controller tracking errors*

3.5 Conclusions

The design and control of robot arms is a process that has traditionally been seen to be sequential in nature. Some of the issues brought forth in this document hopefully demonstrate that the whole process is integral, and one in which such issues cannot be easily separated. Some analogies with the aircraft industry can be drawn, where even today there are those that insist on designing aircraft without any regard for its subsequent control, even in the face of the success of flight unstable aircraft. It is hoped that the concept of a 'control configured robot' with its concomitant benefit to the ability of robotic arms will eventually be adopted by the manufacturers. Areas, for example, that must be considered alongside the control issues are kinematic design for dexterity (possibly task related), mechanical design in order to reduce robot mass/payload mass ratios whilst maintaining rigidity, and the distribution of appropriate sensors in order to allow a full range of capabilities that should be expected from a manipulative device.

3.6 References

1. PIEPER, D., The kinematics of manipulators under computer control PhD Thesis, Stanford University, 1968.
2. BODDY C L, Implementation of a real-time trajectory planner incorporating end-effector collision avoidance for a manipulator arm, 2nd Int. Workshop on Robot Kinematics, Linz, Austria, 1990.
3. YOSHIKAWA, T. Analysis and Control of Robot Manipulators with Redundancy, Robotics Research eds. M. BRADY and R. PAUL, pp.735-747, Cambridge, MA, MIT Press, 1984.
4. BODDY C L and TAYLOR J D, Whole-Arm Reactive Collision Avoidance Control of Kinematically Redundant Manipulators, IEEE Robotics and Automation, Atlanta, May 1993.
5. SERAJI, H Configuration control of redundant manipulators: theory and implementation, IEEE Trans. Robotics and Automation, Vol. 5, No. 4, 1989.
6. SERAJI, H. and COLBAUGH, R. Improved configuration control for redundant robots, Journal of Robotic Systems, 7(6), 1990.
7. ESPIAU, B, and BOULIC R, Collision Avoidance for Redundant Robots with Proximity Sensors International Symposium on Robotics Research, 1986, MIT Press.
8. LUH J., Conventional Controller Design for Industrial Robots - A Tutorial, IEEE Trans. , 1983, SMC-13, (3), pp.298-316.
9. AN C.H., ATKESON C.G. and HOLLERBACH J.M., Model-Based Control of a Robot Manipulator, MIT Press, 1988.
10. MAGNANI G. et al, Modelling and Simulation of an Industrial Robot, IEEE Robotics and Automation Conf., San Francisco, 1988.
11. CRAIG J.J., An Introduction to Robotics - Mechanics and Control, Addison-Wesley, Reading, Mass. , 1988.
12. AN C.H., ATKESON C.G., GRIFFITHS J.D. and HOLLERBACH J.M.,Experimental Evalution of Feedforward and Computed Torque Control, IEEE Robotics and Automation Conference, 1987.

Chapter 4

Intelligent gripping systems

J.S. Dai, D.R. Kerr and D.J. Sanger

4.1 Introduction

In the past ten years, robotic grasping and restraint have evolved into an important field of robotics research, which is due partly to the evolution of industrial automation towards flexible automation, and partly to the progress in the study of fundamentals. Attention has been given to the mechanics of grasps [e.g. 1 to 5], and to the grasping of a broader class of objects, both in precise assembly and in hazardous environments. This has pointed out the need for us to determine the graspability of objects, so as to plan a stable and optimal grasp, and to impart fine motion and force control. The requirement is also heavily dependent on the need for a versatile grasping system with integration between sensors and grippers, together with appropriate software.

Gripping systems and mechanical hands [6] are at present only weakly related to the human hand, which raises the issue of the study of anthropomorphic hands. Anthropomorphic hands make use of human knowledge, relying upon a large database for gripper models and sensors for feedback. However, the use of anthropomorphic hands is restricted in many specialist areas, and they lack general applicability.

Further, human hands focus attention on the issue of sensors in the grippers. Although vision systems [7] have been widely used, the successful execution of grasps still needs the assistance of tactile sensors, which not only sense the contact forces but also sense the geometry of an object, to establish feedback to plan and control the grasp. In addition, high expense and some special environments also hinder the use of vision systems.

Meanwhile, although much work has gone into grasping research, the property of elasticity of the contacts has rarely been used. Thus the real understanding of a grasp, and its force control have been hindered; in spite of grasp mode paradigms [8] and exhaustive heuristic methods [9] being used, many published analyses of grasps have not been towards a general applicability. Thus, a good understanding of grasps including their planning and optimisation are needed in building gripping systems. This is now achieved at the University of Salford.

4.2 Overview of the Salford theories

A new system of theories has been established at Salford to analyse and synthesise the grasp, with the application of screw algebra [10], convex algebra [11] and algebraic geometry, and a set of new approaches has been created to facilitate the use of the theories. Thus a known grasp can be analysed, a grasp of a known and unknown objects can be synthesised and planned.

4.2.1 Theories on restraint and grasping

The minimum number of frictionless point contacts to complete restraint is shown in the following table, which is needed to perform either translational restraint or complete restraint.

Table 4.1   *Minimum number of contacts in restraint*

	Translational	Complete
Planar	3	4
Spatial	4	6

The restraint equation is thus formed to complete the restraint in matrix form

$$\sum_{i=1}^{n+1} f_i \$_i = w \qquad\qquad (4.1)$$

where f_i is the magnitude of ith contact force, $\$_i$ is the screw of ith contact force, w is the external force. By incorporating screw theory, a set of theories [12] of analysing and synthesising the grasp has been proposed, which describes the state of instantaneous kinematic restraint of a rigid body in multiple point contact with its surroundings, and is used for the synthesis of any required state of kinematic restraint or freedom. Based on a known set of contacts, their positions and screws, a further contact can be established to add to the existing set, whose line vector is reciprocal to all the desired freedoms.

As a necessary condition to form restraint either in the sense of partial restraint or complete restraint, the above restraint equation is overconstrained with more contacts than equations. In consideration of frictional restraint, equation (4.1) is in the form

$$\sum_{i=1}^{vn} f_i \$_i = w \qquad\qquad (4.2)$$

where n is the number of contact points, and v is 2 or 3 in planar or spatial cases respectively. A further equation of geometric compatibility [13] of contact points has to be introduced in the form

$$u = [J]^T [\Delta] D \qquad (4.3)$$

where [J] is a Jacobian matrix formed by contact screws $\$_i$, $i=1,...,vn$, **u** is the displacement along vn contact screws, $[\Delta]$ is the matrix operator [14] to exchange both parts of **D**, and **D** is the general twisting displacement of an object. The property of elasticity of fingertips on the contact points has to be introduced in the form [15]

$$f = [K] (\delta - u) \qquad (4.4)$$

where δ is the displacement caused by applying preload on one or more contacts. The three equations, of restraint, of compatibility and of elasticity complete the mathematical model of restraint and grasp, giving a basis for further analysing and synthesising restraint and grasp.

4.2.2   Restraint mapping and screw image space

The analysis and synthesis of grasp is inevitably associated with the geometry of the object. Convex algebra has thus been introduced to deal with the analysis of grasp [16], and the boundless frictionless grasp and stable grasp in the sense of form closure have been described [17]. Restraint mapping has further been proposed [18] as the extension of screw image space. A grasp can thus be mapped into screw image space as a simplex in the form

$$conv(G) = \{\$_1, \$_2, \cdots, \$_{n+1}\} \qquad (4.5)$$

A valid grasp consists of a simplex which contains the origin of the image space. Thus graspability has been defined with a number of specific criteria. Restraint mapping can either map the grasp into the screw image space or the surface of the object into the image space. This facilitates the visualisation of the analysis and synthesis of restraint and grasp, bringing together the screw algebra, convex algebra and algebraic geometry. The screw image space so established [19] reconciles the six-dimensional space and five-dimensional projective space, with its relevant physical Euclidean three-dimensional space and two-dimensional space. A system of new theories has thus been proposed for the application of screw image space. A systematic approach has then been established to analyse any existing grasp, to plan a grasp on an arbitrary object and further to optimise such a grasp. The procedure of this approach can be summarised as

i)   mapping a grasp or an object into the screw image space;

ii)   checking graspability by ensuring the image mapping contains the origin;

iii) planning a grasp on an object with the strategy proposed in the screw image space;

iv)   optimising a grasp in the screw image space to let the centroid of the simplex coincide with the origin.

The optimal grasp thus obtained has isotropic resistance to any arbitrary external disturbance. The augmentation approach [20,21] augments the Jacobian matrix, incorporating all three sets of equations in the solving of the grasp, giving an augmented solution of the restraint equations (4.1) and (4.2). This gives a full understanding of the significance of the effect of the elastic compatibility in solving the redundant grasp [22,23].

       The screw image space and augmentation approach have established the relationship between algebra and associated geometry in a grasp, facilitating the synthesis and prediction of a grasp.

4.3   Need and provision of fingertip sensor system

The implications of the above results suggest that for a grasp to be analysed it is necessary to have a knowledge of the positions of the contact points, and the directions of the normals to the surface of the object at those contact points. If friction is involved, then it is also necessary to know the elastic properties of the contacts or fingers, and also the coefficients of limiting friction at the contacts.

       It is further assumed that the contacts behave as point contacts. To verify these theories, a set of novel fingertip sensors [24] has been developed, based upon the previous work [25] showing how a Stewart platform arrangement could be used as the basis of design of a 6-component force transducer.

       The new fingertip sensor system is capable of allowing the detection of position of contact point relative to the known position and orientation of the supporting frame or fingertip to a satisfactory level of accuracy, typically to 0.5 mm in any co-ordinate direction. The direction of the common normal between object and fingertip can similarly be assessed to a very high level of accuracy, typically to within 1°. The fingertip sensors can be used to discriminate between normal and tangential components of contact force so that the ratio between these components can be assessed, to an accuracy sufficient for most purposes of grasping. Given the levels of uncertainty normally associated with the phenomenon of surface friction, and provided that relative motion can be assured in the tangent direction on the object's surface, the coefficient of limiting friction can also be estimated. Typical errors are less than 1.0%

in the use of the fingertip sensor systems. These were verified using random but known positions of fingertip and the surface of objects.

4.4   Computer software package implementation

To implement the new theories and approaches, and to effectively use the information from the fingertip system, computer software has been developed to facilitate the analysis and synthesis of grasps both in the analytical and experimental work. The software written in C++ consists of a user-oriented package, two libraries of linear algebra and screw algebra, and a set of independent programmes. It is capable of

i)   interrogating contact information, normals, position vectors etc. from the fingertip sensor system;

ii)   giving detailed force distribution from the detection;

iii)   assessing a grasp and giving the diagnosis of the grasp;

iv)   planning a grasp on a set of contact information and further optimising the grasp;

v)   analysing a grasp with different properties such as stiffness;

vi)   predicting preload and force distribution corresponding to any given external force.

The software package is composed of multiple user layers and paths, incorporated in a window interface with pop-up menus, hot keys, help menus and dialogue boxes. Figure 4.1 gives the flowchart of the RASP facility (Restraint Analysis and Synthesis Package).

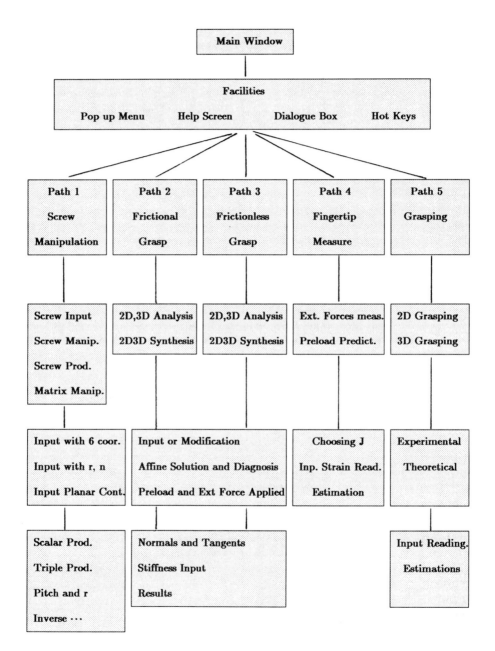

Figure 4.1 *Flowchart of RASP package*

4.5 Overview of the implementation of the gripping system

The implementation of the gripping system consists of three parts: detection of the object with the novel fingertip sensor system; use of the software package which incorporates a new system of methodologies and algorithms; execution of the grasp predicted from the software package. Figure 4.2 gives the flowchart of the procedure in the experiment.

Figure 4.2 *Flowchart of the experiment procedure*

A typical case can be seen from one of the most demanding grasps, that of a horizontal pipe, by implementing the gripping system. The tube is arranged so that its axis is horizontal, with three fingertips arranged in a horizontal plane which is located above the central axis of the tube. The location of this plane is indicated by an elevation angle θ. The grouping of the three fingers is further arranged with its centre either at or horizontally offset from the centroid of the tube. The horizontal offset of the group of the contacts is indicated by a centre distance p. The longitudinal spacing of the contacts is indicated by q. The ratio of p to q is termed the grouping ratio. The whole arrangement is illustrated in Figure 4.3.

By arranging the case study with different grouping ratios p/q and different elevation angles θ, a set of experiments was carried out. A comparison between the theoretical and experimental results gave excellent agreement. The errors were usually of the order of 10^{-2} kgf in magnitude, 10^{-1} degree in direction, and 10^{-3} mm in pitch.

4.6 Conclusion

The gripping system is composed of a new system of theories and methodologies, a novel fingertip sensor system, and a versatile user-oriented computer package. The new system of theories has shown great advantage in grasping arbitrary objects. The implementation of the gripping system has been carried out by a series of experiments on the grasping of practical objects. The theories and their experimental verification are therefore regarded as being complete.

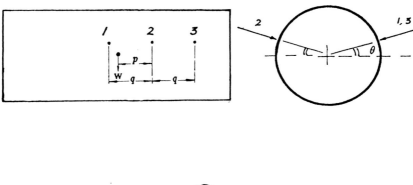

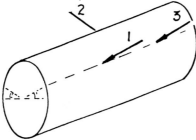

Figure 4.3 *Three fingered grasping of a horizontal pipe*

4.7 References

1   LAKSHMINARAYANA, K., 'Mechanics of Form Closure', ASME Paper 78-DET-32, New York, 1978.

2   SALISBURY, J. K., and ROTH, B., 'Kinematic and Force Analysis of Articulated Mechanical Hands', Trans. ASME, J. of Mechanisms, Transmissions, and Automation in Design, Vol. 105, pp. 33-41, 1982.

3   NGUYEN, V. D., 'Constructing Force-Closure Grasps', Proceedings of 1986 IEEE Conference on Robotics and Automations, Vol. 3, 1986, pp. 1368-1373, Int. Journal of Robotics Research, Vol. 7, No. 3, June, 1988, pp. 3-15.

4   OHWOVORIOLE, E. N., 'Kinematics and Friction in Grasping by Robotic Hands', Trans. ASME, Journal of Mechanisms, Transmissions, and Automation in Design, Vol. 109, pp. 398-403, Sept. 1987.

5   LI, Z. and SASTRY, S. S., 'TaskOriented Optimal Grasping by Multifingered Robot Hands', IEEE Journal of Robotics and Automation, Vol. 4, No. 1, Feb, 1988, pp. 32-44.

6   RAO, K., MEDIONI, G. and BEKEY, G. A., 'Robot Hand-Eye Coordination: Shape Description and Grasping', Proceedings of 1988 IEEE Conference on Robotics and Automations, Vol. 1, 1988, pp. 407-414.

7   JACOBEN, S. C., WOOD, J. E., KNUTTI, D. F., BIGGERS, K. B. and IVERSON, E. K., 'The UTA/MIT Dextrous Hand', Robotics Research, 2nd International Symposium, Cambridge, MIT Press, 1985, pp. 301-308.

8   LIU, H., and BEKEY, G. A., 'A Generic Grasp Mode Paradigm for Robot Hands on the Basis of Object Geometric Primitives', Project Report IRIS#218, 1987, Computer Science Department and Institute for Robotics and Intelligent Systems, University of Southern California, California 90089-0273.

9   IBERALL, T., JACKSON, J., LABBE, L., and ZAMPANO, R., 'Knowledgebased Prehension: Capturing Human Dexterity', Proceedings of IEEE Conference on Robotics and Automation, 1988, pp. 8287.

10 BRAND, L., Vector and Tensor Analysis, Wiley, New York, 1947.

11 ROCKAFELLAR, R. T., Convex Analysis, Princeton, New Jersey, 1970.

12 KERR, D. R. and SANGER, D. J., 'The analysis of kinematic restraint', Proceedings of 6th World Congress on the Theory of Machines and Mechanisms, IFToMM, New Delhi, pp. 299-303, Dec. 1983.

13 KERR, D. R. and SANGER, D. J., 'Grasping using a three-fingered gripper', Proceedings of 26th International Machine Tool Design and Research Conference, UMIST, Manchester, pp. 123-126. Sept. 1983.

14 LIPKIN, H. and DUFFY, J., 'The Elliptic Polarity of Screws', Journal of Mechanisms, Transmissions, and Automation in Design, Transaction of ASME, Vol. 107, pp. 377-387, Sept. 1982.

15 KERR, D. R. and SANGER, D. J., 'The synthesis of point contact restraint of a rigid body', Trans. ASME J. Mech. Trans. and Auto. in Des., Vol. 107, No. 4, pp. 521-525. 1987.

16 XIONG, Y. L., SANGER, D. J. and KERR, D. R., 'Geometric Modelling of Boundless Grasps', Robotica, pp. 19-23, Vol. 10, Jan. 1993.

17 XIONG, Y. L., SANGER, D. J. and KERR, D. R., 'Geometric Modelling of Bounded and frictional grasps', Robotica, pp. 189-192, Vol. 11, Mar. 1993.

18 DAI, J. S. and KERR, D. R., 'Analysis and synthesis of planar grasping in an image space', Proceedings 22nd ASME Mech. Conf., Scottsdale, Arizona, DE-Vol 45, pp. 283-292, Sept. 1992.

19 DAI, J. S., 'Screw image space and its application to robotic grasping', PhD Thesis, University of Salford, Salford M4 5WT, 1993.

20 DAI, J. S. and KERR, D. R., 'Synthesis of Frictionless Grasps in Image Space Using Affine Augmentation', to be presented at the Ninth World Congress on the Theory of Machines and Mechanisms, Milano, Italy, August-September, 1995.

21 DAI, J. S. and KERR, D. R., 'Analysis of Force Distribution of Grasps Using Augmentation', accepted by Journal of Mechanical Engineering Science, proceedings of IMechE, Mar., 1995.

22 SANGER, D. J., KERR, D. R., KUO, J. A. and DAI, J. S., 'Modelling redundant systems', Submitted to IEEE J. Robotics and Automation, Feb. 1993.23

23 KERR, D. R., GRIFFIS, M., SANGER, D. J. and DUFFY, J., 'Redundant grasps, redundant manipulators and their dual relationships', J. Rob. Systems, Vol. 9, No. 7, pp. 973-1000, Oct. 1992.

24 DAI, J. S., SODHI, C. and KERR, D. R., 'The design and analysis of a new six-component force transducer for robotic grasping', Proceedings of the Second Biennial European Joint Conference on Engineering Systems Design and Analysis, Vol. 8, part C, pp. 809-817, London, U.K., July 4-7, 1994.

25 KERR, D. R. 'Analysis, properties, and design of a Stewart platform transducer', Trans. ASME J. Mech. Trans. and Auto. in Des., Vol. 111, No. 1, pp. 25-28, Mar. 1989.

Chapter 5

Force feedback control in robots and its application to decommissioning

R.W. Daniel, P.J. Fischer and P.R McAree

Force control is a central requirement if robot arms are to use tools or interact with workpieces in an unstructured environment. This chapter concentrates on a form of force control called active compliance control. Experimental results are presented to illustrate the difficulties encountered when designing such a force controller on a practical robot. A new theorem on the stability of such a controller is also proved. The paradigm considered is the use of tele-robots' - arms which are operated under human supervision while carrying out complex tasks. The application area considered is nuclear decommissioning but the technology is widely applicable to such diverse activities as bomb disposal and the 'virtual reality' simulators of drug interactions.

5.1   Introduction

Robot force control is a generic term for a technique to modify an arm's servo behaviour when contacting its environment. A good survey of force control techniques is provided in Whitney [17] who grouped the various ways of utilising force information under the headings Stiffness Control, Damping Control, Impedance Control, Explicit Force Control, and Hybrid Control. This chapter will concentrate on a form of Stiffness Control commonly adopted for practical industrial applications and aims to describe the use of such control in the solution of a real world problem, namely 'How does one provide an operator with a high performance system to carry out remote decommissioning tasks?'

Decommissioning is a task which requires the use of tools, such as drills and saws, in a radioactive contaminated environment. A number of systems have been, or are being, developed to accomplish this. The experimental system at Oxford is based on a very high performance parallel robot used as an input device coupled to a remote slave arm, in our case a Puma 560 robot. The design of the input device and its coupling to the remote slave has been the focus of our laboratory's work for a number of years and has involved us in a number of design issues. One possible control strategy is that of force reflection or position-force tele-control. This strategy measures the position of the input device, or master arm, and transmits this to the position servo of the remote, or slave arm. The slave arm moves in response to this command and a force sensor mounted on its wrist is used to drive the master arm motor torques. It is well known that such a system suffers from stability problems on contact with a stiff environment, the most succinct explanation of this phenomenon may be found in Colgate and Hogan's work [6] on the lack of passivity of this strategy.

One may ask why one perseveres with force reflection if the strategy is in some sense inherently 'unstable', or at least non-passive. The main design point is that unless one is willing to pay for esoteric technology, slave arms should be based on standard industrial robots which brings the advantages of robustness and reliability. Traditional master-slave systems rely on using common error as their control strategy, which links the dynamics, and more importantly, the drive friction of the slave, with the force sensed at the master. This may be acceptable for specialist designs using low friction drives and kinematically similar masters and slaves, but prevents the use of standard robots with their associated commercial advantages. The objective is thus to control the behaviour of the slave arm in a manner which permits the use of force reflection and yet still removes the effect of the slave drive friction from the force sensed at the master.

The design problem may be translated into that of modifying the perceived impedance at the master when the slave arm makes contact with the environment. We have achieved this by a combination of force feedback around the slave and signal shaping between the force sensor and the torque fed to the master. The practical result is a high performance master-slave system, which is able to reflect high frequency forces back to the operator with fidelity sufficient for remote tool use and typical operations such as saw blade replacement and insertion into a partially cut slot.

This chapter will describe the advantages of a force feedback loop placed around the slave and some of the problems which limit its performance. Section 5.2 is a brief resume of force feedback strategies following Whitney [17]. Section 5.3 describes why active compliance control needs to be used around the slave and provides a framework for discussing the stability of force control. Section 5.4 provides a brief experimental demonstration of the problems an engineer needs to consider when designing a force controller. Section 5.5 is a new theoretical treatment of active compliance control. Section 5.6 concludes with suggestions for future research.

5.2   Force feedback strategies

A number of force feedback strategies have been suggested. One of the earliest was active stiffness control [16] which was originally formulated as a method for choosing the position gains of the robot joint servos. In this original formulation the joint torque to the motors is given by

$$\tau = J^T K_p J(q_d - q) + K_v(\dot{q}_d - \dot{q})$$

where J is the arm Jacobian, K_p the position gain chosen to achieve the requisite stiffness. This is a simple form of impedance control as developed by Hogan [11]. The philosophy of impedance control is to specify the apparent robot end point dynamic impedance as seen by the environment. The term impedance is used within the context of the classical analogy (as opposed to the mobility analogy) of Bond graph theory, which maps force into effort (electrical voltage) and velocity into flow (electrical current).

In keeping with the fact that most industrial robots have considerable friction, Whitney [17] describes stiffness control, or more properly active compliance control [9], in terms of an inner loop PD joint or Cartesian based servo, with an outer force sensor providing a signal to modify the position demand sent to the robot, i.e.

$$\tau = K_p(q_d - q) + K_v(\dot{q}_d - \dot{q})$$

$$q_d = K_{fp}(f_d - f) + K_{fi}\int(f_d - f)dt$$

where K_{fp} and K_{fi} are the force proportional and integral gains respectively. Often the desired force is set to zero to achieve pure compliance control.

The strategy above overcomes the problem of drive non-linearities at the expense of using a non-colocated sensor (end point force) which can result in phase deficit problems - see for example An and Hollerbach's work [3] on the dynamic stability of force control. It is also known - for example the work by Tae-Sang Chung on Cartesian compliance control [5] - that even a perfectly stiff robot can exhibit unstable behaviour because of the multivariable nature of force servoing, a point that we will return to in Section 5.5.

If a desired end point force is an *a priori* parameter, then explicit force control may be used [17]. Here the input to the controller is a desired force and it is assumed that the arm dynamic interactions have been compensated by an inner loop servo. One way to achieve this is to use resolved acceleration control, which aims to servo the arm in Cartesian space using the arm full dynamic model. Again one is left with the problem of non-collocation plus that of the proven non-robustness of resolved acceleration control [2] as an inner loop servo. Explicit force servoing uses the transpose rather than the inverse of the Jacobian as the feedback matrix, and so does not have the rigid body instability problems of active compliance control as described in Section 5.5, but the problem of how to specify end point forces, rather than an end point compliance or impedance to emulate, is a significant open research question.

A number of early researchers addressed the problem of specifying end-point forces and/or compliance. It seemed self evident that specific tasks exhibit specific degrees of freedom and types of constraint, i.e. inserting a peg into a hole might, at first glance, require position servoing along the peg's axis and force servoing orthogonal to it. A postulated solution to the associated control problem was hybrid control in which some Cartesian directions were position servoed, and their orthogonal complement force servoed <u>directly</u> (without any inner position loop). The original hybrid control paper [15] sparked a flurry of activity in the kinematics community who pointed out that the existence of such an orthogonal subspace as originally defined was not necessarily guaranteed. It was then shown that the original formulation also resulted in kinematic instability for revolute robots in certain configurations [4]. The kinematic instability arose because of interactions between the robot mass matrix and the robot Jacobian and is a simple consequence of not using an inner loop servo. Using a proper inner loop joint servo overcomes these kinematic instability problems [1], although one is still left with the problem of specifying a co-ordinate system for the 'Selection Matrix' used in such controllers for an unstructured environment.

There are more robust techniques for designing force controllers than those described above. One in particular is the class of controllers based on joint torque servoing [12]. As stated above, any practical controller requires a high gain inner loop servo to overcome the effects of drive friction, and this is usually provided by PD loops placed around the joints. Joint torque servos should be placed downstream of the major joint non-linearities and can be tuned to reduce the effects of friction and cogging. They have the added advantage that the loop gains of a joint torque servoed system are not modulated by the environmental stiffness and in principal are more robust when the arm contacts a high stiffness environment [10]. However, the robustness of the joint torque servo itself, rather than the outer force loop, becomes an issue when the arm is in free space. Torque servos are non-collocated, and thus hard to stabilise, and cogging in the joint gears can introduce impulsive torques if the sensor is mounted directly on the output of the reducer and there is little inertia in the transmission; this limits the amount of phase advance or derivative action in the servo loop. These practical difficulties lead us to concentrate on traditional robot servo architectures, and in particular, active compliance control.

5.3   Active compliance control for teleoperation

In this section we will outline the factors which affect the design of a force controller. Our objective is to modify the end point impedance of a robot arm to facilitate remote manipulation under human control, but the same design observations hold for all position servo based force controlled tasks.

An objective of high performance teleoperation is to give a human operator a sense of feel which can aid in the execution of a remote task. One method of achieving this is to use force reflection from the slave to a high performance master input device. However, as stated in the introduction, such an approach leads to stability problems when the remote slave contacts a stiff environment.

A full explanation of the problem is given by Colgate and Hogan [6] using Bond graph models and passivity. We will adopt their general approach but will apply their argument to a simpler linearised model using block diagrams, rather than using Bond Graph effort and flow variables, and will restrict our attention to one degree of freedom (dof).

As a first step, consider the simple model shown in Figure 5.1 of a human using a tool to touch the environment, e.g. using a pencil.

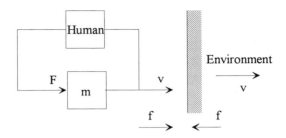

Figure 5.1 *Human using a tool to contact the environment*

The human arm with tool is modelled as a simple lumped mass under control, the environment consists of a complex collection of masses, dampers, and springs lying to the right of a boundary at which energy exchange is observed. It is assumed that net power flow is from left to right. A possible interpretation of the causality of this system may be described by the following sequence of operations:

1. The human applies a force F.
2. The mass m (the arm) accelerates and has an instantaneous velocity and position $v(t)$, $x(t)$ respectively.
3. The tool and environment react with instantaneous force $f(t)$.
4. The reaction and applied forces combine to define the consequent motion of the system.

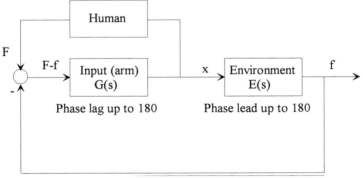

Figure 5.2 *A block diagram of the simple human-tool system*

The block diagram in Figure 5.2 describes the proposed assignment of causality. Note that the input device, or arm, has a force as its input and a position as its output and so has a lagging transfer function, i.e. the arm model has integral causality. For example, a pure mass acts as a double integrator and has 180 degrees of phase lag, the maximum lag possible for a passive (collocated) system. The environment, on the other hand, has a position input and a <u>reaction</u> force output and its transfer function must have a leading phase, i.e. differential causality, a pure mass this time resulting in a 180 degree phase lead. In fact any passive system differential model between imposed position and reaction force has a maximum phase lead of 180 degrees for a pure mass, and a minimum phase lead of zero for a perfect spring. As the total phase lag around the full system loop cannot exceed 180 degrees it can be seen that the system must always be stable, or at worst resonant if there is no damping. The task of the human is thus to inject sufficient damping to achieve the desired performance.

Now consider a force reflecting master-slave system, i.e. the position of the master held by the human is treated as a position demand to the slave, and the force sensed at the end point of the slave is used to specify the master motor torques. Figure 5.3 is a block diagram of such a system.

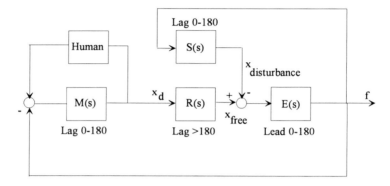

Figure 5.3 *Master-slave system in contact with environment*

M(s) is the master transfer function between applied force and position (assumed to be lagging by up to 180 degrees, i.e. the master is high performance and there are no non-collocated modes between input torque and output position), *E(s)* is the environmental transfer function between robot end point position and reaction force (leading by up to 180 degrees). The robot is modelled by *R(s)*, its free space position demand to <u>end point</u> position transfer function (assumed to be lagging, probably by considerably more than 180 degrees at high frequencies because of stiction and the many sources of flexibility in the arm between the actuator and the end point) and *S(s)*, the robot end point force to end point position disturbance response (assumed lagging by at most 180 degrees). The limit of 180 degrees lagging on *S(s)* arises because the robot controller is assumed to be proportional plus derivative at the joints and so acts as a passive device when viewed from an imposed end point force.

Simple block diagram manipulation results in the modified system shown in Figure 5.4.

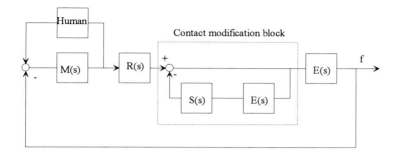

Figure 5.4 *Block diagram reconfiguration of master-slave system*

The stability of the master-slave system can now be seen to depend on 4 transfer functions:

1. *E(s)*: A phase lead representing the reaction of the environment, probably with very high gain for a stiff environment.

2. $(1+E(s)S(s))^{-1}$: representing the interaction between the robot stiffness and the environment reaction, where *S(s)* has a phase lag of up to 180 degrees with low gain if the robot is stiff or high gain if compliant.

3. *R(s)*: The robot end-point position response, probably with very large phase lag with a gain close to one at DC and rolling off with the occasional resonant peak and trough above 1-10 Hz.

4. *M(s)*: The master response with a phase lag up to 180 degrees.

As *R(s)* and *S(s)* refer to the same structure (the robot arm) then they differ only in the dynamics that the actuators introduce before the gear stiction and the zero polynomials which reflect the degree of collocation of the two transfer functions. For this simple one degree of freedom idealised system, let the environment be a simple spring k_e, the

robot free space transfer function $R(s) = \dfrac{n_1(s)}{d_{act}(s)d_{struct}(s)}$, the free space disturbance

response $S(s) = \dfrac{n_2(s)}{d_{struct}(s)}$, where $d_{struct}(s)$ is the arm structure poles, $n_1(s)$ the actuator to

free response zeros, $n_2(s)$ the free space disturbance response zeros (of order two less than $d_{struct}(s)$ as a consequence of collocation) and $d_{act}(s)$ the linearised dynamics of the drive train up to the shaft carrying the link. The transfer function of the slave position demand to slave force is then given by

$$R(s).(1+S(s)E(s))^{-1}E(s) = \frac{n_1(s)k_e}{d_{act}(s)(d_{struct}(s) + k_e n_2(s))}.$$

It can be seen that the environmental stiffness enters as a multiplicative gain in the forward loop as well as modifying the vibration behaviour of the arm.

The transfer function $R(s)$ has considerable phase lag for geared robots and in general a force reflection system constructed as above will be unstable in contact with a stiff environment.   Kim [13] suggested that the problem is caused by the manipulator being too stiff and that the problem may be overcome by using force feedback around the slave and called this approach shared compliance control (SCC). As the structure of the arm vibration modes changes radically with configuration, SCC must be stabilised with a dominant lag. The structure of the force controller part of a shared compliance controller is shown in Figure 5.5.

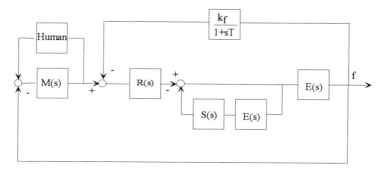

Figure 5.5 *Dominant lag force controller around remote slave*

The dominant lag in the force feedback path generates a phase lead around the force reflection path and so eases the operator's problems in stabilising the system. High gain and a large phase lag are still present in the forward path, and so further filtering is required to provide the operator with a high quality sense of feel.   The design of these filters is outside the scope of this chapter, the interested reader is referred to [7] where further problems which only occur in such master-slave systems are discussed.

5.4   Typical experimental results during the design of a force controller

Space precludes a full design study of a robot force controller. This section aims only to highlight some typical problems encountered by an engineer during the design process.

Inner loop position servoed force controllers suffer from the following stability problems:

• Modulation of the force loop gain by (unknown) environmental stiffness.
• Non-colocation leading to large (linear system) phase deficits near arm resonant frequencies, i.e. $R(s)$ above has a numerator polynomial with many fewer zeros than $S(s)$.

- Drive friction leading to transmission (forward path, non-linear) time delays.
- Joint position quantisation limiting servo gain and affecting the servo dynamic range.

The 'linear' system problems of an unknown loop gain (modulation by the environmental stiffness) and complex (changing) arm mode shapes and modal momentum distribution (leading to non-collocated complex pole-zero pairs) are difficult to design for in themselves. When these are coupled with the large joint friction and loss of joint velocity information on contact through quantisation, the only closed loop controller that works in practice is based on a dominant low frequency pole, the shared compliance controller described in the previous section.

Our illustration will be based on a real industrial robot, the Puma 560 robot used in our remote manipulation cell. Only a single degree of freedom test will be reported - a full characterisation would require the identification of two full six by six frequency response matrices.

The Puma 560 used in the Oxford University telemanipulation test cell is driven using a standard VAL 2 controller but with a high speed, Ethernet based, communication enhancement called ESLAVE. This provides a 278 Hz direct communications link to the robot servo drives, top level control being provided by a Pentium PC. The servo gains in the robot have also been set to optimise performance for force reflection, the position gains are considerably lower than in a robotic application and the integral terms have been disabled.

5.4.1   Experimental method

The robot was moved manually with the brakes released into a typical manipulation configuration. The resulting joint angles are given in Table 5.1. Our objective was to measure the transfer functions $S(s)$, the free space force disturbance response, and $R(s)$, the free space position response. The world Z direction, parallel with joint one axis, was chosen as the example co-ordinate to use. In the first experiment a suspended electromagnetic shaker, providing external excitation to measure $S(s)$, was attached via a stinger to the end plate of an ATI force-torque sensor attached to the robot wrist. In the second experiment, joint 2 of the robot was driven via the controlling PC and ESLAVE to provide excitation to measure $R(s)$ - the robot configuration was chosen to provide a good approximation to straight line motion in the Z direction for small amplitude motion of joint 2. Applied force during the estimation of $S(s)$ was measured using the ATI force-torque sensor. Motion in the Z direction was measured using an LVDT for low frequency position and an accelerometer for vibration. The acceleration response was converted into position by double integration above a threshold frequency. The threshold frequency was chosen by studying the coherence plots corresponding to the LVDT and accelerometer derived transfer functions and keeping the combined coherence as close to one as possible over the identification frequency range. Identification was carried out using a

Tektronix Spectrum Analyser, excitation was via an internally generated 100 Hz baseband PRBS signal. Identification was via a non-overlapped 1024 point FFT averaged over 32 windows. All signals were converted to and from analogue signals appropriate to the analyser via an A to D board in the controlling PC.

Joint 1	Joint 2	Joint 3	Joint 4	Joint 5	Joint 6
-97.6	-18.6	140.9	-2.1	56.8	103.9

Table 5.1 *Joint angles used during the estimation process, degrees*

5.4.2   Results

The threshold frequency for swapping between LVDT and accelerometer derived position was 40 rad/s. Figure 5.6 shows the free space force disturbance response, $S(s)$, the discontinuities in magnitude and phase corresponding to small errors in the calibrations of the two sensors used. Note the strong resonances at 90 and 190 rad/s and the low value of stiffness used in our position servos, 0.007 mm/N. Non-linearities in the structure, together with quantisation, aliasing in the force-torque sensor signal (which is sampled at 278 Hz), destroy the coherence above 400 rad/s - note the noise-like behaviour of the plot.

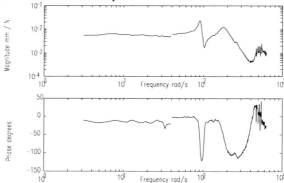

Figure 5.6 *Bode plot of S(s) for a Puma 560, Z direction*

Figure 5.7 shows the free space position response, $R(s)$, along the Z axis. Again there is a discontinuity as the two position sensors are swapped over at 40 rad/s. Note that there is now added damping as the whole drive train is moving and the resonances are not so pronounced - less energy is able to be pumped into the structure from the motors than by the environment at the arm resonant frequencies. Also note the exponentially increasing phase lag caused by time delay. A more detailed investigation, driving the joint servo directly by an analogue signal, revealed that the majority of this lag is caused by a torque dead zone in the drives induced by gear tooth stiction and not computation delay. The forward gain is also much less than one for the very small excitation signal used - the gain increases with size of joint demand. This has a further implication for *S(s)* measured above - it is most probable that the joints remained in a 'stuck' configuration during the identification process, and in any case, the joint motion is below the quantisation level of the encoders.

The coherence function is very nearly one for the LVDT at low frequencies, and this indicates that a describing function analysis would be appropriate. However, the massive phase lag means that any end point position dependent feedback loop must roll off its gain well before 30 rad/s, where the forward path lag drops below 90 degrees. Our SCC feedback loop needed a single pole at 1 Hz .

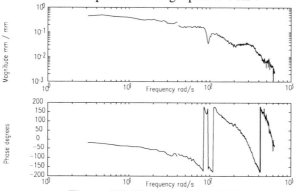

Figure 5.7 *R(s) for a Puma 560*

5.4.3   Conclusion

The results above illustrate the typical data to be expected when designing a force controller for an industrial robot. There is very little that sophisticated controllers can achieve to compensate for such a system, and the performance achieved with a simple lag is close to the limit. The only design parameter left is gain-bandwidth product around the feedback loop - a very similar situation to achievable performance from operation amplifiers such as the 741. The main conclusion to be drawn from this simple experiment is that standard industrial robots are not suited to high performance force servoing. If a high bandwidth active compliance servo is required, then modifications to the servo sensors (joint torque servoing), or to the method of actuation (direct drive) are required.

5.5   Active compliance control stability

This section will prove a new stability result for perfect rigid robots when using joint position servos as inner loops. There is some confusion in the literature about what does, or does not, cause instability within a force controlled environment. For example, the original hybrid control algorithm,

$$\tau = K_p J^{-1} S(x_d - x) + K_v J^{-1} S(\dot{x}_d - \dot{x}) + K_f J^T (I - S)(f_d - f)$$

has both position and force servoed in Cartesian co-ordinates and a diagonal selection matrix S is used to choose which co-ordinates are force controlled and which are position servoed. S is diagonal, each entry being either one or zero, hence I-S is also a diagonal matrix with complementary entries. It may be shown that such a controller is unstable in some configurations on a revolute joint robot [4] because of interactions between the Jacobian and the robot mass matrix. However, such a controller would not be used in practice as it has no joint servos.

The design of force servos involves a hierarchy of problems - this section is only a simple theoretical treatment which holds for an ideal perfectly rigid 6 DOF robot contacting an environment built entirely from springs.

The force controller architecture of our Puma 560 robot system shown in Figure 5.8 is typical of industrial practice.

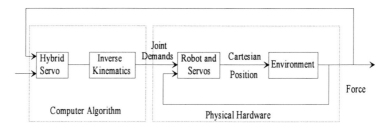

Figure 5.8 *Puma 560 force control system*

There are two methods of providing compliance control for the above system. The first is to use the information from the force sensor to artificially increase the perceived stiffness of the environment. The second is to program the position servoed arm to behave as a set of Cartesian springs.

As the robot is assumed to be rigid, with no actuator dynamics or non-linearities, the dynamics of the arm in contact with the environment are:

$$M(q)\ddot{q} + C(q, \dot{q})\dot{q} + G(q) = \tau + J^T K_e (x_0 - x)$$

where $M(q)$ is the mass matrix of the arm, $C(q, \dot{q})\dot{q}$ is the vector of Coriolis and centripetal torques, $G(q)$ is the gravity vector, K_e is the environment stiffness, x_0 is the position of the end point when just in contact, x is the current end point position of the robot corresponding to q, and J^T is the transpose of the arm Jacobian. We will further assume that the arm is stationary and the gravity term has been factored out, i.e. $C(q, \dot{q})\dot{q}$ and $G(q)$ are both zero incrementally. The linearised equations of motion of the arm with a PD servo are then:

$$M\delta\ddot{q} = -K_p\delta q - K_v\delta\dot{q} - J^T K_e J\delta q$$

where δq is the perturbation vector of joint angles.

It will be assumed that the inverse kinematics are calculated infinitely fast and may be represented incrementally as the inverse of the arm Jacobian. The first approach as discussed above consists of measuring Cartesian force, multiplying it by $JJ^T S$ where S is a gain, or selection matrix with gain for hybrid control, and using this signal in a negative feedback loop as the robot inverse kinematics algorithm position demand. The second approach - making the robot act as a Cartesian spring - consists of multiplying the sensed force just by the gain selection matrix before using in the same negative feedback loop to the kinematic algorithm joint demand.

We will now compare the multivariable root loci of these two approaches as the environmental stiffness is increased, or equivalently, the force feedback scalar gain is increased. The state space models of the two systems using incremental joint position and velocity as states, joint torques as the vector of inputs, and Cartesian force as the output are:

1. Pseudo-stiffening of the environment

$$\begin{bmatrix} \delta\dot{q} \\ \delta\ddot{q} \end{bmatrix} = \begin{bmatrix} 0 & I \\ -M^{-1}(J^T K_e J + K_p) & -M^{-1}K_v \end{bmatrix}\begin{bmatrix} \delta q \\ \delta\dot{q} \end{bmatrix} + \begin{bmatrix} 0 \\ M^{-1}J^T \end{bmatrix}\tau$$

$$f = \begin{bmatrix} K_e J & 0 \end{bmatrix}\begin{bmatrix} \delta q \\ \delta\dot{q} \end{bmatrix}$$

2. Cartesian spring via the inverse kinematics

$$\begin{bmatrix} \delta\dot{q} \\ \delta\ddot{q} \end{bmatrix} = \begin{bmatrix} 0 & I \\ -M^{-1}(J^T K_e J + K_p) & -M^{-1}K_v \end{bmatrix}\begin{bmatrix} \delta q \\ \delta\dot{q} \end{bmatrix} + \begin{bmatrix} 0 \\ M^{-1}K_p J^{-1} \end{bmatrix}\tau$$

$$f = \begin{bmatrix} K_e J & 0 \end{bmatrix}\begin{bmatrix} \delta q \\ \delta\dot{q} \end{bmatrix}$$

The gain selection matrix will be assumed to be a scalar times a unit matrix. Under these conditions, the asymptotic stability of the two feedback strategies above may be analysed by looking at the positions of the finite zeros and the asymptotes associated with the infinite zeros. The zero structure of a multivariable system can be obtained from a MacLaurin series expansion (the Markov parameters) of the system transfer function using the NAM algorithm [14]. For a state space model of the form

$$\dot{x} = Ax + Bu$$
$$y = Cx + Du$$

the transfer function is given by

$$G(s) = C(sI - A)^{-1}B + D$$

and for the systems above, $D=0$ and the series expansion is given by

$$G(s) = \frac{CB}{s} + \frac{CAB}{s^2} + \frac{CA^2B}{s^3} + \cdots$$

The finite zeros are the finite solutions of

$$|sNM - NAM| = 0$$

where N and M are full-rank matrices representing the left null-space of B and the kernel of C respectively, i.e. $NB=0$ and $CM=0$. For both the above systems, $N = [I \quad 0]$

and $M = \begin{bmatrix} 0 \\ I \end{bmatrix}$ so that $NM = 0$, i.e. there are no finite zeros. As the first order Markov

parameter, CB, is also zero for both systems, there are no first order infinite zeros. There are, however, second order infinite zeros for both systems and none of higher order. It is the difference in the patterns made by these second order zeros that differentiates the stability properties of the two strategies.

To obtain the higher order zeros one needs to compute the *projected* Markov parameters for a system. These are obtained by performing an eigenvalue decomposition on each, successively higher order, projected Markov parameter, starting with the unprojected lowest order parameter CB. Projection is achieved by multiplying the Markov parameters successively on the left by the basis matrices spanning the right null spaces of the previous projected parameter and on the right by the left null space bases. The projection process stops when the resulting parameter is full rank for $i = v$. Following the notation of Kouvaritakis [14], let G_i be the ith projected parameter, then

$$G_1 = CB = [U_1 \quad M_1] \begin{bmatrix} \Lambda_1 & 0 \\ 0 & 0_{d_1} \end{bmatrix} \begin{bmatrix} V_1 \\ N_1 \end{bmatrix}$$

$$G_2 = N_1 CABM_1 = [U_2 \quad M_2] \begin{bmatrix} \Lambda_2 & 0 \\ 0 & 0_{d_2} \end{bmatrix} \begin{bmatrix} V_2 \\ N_2 \end{bmatrix}$$

The expansions of the two systems above stop at the second projected Markov parameter, and so they only have second order infinite zeros. In order to predict the stability of the system at very high environmental stiffness one uses the following theorem:

Theorem (Kouvaritakis [14])

If all the projected Markov parameters of an m by m system have simple null structure and $d_{i+1} < d_i$ for all $i = 1, 2, \ldots v -$ then, as the loop gain goes to infinity, the root loci that tend to the *i*th order infinite zeros travel along asymptotes whose angles are

$$\alpha_{j,l}^{(i)} = \frac{1}{i}\left[\arg(\lambda_j^{(i)}) + (2l+1)\pi\right] \text{ for } l = 1, 2, \ldots, i$$

$$j = 1, 2, \ldots, d_{i-1} - d_i$$
$$i = 1, 2, \ldots, v$$

where $d_0 = m$, $d_v = 0$ and $\lambda_j^{(i)}$ is the *j*th non-zero eigenvalue of the *ith* projected Markov parameter.

The second projected Markov parameter for the first case of pseudo-stiffening of the environment is given by

$$G_2 = K_e J M^{-1} J^T$$

and the second case, making the arm behave as a Cartesian spring,

$$G_2 = K_e J M^{-1} K_p J^{-1}$$

For both cases, assuming *n* DOF, the root asymptote angles are given by

$$\alpha_{j,l} = \frac{1}{2}\left[\arg(\lambda_j) + (2l+1)\pi\right] \text{ for } l = 1, 2$$

$$j = 1, n$$

The first case has eigenvalues which are all positive real and so the root loci travel along asymptotes which are at right angles to the real axis. This is as expected, as this control algorithm has identical dynamics to that of contacting a stiffer environment than that present with just position servoing, i.e. a passive algorithm. This algorithm is thus stable for all force gains and environmental stiffness. The stability of the second algorithm depends on a matrix which does not necessarily have positive real eigenvalues. If the eigenvalues are not real, then the asymptotes are not at right angles to the real axis and the system will go unstable for some large environmental stiffness or force feedback gain.

Example

To demonstrate the difference between the two algorithms, consider a two degrees of freedom manipulator, with massless links but with motor inertia contacting a variable stiffness environment.

Let the system parameters be

$$J = \begin{bmatrix} 1 & 2 \\ 3 & 4 \end{bmatrix}, \ K_e = \begin{bmatrix} 1.5 & 0 \\ 0 & 1 \end{bmatrix} \times k, \ M = \begin{bmatrix} 1 & 0 \\ 0 & 5 \end{bmatrix}, \ K_p = \begin{bmatrix} 10 & 0 \\ 0 & 10 \end{bmatrix}, \ K_v = \begin{bmatrix} 6.32 & 0 \\ 0 & 14.1 \end{bmatrix}$$

with critical damping and variable *k*. Figure 5.9 shows such a system configuration.

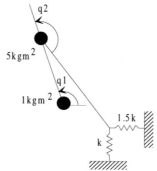

Figure 5.9 *Two link robot contacting variable stiffness environment*

The first approach of feeding force back via the transpose of the Jacobian directly to motor torque results in a second projected Markov parameter

$$G_2 = K_e J M^{-1} J = \begin{bmatrix} 2.7 & 6.9 \\ 4.6 & 12.21 \end{bmatrix}$$

Both eigenvalues are positive real, and so the multivariable root loci will have asymptotes at plus and minus 90 degrees. The root locus of a unit gain force feedback system as the environmental stiffness is increased is shown in Figure 5.10, agreeing with the theoretical predictions.

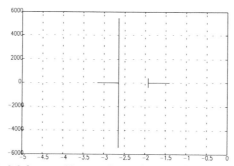

Figure 5.10 *Multivariable root locus for Jacobian transpose feedback*

The second approach, using inner loop position servo based force feedback, has a second projected Markov parameter

$$G_2 = K_e J M^{-1} K_p J^{-1} = \begin{bmatrix} -21 & 12 \\ -48 & 126 \end{bmatrix}$$

the eigenvalues now being $2.5\pm4.87j$ and the predicted angles of departure are

±58.58° and ±121.42°. The multivariable root locus for this system in contact with an increasingly stiff environment with unit force feedback gain is shown in Figure 5.11.

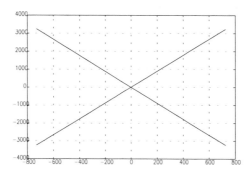

Figure 5.11 *Multivariable root locus of Cartesian spring feedback*

We have thus shown that it is possible to have instability for sufficiently stiff environments even for servoed position based active compliance control and we have given an explicit test for a perfectly rigid robot. Non-ideal robots have a greater phase deficit for the reasons outlined in previous sections, and so are likely to go unstable for lower gains than that predicted by the expression above, but this result does put an upper bound on the supportable force feedback gain. This work may be extended to partially constrained contact, but space precludes any further development.

5.6   Conclusions and future directions

We have presented a brief overview of the robot force control problem in the context of teleoperation. We chose to study active compliance control in detail as it is the control strategy most easily implemented on practical industrial robots.

The initial discussion in Section 5.3 covers most cases of force control. We argued that only simple dominant lag controllers work in practice on position servoed arms because of the non-ideal nature of real robots. Experimental data were presented in section 5.4 to support the assertions of Section 5.3, the main feature to note being the large phase deficits induced by actuator non-linearity. We concluded by showing in Section 5.5 that one has to be careful adopting a simple minded approach to position based force control and that programming the robot to act as a Cartesian spring can result in instability when contacting a rigid environment. The result proved is general for perfect robots and subsumes many of the reported problems of instability reported in the force control literature.

We have concentrated on the problems of designing a force controller for an existing robot. An open problem is how to design an arm specifically for force control. Direct drive arms have the advantage that their vibration characteristics are relatively constant throughout their work-space, whereas geared robot vibration frequencies tend to have an *r*-squared relationship with the distance from the base to the point of contact. Another factor is the intrinsic stiffness of the arm. Force control is generally used in a negative feedback loop to make an arm more compliant. An alternative approach would be to stiffen an already compliant arm with positive feedback. One would then need to compensate for free space vibrations in the arm, possibly using accelerometer feedback. Force control and the control of arm vibration modes are thus intimately related - future research should see results from both areas converging.

5.7 References

1. Alici, G, and Daniel, R.W., Development and experimental verification of a mathematical model for robot force control design. Proceedings of the IEEE/RSJ International Conference on Intelligent Robots and Systems, July 26-30, 1993, Yokohama, Japan.

2. Alici, G. and Daniel, R.W. Experimental comparison of model-based robot position control strategies. Proceedings of the IEEE/RSJ International Conference on Intelligent Robots and Systems, July 26-30, 1993, Yokohama, Japan.

3. An, C.H. and Hollerbach, J.M. Dynamic stability issues in force control of manipulators. Proceedings of the IEEE Conference on Robotics and Automation, Rayleigh, 1987, 821-827.

4. An, C.H. and Hollerbach, J.M. Kinematic stability issues in force control of manipulators. Proceedings of the IEEE Conference on Robotics and Automation, Rayleigh, 1987, 897-903.

5. Tae-Sang Chung An inherent stability problem in Cartesian compliance and an alternative structure of compliance control. IEEE Transactions on Robotics and Automation, Vol 7, No. 1, February 1991, 21-30.

6. Colgate, J.E. and Hogan, N. Robust control of dynamically interacting systems. The International Journal of Control, Vol. 48, No. 1, 1988, 65-88.

7. Daniel, R.W., Fischer, P.J. and Hunter,B. High-performance parallel input device. Proceedings of the SPIE, Telemanipulator Technology and Space Robotics, 7-9 Sept 1993, Boston, 272-281.

9. De Schutter, J. and Van Brussel, H. Compliant robot motion II. A control approach based on external loops. International Journal of Robotics Research, Vol 7, No. 4, Fall 1988, 18-33.

10. P. Elosegui Analysis of the compliant motion achievable with and industrial robot identification, feedback and design. D Phil Thesis, Oxford University, 1991.

11. Hogan, N. Impedance control An approach to manipulation Part 1 - Theory. ASME Journal of Dynamic Systems, Measurement, and Control, Vol. 107, March 1985, 1-7.

12. Jansen, J.F. and Herndon, J.N. Design of a telerobotic controller with joint torque sensors. Proceedings IEEE International Conference on Robotics and Automation, Cincinnati, 1990, 1109-1115.

13. Kim, W.S. Shared compliant control a stability analysis and experiments. Proceedings of IEEE Conference on Systems, Man and Cybernetics, Los Angeles, Nov 1990, 620-623.

14. Kouvaritakis, B. The optimal root loci of linear multivariable systems. The International Journal of Control, Vol. 28, No. 1, 1978, 33-62.

15. Raibert, M.H. and Craig, J.J. Hybrid position/force control of manipulators. Transactions of ASME Journal and Dynamic Systems, Measurement, and Control. Vol 102, 1981, 126-133.

16. Salisbury, J.K. Active stiffness control of a manipulator in Cartesian co-ordinates. Proceedings of the 19th IEEE Conference on Decision and Control, December 1980, Albuquerque.

17. Whitney, D.E. Historical perspective and state of the art in robot force control. The International Journal of Robotics Research, Vol. 6, No. 1, Spring 1987, 3-14.

Chapter 6

Tele-presence control of robots

D.G.Caldwell and A.J.Wardle

Robots are often required to function in environments that are too dangerous or expensive for direct human operation, but, computer control and intelligence are not sufficiently developed to permit them to operate under their own initiative. In these instances teleoperational control forms a popular solution set. For optimum performance under these conditions the operator would wish to control an advanced instrumented robot, comparable in function (where possible) with the human body, integrated into a sophisticated tele-presence system to provide the operator with a full range of motion inputs and sensory feedback data.

This chapter considers the development of input, control and feedback (visual, audio and tactile) systems (man-machine interface) for a twin armed mobile robot to be used in tele-presence applications.

Modules for leg, arm, hand and head motion monitoring have been developed to control the activities. Cameras, microphones and a multi-functional tactile sensing system provide feedback signals that give the operator a relatively realistic impression of the robot's activities. The signals from these sensors can be fed to the operator as direct stimuli, comparable in many instances with normal sensations. All the feedback modules have been made sufficiently light and compact that they can be worn by the operator during normal usage, without restricting motion or comfort. Comparisons of various visual, control, audio and tactile techniques have also revealed the optimum sensory configurations for specific tasks.

6.1   Introduction

Full autonomous operation in complex environments is a primary goal of robotic research, but this is not yet achievable, and sophisticated tasks still need direct human input. In such applications teleoperational abilities are of the utmost importance but accurate control and efficient use are severely hampered by the manipulator design and input drive, system sensing and sensory data feedback.

To improve these facilities recent work has focused on the development of integrated visual, audio and tactile sensing with operator control of advanced dexterous end-effectors and high power/weight actuation systems. These abilities extend the work on teleoperation into a new domain of tele-presence (or tele-existence), where the aim is to convince the operators they are actually living and moving in a remote environment [1-3]. In robotics these activities can, in part, be replicated directly by mimicking three of the primary human senses, namely vision, hearing and touch. In these teleoperated circumstances the operators would wish to input body motions (legs, arm, hand and head) which the robot would duplicate, and receive from the remote sensors full visual, audio and tactile feedback of a quality and form comparable with that normally produced by the eyes, ears and skin [1-2].

As remote manipulators become more sophisticated and the tasks they undertake become more complex three main sensory obstacles to effective widespread use have been identified:

  i) The detection of sensory signals
  ii) The feedback of real-time sensory information to the operator
  iii) The presentation of this information in a form that can be easily detected and processed by the brain as a reflex action, since an excessive need for thought would detract from performance of the primary task.

Tele-operational usage of robotic systems has traditionally concentrated on the control of a single fixed manipulator with feedback limited to force reflection, which can nevertheless be a very powerful sensation [3-4]. However, as video and audio feedback was added to the sensory capabilities of manipulators, the area of tele-presence was developed, but the robot was still usually a static platform with only limited user interface capacity [5]. Recently increased system flexibility has been added through developments into controlled mobile systems [6-7].

In this work the tele-presence robot has three sensory levels; visual, audio and multi-functional tactile, which can be fed back to the operator in a user compatible format. Control of the actions of this twin armed mobile system are regulated by user derived movements of arms, legs, hands and head, that largely correspond to expected human actions. All links are wireless using radio and video communication. Since the robot is of approximately human dimensions this further helps to assist with control making 'hands free' operation feasible.

6.2   Tele-presence robot design

6.2.1   Mobile base

Figure 6.1 shows the tele-presence robotic system. The robot 'body' is mounted on a three wheeled platform with two driven front wheels and a castoring rear wheel. The wheels are driven differentially to give the robot a minimum turning circle within its own width, ie 0.8m. The battery pack uses two 12V 24Ah sealed lead acid rechargeable batteries providing ±12V. These units are mounted over the front drive wheels for stability.

Figure 6.1 *Tele-presence robot*

Motion of the base is controlled by a PWM controller driving two H-bridge motor control circuits. Position and velocity data are available from optical encoders mounted on the motors giving precise closed loop control. The maximum forward and reverse speed of the robot is $1\mathrm{ms}^{-1}$, a slow walking pace.

The robot 'torso' is formed from two aluminium columns for light rigid construction. The inter-column area is the main electronics mounting site and the pillars serve to give a measure of protection for this area.

6.2.2 Robot manipulator arms

The robot has two manipulator arms of different configurations, designed to demonstrate co-ordinated tele-presence control. The first of these arms uses a conventional 6 degree of freedom SCARA type design with electrically driven gripper. This is designed for use during relatively crude manipulation and gripping. The second arm is of an anthropomorphic design with controlled shoulder, elbow and wrist motions, giving greater manipulative capacity and dexterity. These arms will be used for comparative performance studies during teleoperation. The shoulder is moved in three axes (two rotational and one twisting) by three dc servo motors. The position of each is measured using potentiometers and controlled via a microcomputer.

The elbow joint uses pneumatic muscles operated by an air cylinder via piezo-electric valves [8]. The valves allow relatively fast switching (25 ms) to provide pulse width modulation of the air supply to the muscles. The muscles, mounted on either side of the joint (flexors and extensors), are controlled differentially using a high order control algorithm and microcomputer, resulting in rapid and accurate response to inputs. The motion range is from 180° (fully extended) to 90°. The angles of the joints are measured using a linear Hall effect device.

Figure 6.2 *Anthropomorphic arm*

The wrist consists of a two axis joint operated by pneumatic muscles and controlled in a similar way to the elbow. Again, a Hall effect sensor is used to measure its position between ±30° vertically and horizontally from the forearm axis.

Both arms have dimensions comparable with an 'average' adult and are mounted at a height of 1.3m. The SCARA arm can also move prismatically over a height range from 0.3-1.45m.

6.2.3   Robot pan and tilt head

The robot's pan and tilt head unit which provides stereo visual and audio sensing is mounted on the torso at a height of 1.5m. Movement about the vertical axis or pan, and that about the horizontal axis, tilt, is accomplished by dc servo motors operating through planetary gearboxes, with the position of the head being sensed by high linearity potentiometers. The maximum angles of deflection of the robot head are 50° left and right, 20° up and 40° down at a maximum speed in both axes of 60°s⁻¹.

Two camera mounting points are provided 100mm in front of the pan axis. Each camera weighs 650g and to offset this moment, counter-balance springs are employed, as shown in Figure 6.3.

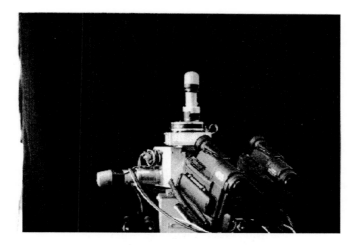

Figure 6.3 *Robot pan and tilt head*

6.2.4   Tactile sensing systems

Touch provides unique information used to prevent injury and to assess various object perceptual attributes such as texture, hardness, discontinuities such as holes or edges, thermal states and movement, including vibration [9].

A multi-modal tactile system has been constructed, Figure 6.4, to replicate the four basic human cutaneous nerve sensations: pressure, vibration, temperature, and pain. Using these parameters contact pressure/force, object texture, slip, hardness, profile/shape, absolute temperature, and thermal conductivity can be measured. Pain is recorded as overload in the 3 prime sensors, although in humans there are dedicated pain sensors.

The robot finger uses three forms of sensing element: PVDF for dynamic sensing (slip and texture), force sensitive resistors (FSR) for pressure (shape and hardness), and a fast response thermocouple (10ms) for temperature (thermal conductivity).

The force sensitive resistor (FSR) is mounted on a compliant rubber seat. On top of this are attached the PVDF dynamic sensor with a 'finger nail' attachment, and a Peltier effect heat pump (power output 0.2W). The thermocouple is mounted directly on top of the Peltier device, which maintains the robot fingertip at a temperature of 40°C (approximately human body temperature) using the thermocouple and a proportional controller with a 5 second time constant. The thermocouple and Peltier are sealed in a resin for protection and the other sensors are attached compliantly to the finger substrate, Figures 6.4a and b.

The FSR is able to sense forces ranging from 100N to 0.1N depending on the bias resistor and substrate rubber thickness and type [10].

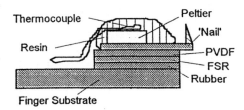

Figure 6.4a *Schematic of finger sensor*

The PVDF sensor is used to determine the roughness of objects. As the nail is moved along a surface, vibrations indicative of the surface roughness are transmitted to the sensor which provides an ac output of up to 100mV depending on vibration amplitude. This signal is filtered to remove unwanted ac interference and amplified.

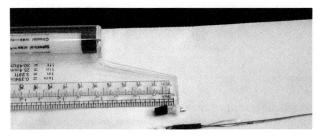

Figure 6.4b *Picture of finger sensor*

If the robot finger contacts any material of different temperature, its temperature varies suddenly and this change is detected by the thermocouple giving an indication of the material's thermal conductivity.

6.3   Operator input systems

It is essential that the operational control of the robot is simple and natural, requiring the minimum of distractive thought which would hamper the effective operation of the system. A number of input channels are required.

6.3.1   Base motion control

To make the operator/robot motion control interface user friendly, foot operation is employed. Movement of the mobile base is regulated by the operator's foot movements via a two axis pedal. This controls the robot's motion in all directions (forward, reverse, left and right). For ease of use and to permit further expansions, the input pedal can be operated using one foot, leaving the second foot free for other tasks. The input system uses position sensing transducers to provide voltages proportional to displacement of the foot from an adjustable centre point.

These signals are transmitted to the robot via a 40MHz 1.5W FM radio link interfaced to the motor drive controller. The speed and radius of turn of the robot are directly proportional to the displacement of the operator's foot. The input pedal can be used effectively while standing but for comfort the operator is usually seated.

6.3.2 Head motion tracking

Pan and tilt of the robot head is to be co-ordinated with the movement of the operator's head. To achieve this objective, a sensor system must be produced to monitor the operator's motion. Motion monitoring systems based on magnetic effects (Polhemus [11]) have been used for this task, as have static position monitoring frames [12]. Unfortunately, these devices operate relative to a fixed frame and cannot easily be moved. In addition, the costs are relatively high. To provide low cost, flexible, and accurate results, the operator wears a helmet on which are mounted two rate gyros (Futaba 153 BB) whose outputs are voltages proportional to the helmet's acceleration in pan and tilt respectively. After amplification and conditioning, these voltage signals are double integrated by computer to obtain the angular position of the operator's head. Gyro and component drift require that the zero acceleration point be reset every ten to fifteen minutes using a foot switch on the free foot. This has proved to be of negligible interference significance and is one of only very few tasks required of the operator which does not involve the robot mimicking human movements.

The head angles are converted to analogue voltages, amplified and filtered before being transmitted to the robot via the 40MHz link. Control circuitry on the robot aligns its head with that of the operator. The performance of the robot head when tracking the operator's inputs is shown in Figure 6.5.

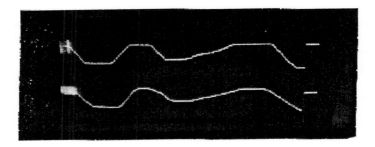

Figure 6.5 *Robot head response*

It can be seen that the robot head follows the helmet with a delay of 0.25s. Although this is acceptable, the delay will be reduced by implementing more efficient integration processing techniques.

6.3.3   *Glove and sleeve input systems*

Two completely different input systems are used to control the robot arms. These provide comparative studies of the effects of controlled tele-presence. The SCARA arm will use a traditional master arm arrangement to drive both the arm and the gripper. The anthropomorphic arm combines gyro, optical and Hall effect techniques in a light, flexible 'data sleeve and glove', as shown in Figure 6.6.

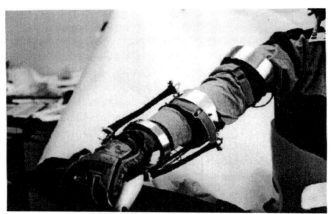

Figure 6.6 *Data Sleeve and Glove*

A flexible tendon is connected from the sensor mounted on one limb across the measured joint to an anchor point on the second limb. The increase in tendon length when a joint is flexed is proportional to the angle of that joint. As the limbs flex, the tendon pulls a shaft out of the optical sensor against a return spring. The end of this shaft is mirrored to reflect IR radiation from a combined IR transmitter/receiver at the end of the tube. The change in path length of the radiation is sensed as a change in output of the receiver. Up to 40mm of movement can be measured. This position is converted into an angle and then transmitted to the robot.

The transfer characteristic of the sensor is shown in Figure 6.7. It can be seen that the output is not linear and in order to calculate a joint angle a look-up table is employed providing repeatable angle measurements to within ±2.5°.

Similar optical sensors are used to measure wrist flexion/extension, and adduction/abduction.

The operator wears an in-house designed data glove, Figure 6.8, to measure the joint angles of the middle and proximal finger joints and control the motion of the end-effector. It employs Hall effect sensors and magnets positioned on the back of each finger joint and three mounted on the base of the thumb. As each joint moves, the signal from the Hall effect varies relative to the joint angle. These signals are processed by computer and the resultant angles used to control the robot hand.

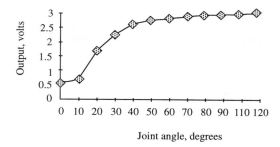

Figure 6.7 *Optical sensor response*

Figure 6.8 *Hall effect data input glove*

6.4 Operator feedback designs

6.4.1 Stereo sound and vision

Visual and audio sensing formed two of the first sensory characteristics possessed by robots and they form a well tested range of systems [14]. In this application two colour CCD cameras with integral microphones, auto-iris and auto-focus are mounted 100mm apart on the robot's head in order to provide the operator with a stereo image.

Each camera has an array of 320000 pixels, a power zoom lens (x8) with focal length variable from 6-48mm providing a field of view of 5° to 15°. Zoom across the full range takes 4 seconds. Wide angle adapters are used which extend the field to 30°. Each camera is powered by a NiCad battery providing in excess of 30min operation between charges.

The outputs, both video and audio from these cameras are transmitted via a 1.3GHz radio link [15] to the operator's station where the pictures are displayed using LCD array displays weighing less than 500g. The operator is able to use the zoom facility to concentrate on areas of interest. This is again operated by foot and uses the radio link previously mentioned.

Prisms and lenses are used to provide an adjustable eyepiece separation and single eyepiece diopter correction. The image angle subtended at the eye is 90.° The sound received from the robot is amplified and fed to two speakers in the helmet. This creates a stereo sound image for the operator.

6.5   Virtual tactile representation (tele-taction)

Tactile feedback (tele-taction) of cutaneous data (pressure, vibration and thermal) is a concept of growing importance in Virtual Reality and tele-manipulation applications [16].

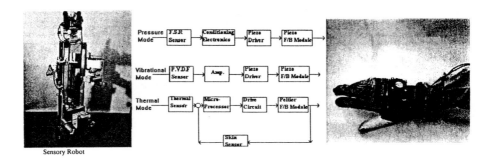

Sensory Robot

Figure 6.9 *Tactile feedback system schematic*

As these tactile feedback devices are to be used by operatives during normal motion, they are required to be light and compact. The units are designed to fit inside a glove (as described for hand control) which could be worn by the operator without restricting motion or comfort. This gives a full tactile closed loop. A schematic of the system is shown in Figure 6.9.

Outputs from various sensory modules are transmitted from the sensory digit previously described to the control glove and the appropriate feedback transducer through conditioning electronics.

6.5.1   Thermal sensation

The thermal feedback unit has been constructed using as a basis a Peltier effect heat pump which weighs less than 10g with overall dimensions of 15mm x 15mm x 3mm. The power output of 15W allows very rapid cooling and heating (20°C/s) of the operator's finger in response to stimuli from the sensor while enhanced sensation discrimination is achieved by mounting a rapid response thermo-couple (response

10ms) on the face of the Peltier module in contact with the operator's skin [9].

The thermal feedback module is fitted within the glove index finger on the back surface of the first (proximal) joint where the density of thermal sensors is high and the skin is thinner. An upper temperature limit of 50°C was set to prevent injury, while the lower limit was -5°C.

6.5.2 Texture/slip sensation feedback

The nerves in the skin such as the Meissner corpuscles detect texture/slip as a series of vibration changes when the object moves relative to the finger [17].

To simulate these vibrational effects a piezo-electric pulse unit was constructed from a PZT (lead zirconate titanate) ceramic disc 10mm in diameter and 1mm thick, mounted on a 1mm metal plate of diameter 15mm for added robustness. The weight of this unit is less than 2g and is driven from a high voltage (160-300V) source converting the electrical signals from the texture sensor on the multi-functional digit into mechanical motion.

For safety and to prevent contamination from dust or moisture, the transducer has been enclosed in a PVC film. The drive electronics are removed from the hand/arm to reduce the mass.

6.5.3 Pressure sensation feedback

The primary use of pressure/force data is in the regulation of manipulative actions. Pneumatic feedback techniques have previously been applied to this domain with some success, although problems associated with this technique are the 'spongy' feel and the need for a pneumatic power pack and associated equipment [18].

An alternative approach is to use the pulsed mode outputs from the texture modules already mentioned. This technique makes use of the learning and adaptive abilities of the human brain. Using this approach, the forces detected by the artificial skin are transferred to the finger as a series of pulses of increasing frequency. This design has the advantage that no new feedback modules need to be incorporated into the glove, reducing system mass and complexity.

The piezo-electric module (used for both texture and pressure feedback) is attached to the inner index finger lining of the sensory glove under the distal pad (the most sensitive pressure and texture sensing site) of the finger.

6.5.4 Pain sensation feedback

No dedicated pain feedback module was specifically developed, but dangerous inputs are reflected to the operator using thermal stimuli.

6.6   System testing

The robot vision, manipulator and motion control systems were tested in a number of ways in order to assess their ease of use and the learn time required for operators to become proficient in using the systems.

6.6.1   Motion control testing

The base control pedals in conjunction with the vision system and radio control circuitry were tested by six operators who had no previous experience of the system. The aim of the test was to determine the ease of use and accuracy of the motion control input system and the visual sensory needs. The response was tested using a series of different visual configurations namely: mono/stereo feedback, colour/monochrome feedback, and standard /wide angle lens. The experimental task involved navigating the robot around and through various obstacles such as benches, tables, chairs, moving objects and doors, in a remote room, as in Figure 6.10. The width of the robot base (63cm) meant that at times the obstacle clearance was less than 5cm. The time taken to complete the course, together with the number of collisions was recorded in order to assess the ease of use of the system and the speed at which accurate control was learnt. Each operator made three attempts at the test. Operator comments and observations were also noted.

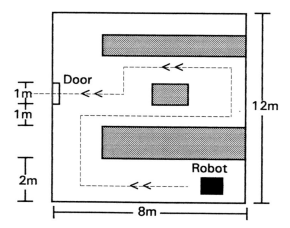

Figure 6.10 *Navigation path*

Initial testing was undertaken with a mono, monochrome, standard field of view configuration. This was used as it is the poorest visual quality. Without exception, all of the test operators found the system very natural to use and within ten minutes could navigate with no or very minor collisions (such as when passing through the narrow doorway). The times taken to complete the task reduced from an average of 5 minutes on the first attempt to less than three minutes on the third attempt. As well as a time decrease, the number of collisions reduced, on average, from seven to just over one from the first to the third attempt. Further improvements with training would be expected.

The visual configuration was then changed to consider the other variables such as stereo vision, colour and wider angle of view. It was found that colour vision did not enhance performance of a purely navigational task. Stereo vision did not affect the speed of completion of the task (if there were no collisions) although it improved depth perception substantially allowing the robot to be positioned more accurately especially when close to an obstacle and hence reducing the danger of collision. The use of a wider field of view was found to both enhance task speed and reduce collisions with all operators reporting a more natural perspective. The operators also advised the use of self centring pedals to ensure a positive stop position.

6.6.2 Manipulator testing

To test the SCARA arm and manipulator when integrated with the vision system a series of tasks was devised. These included:

i) The movement of a cylinder from one point to another
ii) The movement of a cylinder from one point to another around and between obstacles.
iii) The movement of an object around obstacles and placing it at a desired end location.

Performance was measured using the following criteria:

i) The time taken for successful completion of the task
ii) The number of non-catastrophic collisions (defined as not toppling or moving an obstacle to outside its area)
iii) The accuracy of final placement of the object.

The arm was controlled using two double axis joysticks and the time taken for each task was recorded, together with collision occurrences. These tests were initially conducted with the vision system in the stereo, monochrome and wide angle configuration. As can be seen from Figures 6.11 and 6.12, the learn time is very short. As with the guidance system, changes from monochrome to colour have no effect on performance. The use of a standard lens also had little effect provided the objects remained within the field of view at all times. Changes from mono to stereo did affect the accuracy and/or the speed, i.e. if a mono system was used accuracy was possible but greater time was needed.

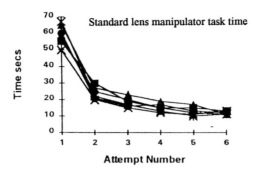

Figure 6.11 *Performance assessment*

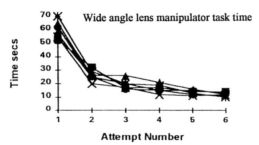

Figure 6.12 *Performance assessment*

6.7   Tactile feedback system testing

6.7.1   *Texture/Slip feedback testing*

Testing of the textural performance by a number of operators (training period 10 mins) revealed that correct identification of a series of surfaces was possible for in excess of 90% of cases. The texture variations were defined according to the spatial frequency and amplitude criteria [9].

The operators reported that recognition of surface texture was relatively simple, and felt 'fairly' human. The differences between natural texture sensations and the fed back signal were felt to be due to:

i) the extremely high sensitivity of the texture sensor, which is beyond human resolution.

ii) the whole finger tip being stimulated as one lumped sensing area rather than many 10s or 100s of sensing points distributed over the skin.

In similar tests for slippage, the operators easily detected motions of 0.5mm or more, but with very small motions (0.1mm) it was difficult to distinguish slip stimuli from a vibration spike. As there is only one sensing point, it was not possible for the operator to say how far the object had moved.

6.7.2   Thermal feedback testing

Thermal feedback tests were conducted with temperatures ranging from 0°C to more than 50°C. When the temperature difference between objects was more than 3-4°C, accurate discrimination was possible, but with smaller changes the recognition ability was reduced to less than 85% correct. These limits are due to the thermal resolution of the human finger. The tests further revealed that combining force data with thermal data increased the correct identification levels to above 90%.

Operators reported that the feedback sensation was lifelike with little thermal lag.

6.7.3   Force/Pressure feedback

Tests of the pressure feedback mechanism were conducted with forces of 2N (0.5Hz), 10N (2Hz), 30N (6Hz), and 60N (10Hz). The feedback pulse frequencies are in brackets. This represents graded changes from the minimum to the maximum detectable forces.

Within these ranges, the operator's force discrimination abilities are good, despite the rather unnatural form of the feedback, see Table 6.1.

Force	Correct Identification
2N	91%
10N	94%
30N	95%
60N	97%

Table 6.1 *Force recognition tests*

When the range of forces was fully variable within the 0.2-60N range, it was found that the operator could recognise the applied force on a gross scale i.e. low pressure, medium pressure, high pressure, and changes in pressure could also be detected, but the users were not able to give exact force estimates accurately (e.g. the force is 32N). This is similar to normal human abilities to estimate applied forces. When the pulse measurements were combined with texture or thermal signals there was no measurable deterioration in performance. The operator simply concentrated on the data required, as occurs in normal usage. The operators also reported that although the sensation was totally different from normal tactile perception of force, the signal was easy to use and to adapt to.

6.7.4   Pain feedback

Pain (danger from force and thermal stimuli) inputs were simulated by rapidly increasing the temperature towards the thermal limit of 50°C. Simulated danger (overload) signals when transmitted to the operators were detected and responded to in an average of 0.9sec.

6.8   Conclusions

The human hand with its complex anatomical structure and network of tactile sensing and proprioceptor elements forms the most versatile end-effector known. Generally, however, the complexity of the design, sensing, feedback and control has mitigated against the widespread use of mechanisms which mimic hand operations, preventing truly efficient tele-manipulator operation.

These short-comings highlight the need for an advanced instrumented robot; comparable in function (where possible) with the human body, integrated into a sophisticated tele-presence system to provide the operator with a full range of motion inputs and sensory feedback data.

Modules for leg, arm, hand and head motion monitoring have been developed to control the activities of a twin armed mobile robot. Cameras, microphones and a multi-functional tactile sensing system provide feedback signals that give the operator a relatively realistic impression of the robot's activities. The signals from these sensors can be fed to the operator as direct stimuli, comparable in many instances with normal sensation. All the feedback modules have been made sufficiently light and compact that they can be worn by the operator during normal usage, without restricting motion or comfort. Comparisons of various visual, control, audio and tactile techniques have also revealed the optimum sensory configurations for specific tasks.

Key areas in which this type of technology could be applied cover a range where environment conditions are too hostile or expensive for human activity. Areas would include: **nuclear, clean room, bomb disposal, explosive handling, space, sub-sea,** and **toxic chemical and bio-hazard handling**.

6.9   Acknowledgements

The authors wish to acknowledge the support given by the EPSRC under grant no. GR/J92637.

6.10   References

1.   BEJCZY A.K., Sensors, Control and Man Machine Interfaces for Advanced Tele-operation, Science, Vol.208, pp.1327-35, 1980.

2.   HAGNER D.G., and WEBSTER J.G., Telepresence for teleoperator systems, IEEE Trans. Systems Man Cybernetics, 1988.

3.   -----, Force Display in Virtual Environments and its Application to Robotic Tele-operation, Workshop S4, IEEE Conference Robotics and Automation, Atlanta, USA, May 1993.

4.   FISHER P.J. and DANIEL R.W., Real Time Kinematics for a 6 DOF Telerobotic Joystick, 9th CISM-IFToMM Sym. on Theory and practice of Robots and Manipulators, pp. 68-9, Udine, Italy, Sept. 1992.

5.   YOKOKOHJI Y., OGAWA A., HASUNUMA H., and T.YOSHIKAWA, Operation Modes for Cooperating with Autonomous Functions in Intelligent Teleoperation Systems,IEEE Conference Robotics and Automation, Atlanta, USA, May 1993.

6.   Causse O. and Crowley J.L., A Man Machine Interface for a Mobile Robot, IROS '93, Yokohama, Japan, pp.327-338, July, 1993.

7.   NISHIYAMA T., SCHULTZ R.J., NAKAJIMA R., and NOMURA J., A Parallel Processing Architecture for a Telepresence Mobile Robot, International Conference Advanced Mechatronics, pp.100-105, Tokyo, Japan, Aug. 1993.

8.   CALDWELL D.G., MEDRANO-CERDA G.A., and GOODWIN M.J., Braided Pneumatic Actuator Control of a Multi-jointed Manipulator, IEEE Int Conference on Systems, Man and Cybernetics, Vol. 1., pp.423-28, Le Touquet, France, Oct. 1993.

9.   CALDWELL D.G. and GOSNEY C., Enhanced Tactile Feedback (Tele-Taction) using a Multi-Functional Sensory System, IEEE Robotics and Automation Conference, Atlanta, Georgia, 2-7th May 1993.

10.   Farnell Electronics, Data Sheet 5524, 1993.

11.   RS Components, Data Sheet 3289, 1993.

12.   YANAGIDA Y. and TACHI S., Virtual Reality system with coherent Kinesthetic and Visual Sensation of Presence, International Conference on Advanced Mechatronics, pp.94-99, Tokyo, Japan, Aug. 1993.

13.   RXM Ltd, UHF Data Sheet, Feb 1993.

14.   EDWARDS P., Robotic Vision Sensing, IEE Colliqium, Digest, 125/81, London, UK, 1981.

15.   Camtech Electronic, Company Literature, 1993.

16.   CALDWELL D.G., BUYSSE A. and ZHOU W., Multi-sensor Tactile Perception for Object Manipulation/Identification, IEEE/RSJ IROS '92 Conf, pp. 1904-11, Raleigh, USA. July 1992.

17.   SCHMIDT R.F., Fundamentals of Sensory Physiology, Springer-Verlag, 1985.

18.   STONE B., Human Factors and Virtual Reality, Comett Workshop on advanced Robotics and Intelligent Machines, pp. 25-27, Manchester, UK, April 1993.

Chapter 7

Sensing and sensor management for planning [1]

P J Probert, A P Gaskell [2] and Huosheng Hu

7.1   Introduction

Most mobile robots or autonomous guided vehicles (AGVs) follow fixed paths either through using an inductive sensor to locate a buried wire or a simple optical sensor to follow a white line. They have, and need, very limited sensing capabilities. In the face of obstacles, they have no option but to stop. Even in a fixed environment, there are drawbacks, including erosion of painted lines, and the expense of altering the work area, especially if buried wires are used. These limitations have seriously hampered commercial applications of AGVs.

Many areas in which mobile robots could make an essential contribution, such as nuclear plant, oil rigs or surveillance demand greater flexibility and a degree of adaptation to the environment. In all of these applications, mobile robots may be required to explore and map environments, and make changes to predetermined plans to cope with uncertain conditions. Such requirements demand that mobile robots have reliable on-board sensing to abstract suitable information from an unknown or dynamically changing environment.

Various sensors can be used in robotic systems to achieve autonomous navigation: for example proximity detectors, shaft encoders, cameras and a variety of rangefinders based on technologies such as sonar and optoelectronic devices. Since every type of sensor has specific areas of operation and certain failure modes, a single sensor can only provide partial information. Therefore to handle general environments, multi-sensor systems become necessary. Many important issues are involved in multi-sensor robots such as establishing a common framework to include different sensors setting up communications between the various sensor systems handling noise, error, and conflict in sensory data, planning strategies for sensor integration.

In this chapter we describe sensors which are being developed for robotics, and discuss how they might be used together to provide a robust facility for real-time sensing, planning and control. The main requirement for robot control and planning is to determine the range of nearby objects - i.e. how far away they are from the robot. In the next section, we describe three categories of sensor for measuring range: those based on sonar, on optoelectronics, and on vision.

[1]This work was supported by the ACME Directorate of SERC
[2]Alexander Gaskell is supported by British Nuclear Fuels plc on a SERC CASE

The most obvious category missing is radar, which is being used, especially by the military, for off-road and outdoor vehicles, but its cost is still high compared with other technologies. We describe how data from a single sensor or from several sensors may be integrated to improve robustness in Section 7.3. We emphasise the importance of quantifying uncertainty so that any decisions made are both informed and have quantifiable risk, illustrating sensor integration methods using an application in obstacle avoidance developed for a factory vehicle guided by a sonar array. In Section 7.4, we describe sensor management, developing the same obstacle avoidance application but now in a multi-tasking, multi-sensor scenario. Finally, we conclude with a brief discussion on current and future research areas.

7.2   Seeing the environment: Common sensors

The type of environment which might be encountered, and its representation for planning, should direct the choice of sensors in robotic systems. As we see in this section, different sensors provide different types of information, and no sensor is completely accurate.

7.2.1   CCD cameras

People rely almost exclusively on vision for finding their way around. CCD cameras are the most obvious electronic parallel and computer vision has been a major research topic for many years. Some impressive results have been obtained using specially built processing hardware, for example guiding vehicles on roadways [1, 2].

However, for more general systems, particularly those in which there is clutter, there are some significant problems which remain to be solved, especially in the image processing:

- picking out the important information from very detailed images, especially in cluttered environments or in varying light conditions

- abstracting range information from intensity images

- achieving a high enough throughput rate.

Although vision systems are still expensive in comparison with the simpler direct range sensors we describe in the rest of this section, the processing and update rate can now be achieved within a frame rate of 25Hz. This is possible largely because of fast parallel processors such as transputers and, more recently, the C4O architecture, together with special purpose vision hardware to provide high bandwidth data acquisition; the perception of vision as a real-time sensor has gone hand in hand with development in processors. In addition, further speed up in bandwidth is achieved through algorithms which restrict themselves to regions of interest in the image rather than the whole image: for example through the prediction of the position of features between frames from knowledge of self-motion [3].

CCD cameras measure intensity; for planning robots need to know range. Stereo range measurement uses the difference in offset of features in the image between the two viewpoints and is now, of course, popularised in 'magic' books of computer generated stereo image pairs (best sellers in the UK when this chapter was written). Although the subject of extensive research, there remain major problems in extracting range from two stereo images. First there is the correspondence problem - deciding which features in one image correspond to those in another. The time taken staring at pairs of stereo images before the picture comes to life suggest that people find this quite hard (indeed some people, such as the first author, never manage it convincingly!) and there are no unambiguous methods for doing it by computation. In addition there is the problem of calibration: converting the offsets in the imaging plane into real range offsets in the world. This is a major problem in vehicle mounted cameras owing to the vibration.

People do not explicitly determine range, or necessarily use both eyes for navigation, so why should robots? Alternative methods of processing exploit the relative motion of features across successive images [4]. One type of motion is described by optic flow, derived from the intensity field in the image, but in practice the computation of optic flow is very prone to noise and error. However imposing additional constraints can make systems more robust and the computation of scene structure for navigation has been demonstrated using these methods[5].

Another way to measure range from CCD cameras is to project structured light, and use relationships between different parts of the image either to improve stereo matching or to determine range directly through triangulation. We discuss triangulation sensors in Section 7.2.3.2.

7.2.2 Sonar Sensors

It is very much easier to detect range directly, usually at the expense of including mechanical scanning and building a bit of electronics. One of the simplest ways to determine range is through measuring the time of flight of ultrasound signals, which are reflected by most common objects. Ultrasound signals can be induced through the piezo-electric effect or through electrostatic forces. Most sensors used in robotics are electrostatic since this mechanism is more efficient for coupling into air. Typical frequencies are between 40 and 100kHz, the higher frequencies being easier to focus but suffering greater attenuation. The same transducer can be used for transmission and reception. Two sonar transducers, mounted on stepper motors are shown in Figure 7.1. They are typical of the type used in many robotics applications.

Sonar sensors give good results when pointed at surfaces which are close to normal incidence. However they have two major problems, both related to the rather long ultrasound wavelength of a few millimetres (long, that is, compared with light). First the beam is hard to focus and typically the beamwidth to the half power points is about 15° - 20°. Secondly, many surfaces appear specular; that is they look like a mirror to the sound wave so that, rather than a portion of the sound being reflected straight back to the source, it is bounced around the room first.

Figure 7.2 shows repeated readings from a single sensor mounted on a vehicle, first at a fixed distance from an obstacle and then as the vehicle moves towards or away from the obstacle.

Figure 7.1 *Two polaroid sonar transducers mounted on servo motors*

The sensor measures the time of flight to the first return of ultrasound energy following a firing pulse. There are several which we can see in these data. First, there is very spiky noise, which is hard to correct with filtering as it is so non-uniform. The spikes which represent erroneously short ranges are probably caused by noise sources in the environment. Those which represent erroneously long ranges are probably specular readings. In addition we see a pronounced overshoot as the reading reaches maximum range in the right hand picture (this is less pronounced when the vehicle moves more slowly).

For feature detection, distortion as well as noise is a problem. Scattering occurs around corners. Another difficulty becomes apparent when a sonar sensor is scanned as the large angular beamwidth causes distortion: for example flat surfaces appear curved. This effect, and specularity, can be exploited and are the basis of work such as that by Leonard and Durrant-Whyte on the representation of sonar readings for feature representation [6]. Rather than thinking in terms of range maps, they have introduced the notion of *regions of constant depth* which are angular intervals across which constant range is measured. For a planar surface this interval corresponds to the beam angular width. Other features similarly have distinctive patterns: for example cylinder, corners and edge targets. Sonar sensors have been used in many vehicles for guidance and for mapping, usually using a two-dimensional grid representation of the world [7, 8].

Modelling the effects of specular reflection has led to their application in localisation [9, 3]. Sonar based navigation amongst obstacles has been demonstrated by several groups, one of the most successful systems being described in [10].

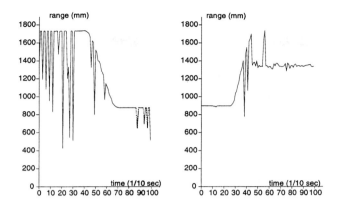

Figure 7.2 *Range readings plotted against time as a sonar mounted on a vehicle moves relative to an object (thanks to Jan Grothusen for the data)*

7.2.3 Optoelectronic sensors

The wide beamwidth of sonar sensors is useful when they are being used as a bumper, or for gathering crude information on the environment. However they have such poor angular resolution that they cannot be used for accurate navigation alone without extensive processing. In addition, as we have seen, they are very noisy, mainly as there are a large number of ultrasound sources in the environment.

The problems arising from the long wavelength of sonar are largely removed by using optoelectronic sources such as light emitting diodes and lasers.

7.2.3.1 Optical amplitude detectors

The simplest sensors project light from a source such as a light emitting diode and simply measure the amplitude of light reflected from close objects onto a photodiode. To gather enough light some type of gathering lens must be used. Figure 7.3 shows a sensor [11] which uses a parabolic reflector for emission and receiving, based on a car headlamp. The amplitude of the reflected beam at the detector depends on range and on reflectivity and these quantities cannot be separated. Therefore amplitude detection will not determine absolute range but is a good way of detecting discontinuities. Results from the sensor are shown in Figure 7.4.

Alternatively amplitude detection can be used alone through suitable modification to the environment. For example a bar code scanner on the Oxford AGV (Figure 7.5) is simply an edge detector; the bar codes are designed so that each presents a different edge pattern. Knowledge of the position of each bar code allows the robot to determine its position to an accuracy of a couple of centimetres.

Figure 7.3   *Optical amplitude detector. Note the large reflector to gather light. This sensor has mounted on it a sonar transducer as well*

Figure 7.4   *Data from the optical amplitude detector shown in Figure 7.3. The scene is made up of cardboard screens. Because the reflectivity of all parts of the scene is similar the data follow the contour well; however recalibration would be needed if the reflectivity were changed. The small triangle shows the location of the rotating sensor*

Figure 7.5 *Laser scanner for reading bar codes for robot localisation*

7.2.3.2   Triangulation sensors

Optical sensors can also be used to measure range. The traditional way to do this uses triangulation, which uses a geometric layout between an active source and a detector such that there is a one to one correspondence between the range at which light from the source is reflected and the lateral offset on the detector. This is shown schematically in Figure 7.6, which shows a beam from a laser source being reflected and focused onto a detector through a lens. Calibration is needed to determine the range from the offset, although early processing can often take place on uncalibrated data.

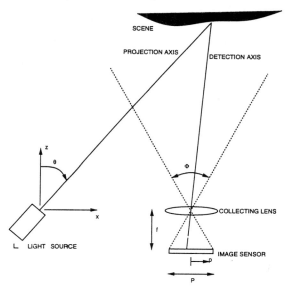

Figure 7.6 *One dimensional active triangulation geometry*

The detector is based either on photodiodes, usually a long two terminal photodiode structure called a lateral effect photodiode (LEP), or on a charge coupled device (CCD) array, such as a video camera. CCD arrays are more sensitive. The advantage of LEPs is the simplicity and high bandwidth of the interface since only the currents at each contact, I_1 and I_2, need be measured and the offset on the detector, p, is given by the difference over the sum:

$$p \frac{I_1 - I_2}{I_1 + I_2}$$

(7.1)

In addition LEPs are suitable as detectors in systems which are modulated and therefore robust to alterations in ambient light.

Triangulation sensors are designed to cover a particular depth of field (which we define as the difference between maximum and minimum ranges to be detected). It is clear from Figure 7.6 that the reflected beam falls off the ends of the detector outside certain ranges. Triangulation can be very accurate for narrow depths of field, especially for close ranges. The reduction in performance with increased depth of field (which is described by a figure of merit called the *triangulation gain)* is a problem in their application in mobile robots where usually quite a large depth of field is needed. As we would expect, accuracy also decreases with lower powered sources.

Measurements on an LEP sensor developed in Oxford by Pears [12] to direct obstacle avoidance show that sub-millimetre repeatability of readings can be achieved at ranges below 1m, increasing to a few centimetres at 2.5m, for reflections from surfaces such as cardboard boxes. The repeatability depends on the fourth power of the distance (rather than on the square as we would expect from an isotropic source) owing to the triangulation gain. The sensor uses an eye safe laser diode and covers a range from 0.4m to 2.5m.

Typical scans are shown in Figure 7.7 for two features: a Cartesian scene consisting of flat surfaces at 90° (boxes) and a cylinder. Information on amplitude is shown as well. In contrast to the results shown in Figure 7.4 it is clear that for some shapes amplitude alone is not a reliable indicator of distance, and, with the laser source, there is significant scatter at corners. However amplitude can be used in conjunction with the range to improve feature detection. First, for the LEP sensor, the amplitude of the signal received (i.e. the *sum* of the currents $(I_1 + I_2)$ - remember that the range is given by the *difference* of currents over the *sum* (equation 1)) provides an estimate of the variance of each range reading [12] and therefore of how much confidence can be placed in each reading. Secondly, certain features have distinctive amplitude profiles, such as the cylinder which gives a very strong return when the incident beam is normal to the surface.

7.2.3.3 Lidar sensors

Another low cost method of detecting range uses modulated light in techniques known generically as lidar (light detection and ranging). Recent developments have concentrated on using phase detection on amplitude modulated carriers in a technique often called AMCW (Amplitude Modulated Carrier Wave). The optical device, light emitting diode or laser diode, is modulated at a radio frequency of a few MHz. The path length is determined in terms of the phase difference between the outgoing and incoming signals at the modulation frequency (Figure 7.8). The high quality of the data has led to the application of such sensors in navigation and mapping [13,14].

The accuracy of AMCW sensors is limited primarily by the accuracy of the phase detecting circuits and the stability of the modulation. It usually falls below triangulation sensors at close range but falls off with the square rather than the fourth power of distance. Typical accuracy is about 1-2 cm at 1m. There is no fundamental limit on maximum or minimum range assuming that the optical layout is well designed.

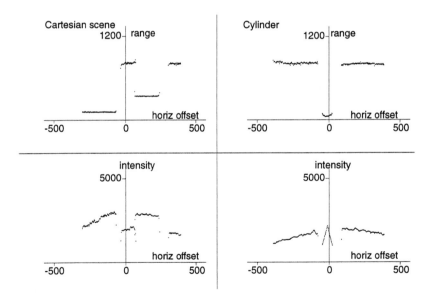

Figure 7.7 *Range and amplitude data from Pears' LEP triangulation sensor. Note the scatter in amplitude at the corners in the Cartesian scene (which was made up of boxes) and the sharp peak in amplitude at the centre of the cylinder*

The high quality of data available from AMCW sensors is shown in Figure 7.9. The sensor which produced these results was designed in Oxford by Brownlow [15]. The electronics was designed for high linearity between range and phase to minimise the need for calibration (less easy than it sounds as many standard amplifiers introduce phase shifts which vary with level). The total cost of the hardware is comparable with a low cost CCD camera, but the cost of processing is at least an order of magnitude less.

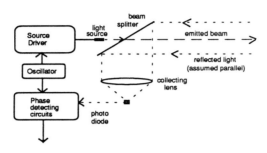

Figure 7.8   *Schematic operation of AMCW sensors*

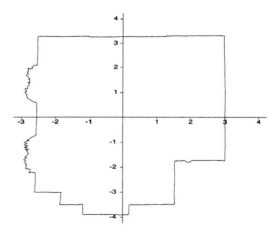

Figure 7.9   *Raw range data from Brownlow's AMCW sensor (15) illustrating the excellent linear characteristic. The data represents internal walls and doors. The two fragmented surfaces on the left are beige, loosely woven curtains. Scale in metres*

Optoelectronic sensors give better noise performance than ultrasound and, a major advantage for processing, exhibit mainly Gaussian rather than spiky noise. The noise behaviour of AMCW and LEP triangulation sensors has much in common. In particular it can be shown that for AMCW sensors as well as for the LEP sensor [14] the variance associated with any particular range reading can be estimated through measuring the *amplitude* signal returned (recall that the range estimate is given by the *phase*). This is a major advantage in the systems which use statistical methods of sensor processing such as those described in Sections 7.3 and 7.4.

7.3   Sensor integration

No matter how good a sensor is or how efficient the signal processing algorithms are, uncertainty in sensor data cannot be eliminated totally. The use of multiple, physically different sensors can help to overcome the shortcomings of each individual device. Further, the integration of redundant data from multiple sensors can reduce uncertainty especially if an individual sensor fails. Typical methods for integrating data from multiple sensors can be categorised as follows:

- a qualitative approach - symbolic descriptions of sensory information are used, such as rule bases [16]

- a quantitative approach - using numerical methods which can incorporate uncertainty, for example through using probabilities [17,18].

Both methods are widely used in robotics research and each has areas of application. However, there are problems with each too: rule bases provide only qualitative information whereas statistical methods may rely on accurate models which are not always possible to formulate. There is a need to integrate both methods to achieve robustness in a changing real world.

7.3.1   *Qualitative approaches*

Qualitative approaches use powerful representation mechanisms to describe complex sensor information and heuristics to guide data interpretation. They are widely used when representation is important and quantification is difficult. They are typified by rule bases, such as expert systems. Rule bases are convenient when a problem can clearly be described in a few statements of the form 'if.. then'. They allow us to build expert information into a system. However they are limited: they cannot handle numbers easily, they become complex very quickly and they do not incorporate uncertainty with any exactness.

A simple example of a rule base was developed by Flynn [16] to combine sonar and optical amplitude data to guide a mobile robot (the sensor hardware was very similar to the sensors described in Sections 7.2.2 and 7.2.3.1). This is a simple, typical example of exploiting the complementary characteristics of the two types of sensor. The sonar was used to provide range information and the near-infrared sensor localised doorways and edges. The integration of data from both sensors was conducted according to simple rules.

Although the rules allow a mobile robot to build a reasonably detailed map of the environment, there are limitations. For example specularity in the sonar sensor and sensitivity to different colours for the optical amplitude sensor are not taken into account. Therefore, there must be considerable risk associated with any robot actions based on this map.

Simple rule bases can also be used to integrate data spatially from a single type of sensor. Hu, Brady and Probert developed rules [19] to interpret data from a three sonar array. The array was to detect unexpected obstacles in the path of a robot in an industrial environment. Using the three sonars together provides more detailed information on the location of obstacles than using three separate readings, as is clear from Figure 7.10 which shows the superposition of the three wave fronts at some distance. Each sonar sensor has a beamwidth of about 30 degrees, so the beams overlap at distances greater than about 0.7m.

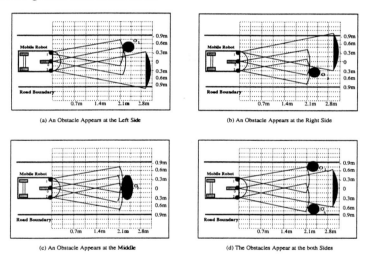

(a) An Obstacle Appears at the Left Side

(b) An Obstacle Appears at the Right Side

(c) An Obstacle Appears at the Middle

(d) The Obstacles Appear at the both Sides

Figure 7.10   *Beam pattern from the three front sonars. The sonars are used to examine a zone ahead of the robot up to 3m in range. Note that the light shaded circles mean that nothing is found in this part of the zone*

The rule base provided information to guide the robot as it followed a predetermined path in a factory with a number of allowable pathways. It provided an estimate of the following states of the robot's current path: (i) the path was clear (ii) it was blocked by a small obstacle (such that a side-step manoeuvre past it can be performed within the pathway) (iii) it was blocked by a large obstacle (which entirely blocks the pathway, requiring the robot to backtrack and find a new pathway). The rules for interpretation of the data from the three sonars are described as follows:

Rule 1 **If** all of three sonars see a clear path, **then** the predetermined path is clear and no action should be taken.

Rule 2 **If** the middle sonar sees nothing in the cone of its acoustic beam, **then** the predetermined path is clear even though the two side sonars may see obstacles, as shown in Figure 7.10(d).

Rule 3 **If** any one of both side sonars sees clearance in the prediction zone and the road width W_r is at least twice the robot width, **then** even though the other two sonars may see an obstacle, there is room for a sidestep manoeuvre (i.e. blocked by a small obstacle) as shown in Figures 7.10(a) and 7.10(b)

Rule 4 **If** all three sonars see something, **then** the path is impassable (i.e. blocked by a large obstacle), as shown in Figure 7.10(c).

In Section 7.3.2 we show how the information produced by these rules can be improved through quantifying the uncertainty in each measurement.

Qualitative approaches have been presented elsewhere. A robotic system presented by Allen [20] demonstrated the integration of vision and touch for object recognition tasks. Vision and touch together provide geometric measures of the surfaces and features that are used in a matching phase to find model objects that are consistent with the sensory data. Stansfield [21] described a robotic perceptual system also utilising vision and touch, but in this case using an active touch sensor. A two-stage exploration is implemented. Vision is first used in a feedforward manner to segment the object and to obtain its position. Touch is then used in a feedback mode to further explore the object. Shekbar *et al.* [22] described how to localise an object using a manipulator end-effector that has been instrumented with both centroid and matrix touch sensors.

Qualitative approaches to the data fusion problem are gaining popularity in robotics because they offer the possibility of capturing the more human-like aspects of decision making. However, more work is needed to cope with the complexity and uncertainty inherent in the multi-sensor systems typical of robotics.

7.3.2 *Quantitative approach*

The problem with rule bases is that they do not handle quantitative information, such as numerical uncertainty on a sensor reading. Although bounds could be placed on uncertainty, only a few categories (e.g. 'small, medium, large') can be used, otherwise the rule base gets much too large for high bandwidth sensing. Since in a number of sensors, such as the optoelectronic ones we described in Section 7.2.3, the uncertainty of any measurement can be expressed numerically in terms of a variance, we would expect to improve the reliability of data interpretation through using this information.

The quantitative approach to sensor integration requires numbers. The information obtained from sensors is quantified so that statistical and decision theoretic methods can be adopted. Through quantitative analysis and mathematical modelling, considerable success has been demonstrated with such approaches. We describe two approaches in this section.

Bayes statistics

Bayes statistics, which use well established mathematical reasoning, can be used to integrate information from various sources (for example from multiple sensors or over time). Unlike the simple sonar rule base we introduced earlier, the Bayes formulation distinguishes between the measurement of a quantity and the true state of the quantity. To account for uncertainty, it reasons using probability density functions. Equation 2 defines the Bayes update rule, which updates an existing hypothesis on a quantity with new information:

$$P(0 \mid z) = \frac{P(z \mid 0) P(0)}{\sum [P(z \mid 0) P(0)]}$$

(7.2)

where 0 is the true state of the system (for example whether a path is clear or blocked) and z is the, possibly incorrect, observation of the state (for example by a sonar reading). In more detail:

1. $P(0|z)$ is the probability density function of 0 following an observation z (the *posterior* density function).

2. $P(z|0)$ is the *likelihood function* which describes how likely the true state is, given the observation that has been made. It accounts for errors in the method of observation.

3. $P(0)$ is the prior information about the state 0.

4. The summation on the bottom line simply normalises the result.

Bayes statistics therefore allow us to combine measurement information with prior knowledge, the beliefs already held about a situation. Sometimes, however, we may have no prior information. For a single reading, this is a serious problem. However if we are taking a number of readings, then we can use Bayes theorem iteratively as follows:

1. At time T start with some arbitrary prior distribution over all states: $P(0)$ (for example assuming that all are equally likely).

2. Take a measurement z. Then assuming we know the likelihood function $P(z|0)$ (we talk about this in a moment) use equation 2 to determine the new distribution $P(0|z)$.

3. Make this the new prior distribution and continue.

Convergence normally occurs over a few cycles even for a grossly incorrect choice of the prior. Of course if the environment changes suddenly, the new estimates of state will not keep up and there will always be a lag in the estimation.

Hu, Brady and Probert [19] demonstrate the use of Bayes statistics to improve predictions made with the rule base described in the previous section for the three sonar array. Recall that the vehicle is travelling along a predetermined path and has to detect and plan a path in the presence of unexpected obstacles. For now, assume that the obstacles are static and that their distribution is random. As we describe above, if an obstacle is detected the path may consist of two states:

$$0 = (0_1, 0_2) = (passable, impassable) \tag{7.3}$$

corresponding to small and large obstacles being present.

The Bayes update rule uses the likelihood function $P(z|0)$ to relate the measurement of state to its true value, and the value of this function determines the relative weightings given to the measurement and the prior information in the update rule. Incorrect values will either make a system unresponsive or over affected by noise. It is clear from Figure 7.2 that the sonar readings are corrupted by spiky noise so significant errors would be expected. Unless a clear physical model can be established (which might be possible for optoelectronic sensors, but not with sonar 5 noise profile) the only way to determine the likelihood function is through taking a large number of readings in known typical environments and analysing the success and failure rates over the states of interest. Details are given in [19].

Measurements show that the probability of missing an object which is present increases as the robot approaches the obstacle closely. This is because there is often no reflection from the object surface itself as it is specular. However there is always a return from an edge, so, at longer distances when the edges are in sight more often the object is more likely to be detected. In contrast, the probability of reporting a non-existent object increases for distant obstacles. This results from the greater effect of environmental ultrasound sources when the signal is low. Unfortunately this means that as the robot approaches an object the sonar sensors are less and less likely to see it - an undesirable property for safe sensing.

Following an estimate of state, the robot must decide what to do. It has two possibilities once an obstacle has been detected: to sidestep or to backtrack. We define formally the robot action space:

$$A = (a_1, a_2) = (sidestep, backtrack) \tag{7.4}$$

In the rule base described in the last section, action and measurement were linked implicitly. However this is incorrect as the action should depend on the *state,* not the measurement. Now that we have established a notion of the true state, we can set up the correct dependency structure. We also use our knowledge of the *uncertainty* of state to improve our chance of taking the best action. We define a *loss function* which quantifies the cost of the action in some common currency such as time or money. For example the loss of performing a sidestep if the path is passable is zero, whereas the loss if the path is impassable may be very high, either because the robot collides with something or because it has lost time in trying to pursue an action which is doomed to failure. We define a *loss function over* all states and actions:

$$L(0,a) \geq 0 \tag{7.5}$$

The *Bayes risk* then allows us to evaluate the expected loss of any action a_j. The Bayes risk of taking action a_j is the expected loss incurred through taking that action, summed over all states:

$$B(a_j) - \sum_i L(0_i, a_j) p(0_i)$$

$$\tag{7.6}$$

where $p(0_i)$ is the probability of 0 being in state i and $L(0_i, a_j)$ the loss associated with action a_j in state 0_i The robot would normally take the action which minimised the Bayes risk.

The Bayes update rule allows the robot to build up a gradual picture of its environment and to build on past knowledge. Reference 23 shows particular cases in which the use of the prior knowledge and temporal integration prevented possible collision from individual incorrect readings. The Bayes risk function allows informed planning decisions to be made, which assess the costs or benefits of various options open to the robot under uncertainty in the state.

The Kalman filter

The Kalman filter, or its non-linear form, the extended Kalman filter, offers an alternative approach to quantitative sensor integration for continuous, rather than discrete, measurement data. The filter uses two stages, each of which predicts an updated state: one based on the previous state and a model of how it changes between iterations, and one from an observation of the new state. Therefore, like the Bayes update rule, it combines both prior expectation with measurements (and indeed is an implementation of the same statistical models).

In the linear case, the Kalman filter is not very *fussy* about how the model is set up, but in the non-linear case, the extended Kalman filter can be very sensitive both to the model and to good estimations of variance. For example, for sonar sensors, abstraction of data to a higher representation than range is needed to build a successful model and to eliminate systematic errors, such as specularity [6].

Kalman filters are used too for spatial integration of data, for example to determine line segments from individual range points. They are particularly well suited to optoelectronic sensors since the *variance* measured range point can be estimated through the *signal amplitude* as we have described in Section 7.2.3 [14].

Kalman filters can been used to integrate data from several sensors. It is often difficult to use them meaningfully to integrate data from different types of sensor, as it is not possible to provide suitable models of each sensor or to find an appropriate common representation of the state. In a system which integrates sonar and optical amplitude detectors (almost identical sensors to those used in Flynn's rule base) Wen has shown how it is best to run Kalman filters *independently* for each sensor prior to sensor integration [24]. The advantage of his scheme over Flynn's is that the map includes information on the variance as well as the expectation of each state variable, so the risk of plans based on this map can be quantified.

7.4   Sensor management in obstacle avoidance

The concept of sensor management broadens out the ideas of sensor integration to include the control and scheduling of sensors. For example a sensor may have a limited field of view, either through hardware or software constraints. Sensors may have to be shared between tasks. In this section we extend the Bayesian framework to include sensor management as well as sensor integration.

The choice of sensor or of the mode of operation of a sensor is often based on minimising some measure of error or the concept of maximising information gain. These methods have been investigated by workers such as Hager and Manyika [17, 25]. We introduce a different measure of choice, choosing not the sensor which provides the maximum information but the one which is most likely to influence higher level decisions. These considerations lead us to the architecture of Figure 7.11, a more general representation of the measurement-state-action paradigm of the last section. The planner plans sensor use as well as robot actions and balances the benefits of sensing and action, to answer questions such as when is it better to delay an action until more sensor information is available (weighing up possible benefits in safety against costs in time).

The planning agent has available a set of sensing actions s_k and a set of external actions α_j. Information on the environment which influences these actions is passed up through a feature detector. The feature detector assembles and maintains a hypothesis on the expectation of various features, integrating information from sensors which it may control. The feature detector uses a Bayesian belief network [26], to integrate data from a number of sources. This representation has been chosen as it allows us to reason about dependence and uncertainty both between states over time and between sensor measurements.

7.4.1   Sensor planning cycle

The planning agent determines the best external action α_j in α_j given the current likelihood of each feature state 0_i from a predetermined, application dependent utility function, $U(0_i, \alpha_j)$. In effect, this utility function expresses the preferences for which action to take in a particular state, and provides the link to the higher level task being carried out. [3] Before making the final decision on an external action, the planner directs

---

[3]The utility function is similar in concept to the loss function described in the last section: minimising loss can be seen as the same as maximising utility. The decision on which to use depends on how the problem is formulated most easily.

the sensing head through sensing actions s_k to improve the estimates of state 0_i either until one is so certain that further sensing actions will have no effect on the utility or until a hard real-time deadline is reached and some action must be taken.

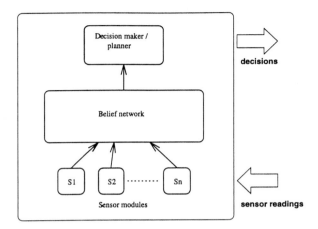

Figure 7.11   *Architecture for sensor planning*

Selection of the sensor actions s_k is based on *expected net increase in utility,* ΔU that would result if s_k were carried out and a different external action selected. The sensor that gives the highest ΔU is selected. Calculation of the value takes into account the time cost involved in carrying out the sensor action (gathering raw data and processing them). Hence the sensing-planning cycle is as follows:

1. The sensors report information which updates the probability density function *P(0)* computed by the feature detector.

2. The planning agent computes new expected utilities,

$$U(\alpha_j) - \sum_i U(0_i, \alpha_j) P(0_i)$$

3. If a real-time deadline is reached, the action α_j with maximum expectedutility U is selected.

4. Otherwise the planning agent looks ahead over possible sensing actions s_k to see whether any measurement of a component of the state might affect the maximum expected utility U_{max}.

5. If so, the sensing action s_k which offers the maximum expected increase on the utility is chosen and the cycle repeats.

7.4.2 Case study

Finally, we illustrate these ideas through our standard example of obstacle avoidance. However, this time, to overcome some of the problems associated with sonar, the vehicle is equipped with a directable optoelectronic sensing head which offers much better accuracy but over a limited field of view. In addition the sensors have to service a low level navigation task (keeping track of walls for corridor navigation). The planner must determine sensor use both between the two types of sensor and between the two tasks.

7.4.2.1 Components of the architecture

Sensor description

The sonar array operates well at long distances and the wide beam means that it can provide information to cover the forward path, but it suffers from lack of angular accuracy and also noise. On the other hand the sensing head is accurate but there is a time penalty if it has to move to view a new part of the scene, for example to locate an obstacle edge. The planner must determine how to balance the trade off between accuracy and time.

In this particular application the sensing head is an LEP based triangulation sensor mounted on a movable platform (Figure 7.12). It provides accurate measurements at short ranges over quite a small angle (40°) in any given position, but the accuracy falls off rapidly with range (see Figure 7.7). Also since optical radiation is absorbed by dark or black surfaces, dark objects are not always detected. The belief network includes the uncertainty between measurement and real state in the likelihood function $p(z|0)$ which the sensors must provide.

Bayesian belief network

The Bayesian network uses the sensors to determine the possibility of a clear pathway to the left or right of an obstacle. Its feature space is as follows:

$$\Omega = [path.left, path.right]$$

Because the LEP sensor can provide much more accurate angular information than the sonar, the states of each path are extended to include a new value: just passable. This implies a situation in which there could be quite considerable time costs in a sidestep maneouvre because the robot would have to go slowly for safety. A finer discretisation of the state space could be included if it was desirable.

$$0_{path} = \{PATH, JUST PASSABLE, TOO SMALL\}$$

For each feature in Ω the belief network holds a belief of state through the probability density function P(0). The network integrates measurements z of 0 from the sonar array and the LEP sensor as they become available, using the Bayes update rule (equation 7.2).

Figure 7.12   *The directable sensor head. Note the mirrors - two to project light into the scene and a small mirror providing a fast scan, which is mounted on an optical galvo projecting below the case*

Utility function

The general structure of the utility function used for driving the selection of sensors and actions for obstacle avoidance is shown in Figure 7.13. In effect this says that the worst case - i.e. attempting to go through a blocked path - has minimum utility, whereas going through a clear pathway has maximum utility. The intermediate case where the path is just passable has a utility, c, that lies between these, the exact value of which is determined by the amount of risk that can be tolerated. The cost of backoff, k, is constant for all cases and can be determined dynamically by considering any alternative path information that may be available from a higher level path planner.

PATH SIZE	SMALL	JUST	PASSABLE
SIDE-STEP	0	c	1
BACKOFF	k	k	k

Figure 7.13 *Utility function for obstacle avoidance*

7.4.3   Results: sensor control

A simulation environment was set up to demonstrate the sensor management and decision structure in the obstacle avoidance task. Typical screen output is shown in Figures 7.14a-d to demonstrate how the LEP sensor is used as the vehicle approaches an obstacle. The vehicle is moving along a narrow corridor following a predetermined path, and observes an obstacle near a junction. Each picture has three parts. One is a graphical representation of vehicle position and an inset showing the LEP bead position. Another, consisting of four bar charts, shows the current belief of any state 0_i configuration. These are given, in left to right order, as the belief values for TOO SMALL, JUST PASSABLE, PASSABLE for the left gap (above) and the right gap (below). The graph to the left shows the current estimates, the one on the right shows the predicted belief for the next timestep. In all examples here we assume static obstacles and so the predicted beliefs are the same as the current estimates. Finally the last part shows the decision trace up to this moment.

In Figure 7.14a, the first decision was to use the LEP scanner to look at the left side wall for corridor navigation; then, when the obstacle was reported by the sonar, to look at the right hand pathway. The grey mottled area on the left hand wall shows the data acquired in the first action. Examination of Figure 7.14a suggests why the next action will be to look right. The left gap has a very high probability of being TOO SMALL, whereas the right gap has quite a high probability of being PASSABLE. Decreasing the uncertainty in the size of the right pathway is likely to give a greater gain in expected utility despite the time cost involved in rotating the head, so the choice is made to look to the right.

Figure 7.14b shows the data acquired by this sensor action in the new mottled area on the right hand wall and a few points (which may not be seen in the scale of the figure) on the right hand end of the obstacle. It also shows the effect of the action on the states 0. The left hand side is unchanged as no new data were acquired. On the right hand side, on the other hand, there is now a high probability that the path is JUST PASSABLE. This implies that there could be significant risks in taking the side-step action. The choice of utility function that quantifies the extent to which the risk might be offset by savings in time from side-stepping rather than backtracking.

Due to the risk of side-stepping into a JUST PASSABLE path, the next action is a final check on the left hand side. The data acquired are shown in Figure 7.14c. The vehicle now has information from the LEP sensor right across the obstacle and extending to the left hand wall. The combination of sonar information and very accurate close range LEP data is sufficient to set the likelihood of the left path being too small as 1.00 (in fact it must be less than this but is rounded numerically). There are now no further sensor readings that will make any difference to the action, and the choice of utility functions is such that the vehicle chooses to side-step to the right of the obstacle. It is worth noting that the LEP sensor continues to be used but, since the obstacle avoidance has chosen an action, the navigation function now has priority on the sensor and continues to locate the original wall, prior to its use in navigating around the obstacle.

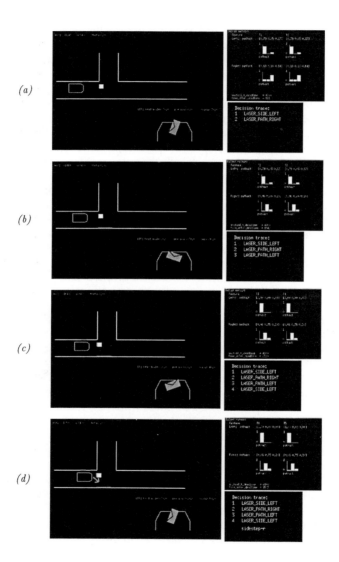

Figure 7.14 *Sensor management as the robot moves up to an obstacle*

Computational complexity

The real-time nature of the problem requires careful consideration of the time taken to make the decisions that control the sensors. Within the architecture presented here there are two areas that contribute to the computation time: updating the state hypothesis by propagating new sensor information in the Bayesian belief network, and the calculation of the expected value of a sensor action s_j. Detail of the belief network has not been described, but suffice it to say that the propagation time is dependent on both the number of nodes and the size of the state spaces of the nodes. Calculating the expected value of a sensor involves using current beliefs and a statistical model of its behaviour to try and predict the reading that it will give. Clearly, this must take less time than taking the actual reading to be of any use. There is a trade-off between accuracy of predictions (and therefore improved decision making) and the time taken to calculate. By careful design of the network, computation times have been kept small enough to succeed in the time pressured obstacle avoidance application.

7.5 Conclusions

The last ten years have seen an explosion in the development of sensors to direct robot planning. However there are still no systems which can handle a wide variety of environments robustly. Indoors the main difficulty is clutter, outdoors ambient conditions are tremendously variable in terms of light conditions and atmospheric conditions such as mist and fog.

Much remains to be done. Sonar sensors have been modelled extensively but there has been little progress in improving the hardware. Optoelectronic sensors are a much more recent development - with particular attraction in the well defined noise characteristics. Vision, which we have mentioned only briefly in this paper, is coming of age as a real-time sensor, but will only compete seriously with more specific range sensors when similar robustness to clutter and changing environments can be demonstrated. If prices fall sufficiently, radar would be an exciting addition to the robot's battery of sensors as it offers excellent long range performance in many conditions.

In this paper we have emphasised statistical methods for sensing and sensor integration. This is for two reasons. First they allow us to incorporate uncertainty in measurement and to base actions on an estimate of the true state of the environment, not on sensor-dependent observations. Secondly, defining features in terms of a likelihood function allows us to quantify average risk and failure. Given the impossibility of achieving complete accuracy this is a crucial requirement in any real industrial environment. However the estimates of risk are only as good as the underlying statistical models on which they depend, and further work is needed to improve modelling methods and to incorporate qualitative as well as quantitative methods in a meaningful way.

The development of further sensors, both hardware and reliable processing algorithms, and their integration into reliable planning, will remain an active research field for many years to come.

7.6 Acknowledgements

We thank fellow members of the Robotics Research Group for providing us with sensor data: Nick Pears and Mike Brownlow (who developed the LEP and AMCW sensors described here) and Jan Grothusen for the sonar data.

7.7 References

1.   GRAEFE, V. Dynamic vision systems for autonomous mobile robots. In RSJ/JEEE International Workshop on Intelligent Robots and Systems, pp 12-23,1989.
2.   GRAEFE, V. and JACOBS. U. Detection of passing vehicles by a robot car driver. in RSJ/IEEE International Workshop on Intelligent Robots and Systems, pp 391-396,1991.
3.   DU, F. and BRADY, M. A four degree-of-freedom robot head for active vision. International Journal of Pattern Recognition ond Artificial Intelligence, 8(6), 1994.
4.   BAKER BOLLES and MARIMONT. Epipolar-plane image analysis: an approach to determining structure from motion. International Journal of Computer Vision, 1 1,1987.
5.   BRADY, J.M. and WANG, H. Vision for mobile robots. Phi. Trans. R. Soc. Lond, pp 341-350,1992.
6.   LEONARD, J.J. and DURRANT-WHYTE, H.F. Directed Sonar Sensing for Mobile Robot Navigation. Kluwer Academic Publishers, 1992.
7.   CROWLEY, J.L. World modelling and position estimation for a mobile robot using ultrasonic ranging. In IEEE Conference on Robotics and Automation, pp 674-680,1989.
8.   ELFES, A. Sonar-based real world mapping and navigation. IEEE J. Robotics and Automation, RA-3 3, 1987.
9.   KUC, R. and SIEGEL, M.W. Physically based simulation model for acoustic sensor robot navigation. IEEE Transactions on Pattern Analysis and Machine Intelligence, 9 6:766-778,1987.
10. BORENSTEIN, J. and KOREN, Y. Noise rejection for ultrasonic sensors in mobile robot applications. In RSJ/IEEE International Workshop on Intelligent Robots and Systems, pp 1727-1732,1992.
11. HU, H. Distributed architecture for sensing and control of a mobile robot. Technical Report OUEL 1836/90, Dept. of Engineering Science, University of Oxford, 1990.
12. PEARS, N.E. and PROBERT, P.J. Active triangulation rangefinder design for mobile robots. In Proc. IEEE Workshop on Intelligent Robots and Systems, pp 2047-2052,1992.
13. COX, I.J. Blanche: an autonomous robot vehicle for structured environments. In IEEE Conference on Robotics and Automotion, pp 978-982,1988.
14. ADAMS, M.D. Optical Range Data Analysis for stable target pursuit in mobile robotics. PhD thesis, Oxford University, 1992.
15. BROWNLOW, M. An Optical Time of Flight Rangefinder for Mobile Robot Navigation. PhD thesis, Oxford University, 1993.
16. FLYNN, A.M. Combining sonar and infrared sensors for mobile robot navigation. International Journal of Robotics Research, 7 6, 1988.
17. HAGER, G. and MINTZ, M. Computational methods for task-directed sensor data fusion and sensor planning. International Journal of Robotics Research, pp 285-313,1991.
18. DURRANT-WHYTE, H. Integration, Co~ordination and Control of Multi-Sensor Robot Systems. Kluwer Academic Publishers, 1988.
19. HU, H., BRADY, M. and PROBERT, P.J. A decision theoretic approach to real-time obstacle avoidance for a mobile robot. In Proceedings of IEEE/RSJ International Conference on Intelligent Robots and Systems, Yokohama, Japan, July 1993.
20. ALLEN, P.K. Integration vision and touch for object recognition tasks. (6):15-33, 1988.
21. STANSFIELD, S.A. A robotic perceptual system utilizing passive vision and active touch. (6):138-161, 1988.

22. SHEKHAR, S., KHATIB, O. and SHIMOJO M. Object localization with multiple sensors. (6):34-44, 1988.

23. BRADY, J.M., HU, H. and PROBERT, P.J. A decision theoretic approach to real time obstacle avoidance for a mobile robot. In Proceedings of the RSJ/IEEE International Workshop on Intelligent Robots and Systems, 1993.

24. WEN, WU. Multi-Sensor Geometric Estimation. PhD thesis, Oxford University, 1992.

25. MANYIKA, J. PhD thesis, Oxford University, 1993.

26. JUDEA PEARL. Probabilistic reasoning in intelligent systems. Morgan Kaufmann, 1988.

Chapter 8

Robotics in the nuclear industry

P.E.Mort and A.W.Webster

The nuclear industry relies on telerobotics for inspection, refurbishment and to carry out basic plant operations, in radioactive facilities. A wide range of remote handling systems have been developed to carry out varied tasks. The remote handling systems can range from simple grab and release mechanically linked systems to complex hydraulically/electrically linked systems with special features such as programmability, resolved motion and collision detection. The driving forces to continue development of remote handling systems come from the need to make the plant more efficient, carry out complex tasks cheaply and quickly and to satisfy new regulations. This chapter is based on systems used in British Nuclear Fuels plc (BNFL).

8.1   Introduction

There are many manipulator systems throughout industry which are termed either robots or manipulators, which can become confusing to the reader as to what constitutes a robot or a manipulator. The next paragraph is the standard definition of a robot [1].

Robot, self-governing, programmable electromechanical device used in industry and in scientific research to perform a task or a limited repertory of tasks. Although no generally recognised criteria exist that distinguishes them from other automated systems, robots tend to be more versatile and adaptable (or reprogrammable) than less sophisticated devices. They offer the advantages of being able to perform more quickly, cheaply and accurately than humans in conducting set routines. They are capable of operating in locations or under conditions hazardous to human health, ranging from areas of the factory floor to the ocean depths and outer space.

About 30 years ago the main form of remote handling equipment used in the nuclear industry were devices known as tongs, see Figure 8.1. These were mounted in a spherical joint where the shaft slid through the centre, with a gripper on the end of the shaft carrying out the task. The problems associated with this device include limited workspace, restricted viewing of task and the level of protection against the activity in the cell/glove box. To overcome this problem a link mechanism was designed allowing the gripper to be more clearly seen and offering more protection to the operator. These are now known as master slave manipulators (MSM) and over the years they have been developed to give greater control over the gripper and increased range of its arm.

MSMs are limited in their reach, payload capacity and have no ability to be programmed. These devices therefore rely upon the skill of the operator to perform tasks efficiently. To overcome this problem a range of remote handling systems has been developed. A system comprises a manipulator, a control system and an input device (sometimes a master arm but more commonly joysticks). In most cases the manipulator is sold with a control system and an input device with limited capabilities. Depending on the task a more complex control system may be required for the manipulator.

To enhance the operation of a remote handling system or to give the system robotic type capabilities special controllers can either be coupled to existing control systems or replace completely the existing control system. There are a number of companies involved in this type of hard/software these being ARRL, (BNFL's RMRC) Thumall and AEA Technology Marwell. At a cost, the manipulator manufacturer will add functionality required by the customer to their controller in most cases. ARRL have used their controller to drive a PUMA and a RRC arm and they are in the process of porting their system to a Schilling hydraulic manipulator. Thumall control system has less functionality than the ARRL control system, but their system replaces the mini-master on the Schilling hydraulic manipulators G7F and TITAN II. Harwell have produced a control system to drive the PUMA robot and, by adding a force transducer at the end-effector, allows force feedback via a force feedback joystick device called the bilateral Stewart platform (BSP) which development was supported by BNFL.

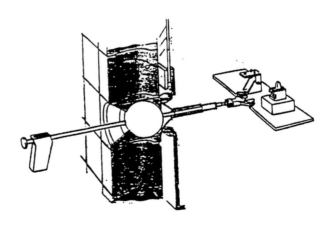

Figure 8.1: *Typical sealed tong installation*

BNFL have identified that the future remote handling challenges associated with operating and upgrading existing plants and decommissioning redundant facilities will be in less well-structured, cluttered and ever changing environments [6,10,11,19]. From experience it is clear that equipment for such tasks will need increasing levels of autonomy and user-friendly operator/equipment interfaces which are now classified collectively as advanced robotics and telerobotics. It is recognised that internal skill and expertise will have to be built up to support the development and application of these advances.

Since the economic pressures on the industry are greater than ever, every effort is being made to create improvements in efficiency. The company initiated a co-ordinated remote handling and robotics programme (RHRP) [12] in 1990 to address the enabling technologies for current and future business needs. This chapter reviews existing and future robotics in BNFL.

8.2 Remote handling systems

This section covers manipulators that have either been designed from scratch for a particular task or which are based on a standard off-the-shelf system which has been modified to carry out a task. These systems have each been developed for particular projects.

8.2.1 DIMAN 1 and 2

DIMAN 1 was designed to be inserted down a 9m long 150mm diameter pipe into the north dissolver vessel where irradiated Magnox fuel rods are dissolved in nitric acid. The vessel is 2.2m in diameter and 5m high and during access the radiation levels are around 1Gy/hr. The purpose of the manipulator is to provide improved visual examination of the vessel and its internal features by means of a REES rotating prism camera mounted on the end of a telescopic boom. Actuation of the degrees of freedom system is via electric motors (at the deployment level) and by solenoid valve operated water hydraulics and pneumatics on the manipulator arm. Positional data are not provided and for control the operator relies entirely upon visual feedback from the on board camera and another monitoring camera inserted separately into the vessel.

DIMAN 1 was shown to be a versatile device by being modified twice on short time scales to satisfy the urgent requirements of another plant. It was designed and built in 1984/85 and was first operated on the plant in late 1985.

Upon the successful completion of DIMAN 1 the need for visual inspection data on the north dissolver vessel was extended to include certain welds which connect process pipe work to the vessel.

DIMAN 2, see Figure 8.2 took the proven concepts of DIMAN 1 and extended them to provide additional degrees of freedom with improved control. On DIMAN 2 positional data are provided by rotary and linear potentiometers located at the joints and extensions giving analogue outputs. The readings from the potentiometers are

Robotics in the nuclear industry

overlayed on the main CCTV camera monitors thereby providing both visual and positional feedback to the operator.

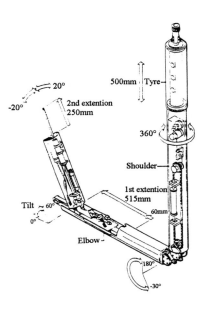

Figure 8.2 *DIMAN 2*

The prime inspection device carried by Diman 2 is a 12.7mm diameter, 250mm long endoscope which is equipped with a built in tungsten light source and is directly coupled to a REES CCTV camera. To facilitate the required linear deployment into the branches connecting to the top dome of the vessel, a special combination of four revolute and three linear degrees of freedom was chosen. Solenoid valve operated water hydraulics actuate the five manipulator movements and two electric motors power the actuators at the deployment level. A view of the endoscope tip is provided by a second camera mounted adjacent to the endoscope with a separate light source comprising a ring of light emitting diodes. This view is used for general positioning up to and entering the process pipe branches.

DIMAN 2 was designed and built in 1986 and entered service in 1987. It continues to provide an excellent inspection service during annual and special plant shut-downs.

8.2.2   Rodman [18]

Rodman was the first truly robotic machine designed for the use in the reprocessing plant. The design brief for Rodman included the following tasks.

(1)   To recover and feed through to the dissolver irradiated uranium fuel rods misplaced during the normal discharge operation from the transfer flask.

(2)   To assist in maintenance scheduled plant shut-down by carrying out decontamination of the cell internals.

(3)   To allow remote removal and replacement of a critical component in the seal valve associated with the transfer route to the dissolver.

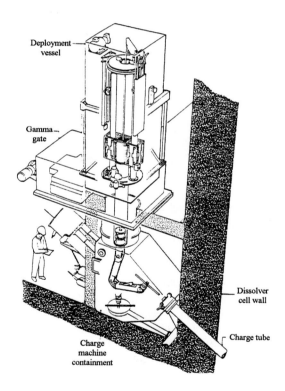

Figure 8.3 *RODMAN*

The Rodman manipulator is a seven function fully gaitered robotic arm with both open loop teleoperation and a teach and repeat capability. In addition, due to the dimensionally restrictive nature of the facility internal features, a continuous collision prevention routine is incorporated into the software. This provides an audible and freeze facility if any part of the manipulator breeches a 50mm nominal inner 'no go' zone. Rodman was design and built in 1986/87 and brought into service in 1989. where it has successfully contributed to improved plant efficiency and lowered the dose uptake of maintenance workers. Figure 8.3 shows Rodman in operation replacing a magnox fuel rod.

8.2.3 Rediman

Calder Hall and Chapelcross Reactors were designed, built and brought into operation during the period 1950 to 1959. To ensure the integrity of the reactor an annual ultrasonic inspection, weld and plate photography, chargepan inspection [17] and in core video work is undertaken during the reactor shut-down for refuelling operations. Ultrasonic inspection of 12m of seam welds in each reactor to meet the Nuclear Installations Inspectorate's (NII) key requirement was first scheduled for 1991. Inspection was to be carried out in three out of eight reactors
REaction DIvision MANipulator or REDIMAN [16], a water hydraulic arm capable of deploying a variety of purpose designed end-effectors to any position between chargepan and top dome in the outlet region, was designed and developed for these tasks. Major units of the system, shown in Figure 8.4 are:

a) the manipulator arm,
b) the slew drive unit,
c) the liner or the deployment tube,
d) the deployment frame, and
e) the control system.

The articulation of the arm is controlled by the operator from three joysticks situated at the control station. The control of the arm is further assisted by provision of two CCTV cameras located at the shoulder and on the lower arm. The manipulator and slew controls are open loop systems with hydraulic and electrical controls through solenoid valves and relays respectively. The computer system is not used for the arm movement control but provides the operator with arm positional information calculated from the joint potentiometers and displayed in the form of graphical mimic and joint angles. It also monitors the manipulator control commands and its response.
Despite minor operational problems, the system has proved very successful. It has been easy to use and maintain the system. The major benefit of the system has been that it can carry out the inspection tasks agreed with the NII to support the case for an extension in the operational life of these reactors. The system has proved capable of completing the inspection task in 10 to 12 days and with minimum disruption to other operations. Also the same system has been used at geographically separated sites Calder Hall and Chapelcross demonstrating its versatility. The system has exceeded its design requirements and is being programmed for use in subsequent inspections. REDIMAN successfully uses some design and control features from RODMAN.

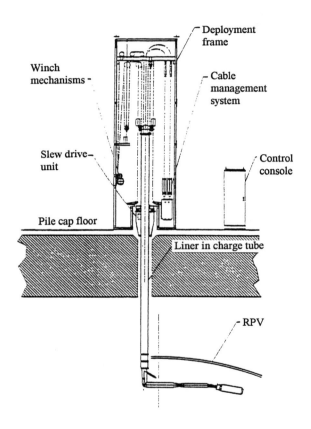

Figure 8.4 *REDIMAN*

8.2.4 *Repman*

Repman, Figure 8.5 is an advanced 'state of the art' robotic manipulator system for use within a highly active enclosed process vessel. It is a hydraulically powered machine using a specially formulated hydraulic medium consisting of 95% pure water. It is capable of carrying tooling packages weighing up to 15 kg and positioning them by taught routines. The manipulator is used for visual examination, ultrasonic wall thickness measurements, localised spot face and tube bore machining, insertion and welding of pipe sleeves and blanking flanges and post weld inspection.

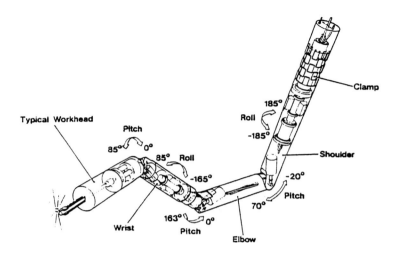

Figure 8.5 *REPMAN*

8.2.5   *Pipeman*

The Pipeman system was set up to clear the pipes in the way to allow Raffman to carry out its task. The equipment to carry out these clearance operations remotely was successfully used in April 1992 during the plant shutdown. The equipment consists mainly of

(a)   seven different work heads,
(b)   a manipulator to manoeuvre the work heads inside the cell,
(c)   a deployment system which deploys both the manipulator and work heads through holes or penetrations in the concrete shielding and into the cell.

The entire project from start to finish (i.e. from initial identification of the problem to the actual clearance exercise on site) took the comparatively short time of only 18 months. All the equipment was designed with safety as a prime consideration and was provided with extensive emergency recovery facilities.

    The deployment system was located inside a shielded enclosure where man access was possible. The deployment system consists of an adjustable framework which supports and deploys two tubes (each approximately 7 inches in diameter and 5 feet in length).

    The upper tube houses the hydraulic manipulator and its linear and rotary drives. The lower tube, known as the tool tube, is used to house and transport the appropriate work heads into cell. Both tubes are driven by the deployment system into two holes (800mm vertically apart), previously drilled in the concrete shielding, to present the manipulator and work heads inside cell. The manipulator is then remotely driven by the operators to pick up a work head from its parking arrangement on top of a tool tray in the tool tube, deploy it to the appropriate work site in cell and then return it to the tool tube once the task is complete.

The bottom half of the deployment frame houses the cable management system for the various work head umbilical. Once a particular task in cell is complete, the two tubes are withdrawn from the concrete shielding, a gamma gate is closed and man access to the deployment frame is then possible in order to change or adjust a work head for the next task. The shielded enclosure, in which the deployment frame was located, ensured containment when the gamma gate was open.

The hydraulic manipulator chosen for this particular application was a modified Schilling HV5F. The major modification was to the manipulator control system, whereby the slave arm is now rate controlled and the master arm controller has been replaced by a series of toggle switches, which are used to operate each joint individually. This was considered to be a more controllable means of operating a high powered manipulator around congested pipe work inside an active cell.

The work heads that were used as part of the operating envelope clearance task are grinder, double cropper, single cropper, pipe pusher, pipe spreader, pipe bender and pipe puller.

8.2.6   Raffman

The Raffman project [2-4], Figure 8.6, started in 1988 and has the specific aims of being able to remove a section of 75mm NB pipe within the working dissolver complex and to substitute a diversion pipe to direct a liquor stream to an alternative part of the plant. This activity is an essential trial part of the changeover to operation of the newly refurbished south dissolver complex which will take place in the near future. During the task the radiation levels in the cell could be up to 5 Gy/hr. To achieve the task a GEC Alsthom advanced slave manipulator is used to deliver tools to the work site through a penetration in the cell wall. Tools are located on racks on both sides of the manipulator. Umbilical cables provide power and control to the tools. A cable management system located under each tool rack automatically feeds out and takes in cable as required. Wherever possible conventional on-site machining tools have been modified for remote use. A large number of CCTV cameras are installed in the cell and in the corridor to provide feedback to the operators. Joint by joint, teach and repeat and resolved motion control are available to drive the manipulator. It is also equipped with a simple collision detection system. Figure 8.7 shows a schematic diagram of Raffman entering the cell.

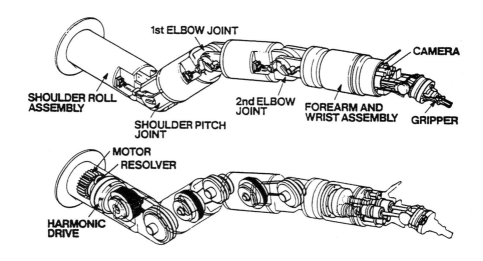

Figure 8.6 *RAFFMAN*

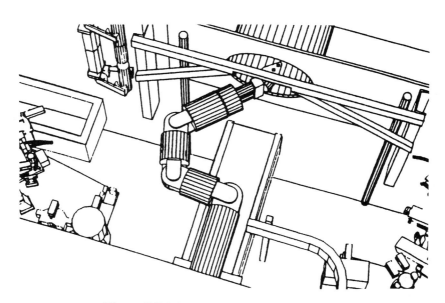

Figure 8.7 *Schematic of Raffman in Operation*

8.3   Future remote handling systems

The remote handling tasks in active plant can vary from simple repetitive tasks to one off complex tasks. Where the tasks are simple the remote handling system does not require lots of special functions, but should be able to gain functionality as required. The more complex tasks may require the development of special equipment and/or software and this should be easily interfaced to the existing systems. This suggests the remote handling equipment requires to be in a modular form both for the manipulator and the control system. The basic system should be cheap, simple and easy to set up requiring no or little specialist knowledge. It should be able to operate in active, dusty or underwater environments. The services required should be minimum to reduce the problems associated with umbilical management.

The layout of the control system should be such that the operation of the equipment is intuitive and is standardised to allow trained staff easy transfer to different remote handling systems. Again a modular set of input devices would enable a control console to be put together depending on their needs. The next section expands what has been discussed above and introduces new areas which require further development.

The experience gained from the remote handling robotics systems described in the above sections has shown that further research and development is required to assist and carry out complex tasks in an active environment [14-15]. The remote handling robotics programme set up at BNFL is aimed at developing the enabling technologies to solve the described problems. The next section describes some of the issues that the programme at BNFL is addressing.

8.3.1   Manipulators

8.3.1.1 Flexible manipulators

As the need for manipulators with higher payloads and greater reach increases, the flexibility of the arm has to be taken into account. This requires a whole host of new developments in the areas of dynamic control of the arm to maintain position, accurate tip location, operator control and much more. These systems would be used where long reach is required such as lifting objects out of tanks or deep caves.

8.3.1.2 Redundant manipulators

There are already a number of redundant manipulators on the market, but their control systems do not make full use of redundancy in areas such as collision avoidance and task optimisation. Kinematic redundancy is where the internal configuration of a mechanism can be changed without changing the orientation of the end-effector. There is also the area of hyper redundant manipulators, sometimes referred to as snakes, which are made up of many links, sometimes up to twelve. These systems are being used in inspection and repair again more development in control and task planning is required. A problem with these redundant systems is that the operator can become mentally overloaded with the number of tasks required to operate them. Research and development work driven and carried out by BNFL through the remote handling and robotics programme is addressing the problem.

This type of robotic equipment will enable the operator to perform what standard six degree of freedom systems cannot do. By utilising the redundancy of the manipulators objects can be avoided, but the end-effector can still achieve its goal position [5,7,13]. The joints can be optimised to reduce the load in each joint, minimise joint movement and increase manipulability.

8.3.1.3 Modular manipulators

A modular manipulator is a device which can be put together from a series of building blocks to achieve a given task. A number of manipulators have already been developed with this concept in mind such as the ARTISAN, Figure 8.8, by AEA Technology Harwell. Manipulators with this capability can be reconfigured to solve other tasks, reducing the need to design one-off specials for particular tasks.

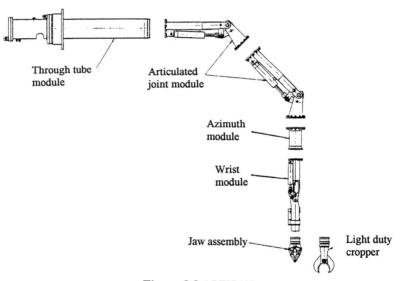

Figure 8.8 *ARTISAN*

A control system with the same concept would allow the engineer to configure the system to drive a manipulator with the required functions for the task. After the task has been completed the control system could be used to drive other manipulators by reconfiguration.

8.3.2   Tooling

The tools a robotic device carries to a site to carry out a task have either to use umbilical connections to power the tool or the robot end-effector has to supply the necessary services for the tool. If umbilicals are used complex cable management is required when moving the tool in and out of active cells. Supplying services to the tool through the end-effector would reduce the problems associated with umbilicals, but would bring in many problems with the design of the manipulator.

A universal interface at the end-effector would allow the problems associated with umbilicals to be a thing of the past. If the services required to drive the tool were standardised, and tool manufacturers used this standard, the cost of tools would be reduced and the development time for a special tool would also be decreased.

8.3.3   Input devices

An input device is a system or object that sends and possibly receives signals from a control system used to move a manipulator. At present there are many different methods available such as joysticks, a kinematically similar master (EL-MSM), mini master (TITAN 7F), force feedback joysticks, spaceball - six axis force sensitive joystick, ARRL Commander - six axis isotonic, Harwell telerobotic controller (HTC) and bilateral Stewart platform.

A new form of input system will be via a computer simulation where an off-line programme is developed in a simulation and down loaded into the robot's control system. Further benefits from the use of simulations and the 3-D information are described in the next section.

8.3.4   Simulation and modelling

Simulation packages have been used for some time within BNFL to improve the design process. Two of the most recently completed complex manipulator systems utilised modelling packages to verify the design of equipment but, as previously reported in the sections on Raffman and Repman, the modelling work did not adequately consider how the equipment was going to be operated. The reasons for this were the inexperience of the design authority in the use of the relatively new modelling packages, the computer graphical power available, and the simulation package used. Major advances have been achieved in each of these areas in the last five years and we now have computer hardware and software capable of real-time stereo viewing performance with the images exhibiting self shadowing, reflections and texturing (which are all used by operators as cues as to where they are in the remote environment).

8.3.4.1 Real time world model updating

World model updating is seen as being the ability to determine the location, shape and size of objects within the working environment and to transmit these to and incorporate them in a computer based geometric model linked to the operational equipment. To have this in real time implies an ability to carry out all operations within a cycle time of the order of 1 sec.

This is seen as an area being critical to the successful application of the technology in the nuclear working environment where it is essential to see changes to the work place as they are being made. As mentioned earlier this applies increasingly within the decommissioning area where the whole reason for the operations being carried out is to change the shape, size and location of objects. In effect taking what might have been a structured environment and changing it into an unstructured one.

Another, though less utilised reason for having this capability is to allow intervention to take place into either previously uncharted territory or into areas where the nature of the events leading to the intervention have themselves caused the environment to become unstructured.

8.3.4.2 Dynamic and random event modelling

Another more esoteric but nevertheless important area for concern involves the modelling of random events for such things as umbilical cable handling. Whilst present graphical systems can model the detailed equipment design down to as fine a detail as one is normally prepared to go, modelling the movement of cables and flexible structures is an area where the technology is not advanced.

The scope for handling dynamic performance is usually limited to pre-defined events using fairly basic dynamic theory. Unfortunately life does not always comply neatly with the theory and where problems have been experienced in the past they have usually been with the random events relating to such things as cable handling and their ability to take up positions not intended by the designer. Such excursions can and have led to cable snags, pinches and, in extreme circumstances, loss of power or control.

8.4　Conclusion

There are many manipulator/robotics systems available on the market, but in some cases the controllers are still in a primitive state and require further development to be utilised fully. Where systems have not been commercially available, one-off designs have been produced and used successfully. These remote handling systems have been expensive and in some cases have required specialist training to operate them. The experience gained from these projects, has brought about the setting up of a research and development programme to facilitate the enabling technologies to solve today's problems in remote handling.

A future remote handling system may consist of a number of building blocks both for the hardware and software. When a task has to be carried out an initial analysis of the problem through the use of simulation would determine the blocks required for the hardware and software. The simulation may replace the need for full scale trials, but the author feels there will still be a need to use full scale mock-ups to commission the equipment. After commissioning and staff training the remote handling system would be used to complete the task. If the task was a one-off, the remote handling system would be decontaminated and disassembled into its building blocks ready for the next task.

8.5   References

1.   McKERROW, J., Introduction to Robotics, Addison-Wesley Publishers Ltd pp.l-52, 1990.
2.   JONES, E.L., Remote Handling Developments For Inspection And Repair Of Highly Active Reprocessing Plant, Remote Techniques for Inspection and Refurbishment of Nuclear Plants, BNES, London 1988, pp.43-48, 1988.
3.   JONES, E.L., Remote Handling and Robotics in the BNFL Sellafield Reprocessing Plants, Proceedings 38th Conference on Remote Systems Technology, v2, 1990, pp.31-36, 1990.
4.   JONES. E.L.; and WEBSTER, A.W., Remote Diversion of a Highly Active Process Line. Remote Technology for the Nuclear Industry, IBC Technical Services Ltd., 10-11 December, 1990.
5.   BODDY, C.L., Implementation of a Real-Time Trajectory Planner Incorporating EndEffector Collision Avoidance for a Manipulator Arm, 2nd International Workshop on Advances In Robot Kinematics, Linz, Austria, 1990.
6.   WRIGHT. E.M.; and COLQUHOUN, A.P., Dose Reduction During Remote dismantling of Pile Chimneys, Proceedings, 38th Conference on Remote Systems Technology, 1990. v2, pp 101 4, 1990.
7.   BODDY, C.L. and WEBSTER, A.W., Introducing Autonomy To Robotic Manipulators In The Nuclear Industry, International Conference Engineering Solutions to the Management of Solid Radioactive Waste, Manchester, UK, 1991, pp.l37-l44, 1991.
8.   JONES, E.L., Remote Diversion of a Highly Active Process Line. Site Preparation and 4th Trial, Remote Technology for the Nuclear Industry, IBC Technical Services Ltd., 2 - 3 December, 1991
9.   SERAJI, H., LONG, M., and LEE T., Configuration Control of 7 dof Arms, Proceedings IEEE International Conference Robotics and Automation, Sacramento, California, 1991.
10. WALTERS, C.L.; WRIGHT, E.M.; and LENG, J.H., Remote dismantling of Nuclear Facilities in the United Kingdom - Experience to Date, The Waste Management 1991. Tuscon, Arizona, USA, 1991
11. WRIGHT. E.M., The Remote Dismantling of the Windscale Pile Chimneys, IMechE, 1992, pp.203-208, 1992.
12. GARLICK, D.R., Remote Handling Developments For New Plant And The Birth Of The BNFL 'REMOTE HANDLING AND ROBOTICS PROGRAMME', Computing and Control Division Colloquium on Robotics in the Nuclear Industry, IEE, London, 1992, Paper 1 pp.1-4.
13. BODDY, C.L., and TAYLOR, J.D., whole-Arm Reactive Collision Avoidance Control of Kinematically Redundant Manipulators, Proceedings IEEE Robotics and Automation Conference, Atlanta, USA, 1993.
14. JONES, E.L., Towards Tomorrow's Solutions, Remote Techniques for Nuclear Plants, BNES, London 1993, pp.24-28.
15   WEBSTER, A.W., and MISTRY, G.C., Telerobotics Experience Within BNFL's Operations, Remote Technology for the Nuclear Industry, IBC Technical Services Ltd., 13-14 January, 1993.
16. BITHELL, S.J., and HOWARD, S.R., Operational Aspects of the Calder Hall and Chapelcross

Pressure Vessel Ultrasonic Inspections. Remote Techniques for Nuclear Plants, BNES, London, 1993, pp.83-89.

17.   PENNICK, A.M., Ultrasonic Inspection of the Calder Hall and Chapelcross Reactor Pressure Vessel. Remote Techniques for Nuclear Plants, BNES, London, 1993, pp.145-151

18.   WEBSTER, A.W., and BODDY, C.L., Introducing Autonomy To Robot Manipulators In the Nuclear Industry, IMechE, 1991, pp.137-144.

19.   WRIGHT, E.M., and COLQUHOUN, A.P., Dose Reduction During Remote Dismantling of Pile Chimneys, Proceedings, 38th conference on Remote Systems Technology, 1990, v2, pp.10-14.

Chapter 9

Robots in surgery: a survey

B.L. Davies

9.1 Introduction

Some of the earliest robotics activity in surgery used small industrial robots for neurosurgery in an attempt to treat brain tumours. In April 1985, the department of Radiology at The Memorial Medical Center, Longbeach, California, carried out clinical trials of neurosurgery using a Unimate Puma 200 robot arm for stereotactic biopsies on the CT scan table [1,2]. Here, however, the robot was only used to hold a fixture at the appropriate position and orientation so that the surgeon could manually insert the biopsy needle into the patient. Subsequently a group at the Hospital for Sick Children in Toronto, Canada also used a Puma robot in a similar way to successfully remove deep seated brain tumours from a group of children who had not responded to conventional surgery[3].

In both the above cases, the robot was used only as a device to hold a fixture. However it was not until April 1991, when the author's group at Imperial College used the robot clinically for removal of the prostate, that the first true robotic operation was performed to use the robot to actively insert the cutting device into the patient and remove quantities of tissue [4,5].

The term 'robotic surgery' is usually regarded by researchers to mean a motorised reprogrammable computer controlled device which can carry a series of sensors and tools as an aid to diagnosis, therapy or surgery. However many clinical groups also use the term 'robot surgery' for the area which more properly should be considered 'computer assisted surgery'. The majority of applications to date have been in this area, partially because they are regarded as safer than active robot systems. For this reason a preliminary discussion of Computer Assisted Surgery is given below.

9.2 Computer assisted surgery [CAS]

CAS [like robotics] generally involves four stages: imaging, modelling, datuming and tracking motions. However, whilst robots are motor powered, CAS systems are moved manually by the surgeon. Medical imaging systems are conventionally used pre-operatively to provide qualitative images of the patient's condition. For CAS systems, these images are usually more sophisticated and involve the provision of quantitative data. They range from the relatively cheap ultrasound [US] and X-ray images through computer tomography [CT] to the more costly magnetic resonance imaging [MRI]. These images are generally of a 2D nature and so are often used in conjunction with a CAD based modelling system. The CAD system takes positional data from multiple 2-D images [e.g. CT slices] and builds them into a 3-D form, so that they are easier to interpret, as an aid in both diagnosis and pre-operative planning of the surgical procedure. The cost of 3-D modelling systems can rise rapidly with increased accuracy, resolution and ability to provide multiple views and to rotate the 3-D image quickly to provide new viewpoints. This is currently an area of rapid growth, primarily focused in computing and medical physics groups. A major challenge is to bring the quality and accuracy of imaging and modelling into the area of physical interventions. There has been considerable work in the past on imaging and modelling systems. This chapter concentrates more on the intervention aspects.

The above images and models may also be used simply as a qualitative guide for the surgeon intra-operatively as well as in pre-operative planning. However, much greater advantage can be gained if the model position data can be linked to motions of the surgeon's tools [e.g. biopsy needles]. This requires two preconditions. Firstly, the tool position must be capable of being sensed and recorded in 3-D space. Secondly, the coordinates of the tool sensing system must be capable of being registered to the patient's current anatomy and then back to the referencing system for the pre-operative imaging and modelling. The accuracy of each link in the processes of imaging, modelling, registration and tool coordinate measurement is critical to the overall accuracy and success of the operation. To improve the accuracy of the registration process, pre-operative imaging often takes place with markers which can be used as intra-operative datuming markers. To ensure that the markers are not affected by motion of soft tissue, they are often anchored to boney landmarks in a pre-operative process requiring local anaesthetic. To avoid this additional trauma, workers have recently attempted to locate anatomical features pre-operatively, to form a surface mapping of the anatomy which can be registered with the same surface that is identified from the pre-operative model[6]. Good accuracies [around 1mm] have been claimed for this technique.

To track the motion of tools and bones; two types of systems have been used: transmitter/receiver remote sensors and linked manipulators.

A. Transmitter/receiver remote sensors

The most common is an optical technique, in which the position and orientation of a number of markers [e.g. LED] are attached to the tool and monitored by a camera system of three or four cameras. These tend to be very expensive to obtain the required accuracy over a large field of view and have the problem that the LEDs in the region of the target area may be obscured by the surgeon or tasks [7]. Even though a larger number of cameras help, the problem can still remain. Alternative cheaper methods use electromagnetic systems. These involve the detection of the position of a series of magnets by a receiver system. The inductance effect of such devices is distorted by metal objects, which can be difficult to compensate for by using calibration techniques. Thus most of these systems use plastic components, but doubts remain about clinical robustness in an environment which may use many metallic devices.

B. Linked Manipulation Arm

The use of a linked manipulator arm, usually of the open chain [anthropomorphic] type that gives a large range of motion, is used to avoid the inaccuracies associated with remote sensors.   Measurement of the joint angles, using encoders or potentiometers, allows the position and end point orientation of tools to be determined.   The quality and cost of both the measurement system and the arm structure is highly variable. Some arms are very cheap, small and light, and give only a poor positional accuracy [8,9]. Others are large, robust and with very costly measurement systems[10]. The surgeon can position the hand held tools so that a tracking cross on a computer display (representing the tool location) can be matched to the appropriate target on the computer model. Thus the surgeon is responsible for keeping track of the target, although the information on which he relies is dependent on the accuracy, integrity and safety in-built into the system by the engineers. In order to allow the surgeon to let go of the tools, manipulator systems are often supplied with electromagnetic brakes to lock the arm in position, allowing tools to be changed and giving improved safety, e.g. hazardous radioactive seeds can be inserted for therapy or X-rays can be taken from a safe position. It can be seen that the simple addition of prime movers to the joints of the manipulator arms results in a fully powered surgical robot.

9.3   Powered robots for surgery, diagnosis and therapy

Powered surgical robots differ from manipulator arms used in Computer Assisted Surgery, simply in the addition of motors, gearboxes and motor controllers. The joint motions are directly controlled from a computer program, rather than being positioned by the surgeon as in Computer Assisted Surgery. The advantage is that the motions are constrained by the computer control system to give high accuracies with predictable velocities and accelerations without overshoot. Repeated and incremental motions can be performed without difficulty, in addition to all the benefits of CAS. The complexity of a robotic procedure lies, not so much in the robot motions, as in the total system requirements for patient clamping and datuming. The following list shows the typical procedure for knee surgery.

Robotic Systems for Surgery

a).   Pre-operatively image patient and create 3-D model
b).   Plan operative procedure
c).   Fix and locate patient on table
d).   Fix and locate robot (on floor or on table)
e).   Input 3-D model into robot controller
f).   Datum robot to patient
g).   Carry out robot motion sequence
h).   Monitor for patient motion
i).   Remove robot from vicinity
j).   Release patient
k).   Check quality of procedure
l).   Re-clamp patient
m).   Re-position and datum robot to patient
n).   Repeat robotic procedure

An additional unexpected benefit that has emerged from robot surgery is that the repeatable motions allow a uniform procedure, independent of the surgeon. This has resulted in standard shapes being resected, etc. allowing a scientific basis for experimentation, leading to further insights into the efficacy of particular surgical procedures. The further potential benefit from robots is that, once the preliminary clamping and datuming activities are over, the actual motion sequence can be carried out rapidly, leading to a reduced time for the actual surgery. The main difficulty with powered robots is that they imply greater safety problems.

9.4   Safety issues

The Health and Safety Executive (HSE) recommendations on industrial robots state that the robot should be used inside a cell from which people are excluded. This would make the use of surgical robots impossible. Thus, there are no official guidelines for safety of surgical robots. Discussions with the HSE and DTI suggest that the robot and medical communities should make recommendations which they and the general public will discuss. The author has made a number of suggestions to this end [11-13]. One of the major problems is that the more safety systems are incorporated, the more complex and costly the system and the less likely it is to be justifiable. Apart from some aspects of neurosurgery, the use of robotic systems are seldom seen as life saving. They generally give a more consistent, repeatable result that, even used by a junior registrar, can give results as good as the best consultant. They are therefore used to improve quality, reduce time and occasionally to reduce hazards [e.g. exposure to X-ray] for both patient and surgeon. In many of these activities the justifiable increase in cost and complexity of a robotic system is limited and, given that no system is absolutely safe, how safe such a system should be is open to question. The use of robotic systems for therapy in treating otherwise inaccessible cancerous tumours is in its early stages. This potentially life saving activity can also change the arguments since it would be senseless to suggest safety constraints so severe that the robot would be too complex and it could not be used to save lives! Table 9.1 shows that as systems proceed from hand held tools towards complex robotic systems, the control is taken out of the hands of the surgeon and resides in the systems provided by the engineers. Medical personnel are often unaware of the full implications and features incorporated in the more automated systems. Thus, they tend to favour the simpler CAS systems over which they feel they have more direct control. However, if full safety systems have been implemented in robots, there is much less reason for error than in more manual systems where motions are totally unconstrained. Where potentially lethal cutting systems are held manually without constraint, the result can be less safe than using a robot. The questions of how safe should robots be and what safety features should be incorporated is one which the author is seeking to promote by setting up a UK forum for Robotic and Computer Assisted Surgery under the aegis of the IEE and IMechE.

Table 9.1 *Levels of complexity of systems in surgery*

Code   Type of System
1.   Hand held tools. Surgeon holds/moves tools freehand using only human innate sensing (touch, vision etc).
2.   Hand held tools with a spatial location system. Freehand held tools, but surgeon can track a target using e.g. cameras + LEDs on tool, or magnetic field source with sensors.
3.   Tools are mounted on a manipulator arm which is associated with a spatial location system. Arm is moved by surgeon, which to some extent constrains his sense of 'feel' and freedom of motion. Joint motions are usually monitored.
3.1   As for 3, but target location is updated with patient movement intra-operatively. Sometimes a second (passive) arm is strapped to patient to monitor patient motion, on others a number of markers are tracked by an external camera system. Target location on a quantitative model is updated in real time to allow surgeon to track target with tool on arm.
4.   Tools are mounted on a manipulator arm which is associated with a spatial location system and powered brakes (passive). The addition of powered brakes to arm permits arm to be locked in position, e.g. to permit long-term treatment at target location, or to permit surgeon to move away to fire X-rays safely.
4.1   As for 4 but the arm is used actively to insert/move tools.
4.2   As for 4 but the arm adapts to patient/organ movement intra-operatively.

5.   Tools are mounted on a powered robot arm equipped with position measurement (used passively). Tools can be moved using the powered arm, either actively (as in 5.1) to enter the patient using the robot, or passively (as in 5). In the latter case the arm acts as a stationary jig which locates the tools so that the surgeon can manually insert them into the patient. In the active case, sometimes the robot is locked in position whilst a special purpose additional single axis moves into the body. In others it is the complete robot which moves to interact with the body. The ability to adapt, on-line, to patient motion (as in 5.2) risks the possibility of errors which cannot be trapped at a planning stage.

5.1   As for 5, but robot is used actively to insert/move tools.

5.2   As for 5, but robot adapts to patient/organ movement intra- operatively.

6.   Tools are mounted on a powered robot arm equipped with force and position control (used passively). The addition of force control permits the robot to switch between position control and force control so that the robot can yield to a given force level in prescribed locations

6.1   As for 6 but robot is used actively to insert/move tools.

6.2   As for 6 but robot adapts to patient/organ movement intra-operatively.

7.   Tools are mounted on a powered robot arm equipped with force and position control with input systems from a 'master' telemanipulator to control the powered robot as a 'slave' system. The use of data gloves and other tactile/force 'feed in' systems to the operator can enhance the operator knowledge of what is occurring at the 'slave' system. Use of master/slave telemanipulator permits 'tele-presence' with a data glove and 'virtual reality' to both control the slave and feed data back to the surgeon at the 'master' input. The addition of a master system as an input, additional to the normal computer control, increases complexity.

7.1   As for 7 but slave manipulator is used actively to insert/move tools

7.2   As for 7 but slave manipulator adapts to patient/organ movement intra-operatively.

9.5   A brief survey of international work

9.5.1   USA and Canada

One of the most advanced robotic surgery systems is the 'Robodoc' hip surgery robot by Integrated Surgical Systems [ISS], California [14-15]. The robot used is a 'scara' style industrial robot, specially adapted to the task by Sanko Seiky of Japan in a joint project. The robot has a number of safety features, including force sensing at each of the joints as well as a six axis wrist force sensor. As soon as any excess forces are felt, the system interrupts and gives the surgeon a message on the display screen, asking for the sequence to be restarted. Work to date has focused on accurate machining of a recess in the hip bone to take a prosthetic implant. Considerable attention is paid to building up 3-D models from preliminary CT scans, the use of landmark pins and clamping of the patient. The system is claimed to be generic and it is hoped to use it for knee and spine orthopaedic tasks. Early experiments on dogs quickly gave way to its use for veterinarian surgery. First human trials took place in November, 1992 and 20 patients have been treated in trials carefully monitored by FDA. A further trial of 150 patients by robot and 150 by conventional surgery is planned for three clinical centres, in an attempt to show the clinical benefits of the robot.

IBM Watson Research Center, N.Y., conducted the early work on the above Robodoc system under the direction of Dr Russ Taylor[16] in conjunction with the University of Davis, California. Current projects are concerned with craniofacial reconstruction using a series of special purpose robots that combine active and passive systems to position tools about a remote centre[17]. A major focus of the work is the provision of a generic computing software and hardware facility that will take information from imaging systems, develop 3-D models and allow pre-operative planning as well as a user-friendly inter-operative human computer interface.

The Department of Radiology, Memorial Medical Center, Longbeach, California has undertaken some of the earliest work on neurosurgery and 3-D modelling of the brain. A Unimation Puma 200 robot was used in preliminary clinical trials for neurosurgery in 1985 under the direction of Dr Kwoh[1]. The purchase of the Unimation Company by Westinghouse resulted in a decision to cease support for all medical robot work. The current owners of Puma robots, Staublie Automation, have written to the author saying that they do not approve of the use of their robots for medical projects as the robots were not designed for use next to people. The Longbeach group has ceased to work in robotic surgery.

The Department of Mechanical Engineering, North Western University, Chicago has been carrying out laboratory studies into robot systems for knee surgery under the direction of Dr M. Peshkin and surgeon Dr D. Stulberger [18]. The project has been aimed at the machining of knees for accurate fitting of prosthetic implants. A series of studies have been concerned with 3-D modelling of CT scans, use of landmark pins, patient clamping and inter-operative datuming of a Puma robot using a six axis force sensor on the end of the robot. The robot then holds a fixture in the appropriate accurate positions in order to allow a surgeon to hand drill the bones for locating holes and associated pins to mount the conventional fixtures. In the next stage, it is planned to hold the fixtures in the robot gripper, so that the surgeon can then cut the shapes for the prosthesis in the bone using a conventional oscillating saw.

The Stanford Research International, [SRI] Stanford, California has been carrying out work into concepts which have been termed 'virtual reality' applied to minimally invasive surgery under the direction of Dr Phillip Green in collaboration with the surgeon, Dr Richard Sattava [19]. The intention is to use a teleoperator master/slave system with force control. Forces experienced by the slave unit are fed back to the master. The addition of a head-up computer generated display of the target area, with a model of the slave end-effector, completes the concept of a virtual reality system. The benefit of the system is that it allows the feedback of amplified forces so that large motions and forces at the master can result in smaller motions and forces at the slave. This is useful for accurate control, particularly for micromotions, such as in eye surgery or in vascular surgery. A further proposed benefit is that the master can be many miles from the slave, so that a master can be in a city whilst the slave can be transported to a Third World country area. A further interest is for a system for the battlefield which will enable soldiers to undergo emergency operations in the field whilst the surgeon is at a safe remote region. The system has been much discussed as a result of a video showing a simulation of the 'slave' being used to sew tissue. However, only one detailed paper has been published as the SRI group have wished to obtain patents prior to revealing details.

The Department of Electrical Engineering, Washington State University has been concerned with force control and micromanipulation under direction of Dr Blake Hannaford. The potential for force control in surgery can be seen in SRI's virtual reality work but also has significance in many datuming and locating systems. The need for very fine control of forces and of micromotions is at the heart of many systems. The Washington group have demonstrated the ability for the slave to sense a single strand of cotton wool and feed this back to the master unit as a tangible force. The sensor unit is constructed from the head of a compact disc drive. A prototype six axis micromanipulator has also been constructed [20].

The Hospital for Sick Children in Toronto, Canada under the direction of Dr James Drake, has used a Puma 200 robot to aid in removing deep seated tumours from a group of children. The group had all had prior conventional treatment which was unsuccessful in accurately targeting the whole tumour. The use of the Puma to hold a fixture resulted in accurate placement of tools by a surgeon to treat the tumour [2]. The group has subsequently gone on to treat tumours as well as epilepsy and vascular problems in neurosurgery using the ISG viewing wand. Using CT and MRI images the tumour is outlined by the surgeon and the boundary fed into a computer to generate the 3-D model on the ISG system, together with the probe direction. After pre-operative planning to determine the optical track, the ISG passive arm is brought into use to show the current probe position superimposed over the model showing the desired track. Alignment of the two allows the surgeon manually to follow the desired track to insert the tool. This system has also been applied in the UK by Dr D. Sandeman at the Department of Neurosurgery, Bristol [21].

9.5.2   UK

The author's group at Imperial College has been concerned with robots in rehabilitation for some time which has given useful insites into their safety requirements and which has also helped in the development of surgical robots[22]. Starting in 1987, preliminary feasibility studies were conducted into prostate surgery using a Unimation Puma 560 robot that was modified to carry two additional frameworks to provide the necessary motions [23]. This led to the development of a special purpose frame to give the required motions with the minimum degrees of freedom. This was manually powered and was tried clinically on 40 patients with good results [24]. Having proved the kinematics, the system was powered under computer control and applied clinically on five patients [25]. A current EPSRC grant for laboratory studies has given the opportunity to use an ultrasound probe attached to the frame to directly measure the gland size at the start of the procedure and interchange this with a diathermic cutter. Other cutting modalities are also being investigated. The group is also researching the use of force control strategies, for prosthetic implant knee surgery, which will allow the surgeon to hold a cutter on the end of the robot and machine the knee bones within software constraints provided by the robot. Low force control is provided within the defined region to allow the surgeon to feel the cutting forces. Towards the edge of that region, the robot gradually switches into high gain position control [26]. The group is also investigating the use of a low cost passive arm, using a 4-bar linkage with potentiometers, to insert biopsy needles into the kidney. The use of calibration objects turns a qualitative image from a low cost X-ray C arm into a quantitative one which can provide 3-D coordinates for the passive arm. The arm can be locked to allow tools to be changed or remote X-rays to be taken[27]. Other work is concerned with force sensing and micro motions for 'virtual reality' surgery. A further group under Professor C. Besant has implemented a substantial and accurate arm for guiding a drill held by the surgeon to drill the spinal pedical. A second arm monitors motion of the spine and updates the model of the current target position.

A group at Loughborough University under Professor Hewitt and Dr Bouazza-Marouf has developed a prototype arm which combines passive and active motions for the drilling of the femur in fracture repairs [28].

Another group at AMARC at Bristol University, under Dr P. Brett, has investigated a 'prototype drill' instrumented for force, which is aimed at performing ear stapedotomies. Useful information about the very small forces has been obtained [29]. Dr R.O. Buckingham, also at AMARC, has conducted preliminary studies of the kinematics of mechanisms for intra cranial surgery [30]. In the Neurological Sciences Department, Dr. D. Sandeman has clinically applied the ISG viewing wand [21].

A further group at Hull University, Computer Science under Professor Phillips, in conjunction with Hull Royal Infirmary is concerned with modelling systems for surgery. Current work is focussed on insertion of a 'nail' for femur fracture repair. A passive arm is being purchased.

One of the few companies in the world concerned solely with medical robots is Armstrong Projects who have a motorised device to help in moving the laparoscope in minimally invasive surgery. They also have a Scara style passive arm whose end vertical axis is counterbalanced and which can carry surgical tools [31].

9.5.3   *Japan*

Considering how much Japanese research has been carried out on industrial robots, there is surprisingly little activity applied to CRATS. Whilst much of the activity could potentially be applied to, e.g. micromotion manipulators for eye surgery, or force control for virtual reality surgery, little has been demonstrated. One of the exceptions is a group at Tokyo University, Department of Neurosurgery who report in

1987 the use of a passive arm for aiding in neurosurgery on twelve patients[32]. An accuracy of around 5mm was reported using CT slices 10mm thick. The six joint arm was specially built from aluminium and used high resolution potentiometers with twelve bit A/D converters.   Beads were imaged in pre-operative CT scans.   Their locations picked off the computer display to allow comparison and datuming with the same beads checked intra-operatively by the arm pointer. The other group working on applications is that of University of Tokyo, Faculty of Engineering (in conjunction with two Tokyo hospitals) who have a background of surgical simulation and 3-D modelling. Recent system developments include a prototype for laser treatment of liver cancer. The same prototype is being modified for neurosurgery to fit into a CT scanner. A further system is being developed for corneal microsurgery using UV pulsed lasers and an x-y galvanometer for scanning. A He-Ne laser is used for scanning, whilst an Excimer laser is used for cutting [33].

9.5.4 Europe

One of the most established research groups in Europe is that of University of Grenoble who have made a speciality of imaging and 3-D modelling for surgery. An early project applied this to head surgery by using a standard industrial robot which had been modified with a very large gear reduction at each joint.   The robot could thus move slowly to position a fixture, held in the robot tip, next to the head. The task was observed by the surgeon who could push a handheld pendant stop button in emergencies.   The surgeon then manually inserted biopsy needles into the fixture. The robot was thus acting purely as a positioning/orientating device which was locked off, during the intervention, for safety reasons [34].

Subsequent work has concentrated on passive devices, primarily concerned with tracking tools using an Optotrak system which used three cameras to track up to 256 light emitting diodes. A major aspect of this work is the ability to datum tools to anatomical features by building up a surface which can be matched with the pre-operative equivalent in a 3-D model [35]. An interesting recent concept is that of a passive constraint arm which allows a surgeon to move only in a predefined direction, all other directions being inhibited by a braking mechanism [36]. The difficulty is that of finding brakes with a fast enough response so that the system does not require excessive viscous damping to slow it down. The concept is the passive equivalent of that used at Imperial College for active knee surgery.

Another French group is that at University of Lille who are concerned with a micro telemanipulator, originally used for ocular vitrectomy but more recently for radial keratomy, which has been tried in phantoms [37].

An active research group in ARTS is that of Professor P. Dario in University of Pisa, Italy.   They have instituted a special purpose laboratory for orthopaedic surgery studies which is equipped with a number of Puma robots for evaluation of procedures. A number of concept evaluations have been conducted concerned with force control scalpels, implantable neurons and computer assisted orthoplasty [38]. Also in Italy, Dr Giorgi at the Institute Neurologica, Milan, has had a number of years of clinical experience in the use of passive devices in neurosurgery. He has recently developed a five axis device which incorporates encoders and electromagnetic brakes which has been used for phantom studies [39].

A further neurosurgery group who have made a recent breakthrough is that of Professor C. Burckhardt, of University of Lausanne, who has recently reported the clinical use of an ambitious special purpose powered robot for neurosurgery in association with a CT imaging system [40]. The system employs a series of single axis motion tools which are locked into position by an automated carousel which then advances the tool into the region of the patient's head where it is datumed to a stereotactic frame. These actions can take place within the CT machine to ensure intra-operative imaging.

A system for ENT surgery has been specially developed by a group at Aachen Hospital, Germany, under Dr R. Mosges [41]. This is a floor standing passive anthropomorphic arm, with very accurate encoders, which has a large reach and is counterbalanced for gravitational forces. The arm has been used for facial reconstruction surgery. It employs a series of bead markers for pre-operative imaging associated with stencilled marks which are used operatively to datum the robot. The markers are placed over bony prominences to minimise local motions of the beads.

9.6   Conclusions

The ability to image and model structures for diagnosis, therapy and surgery has outstripped our ability to perform physical interventions. In an attempt to rectify this situation, the last nine years has seen an upsurge in activity using a range of robotic devices. Some groups have concentrated on providing simple tracking devices on standard tools. Others have placed the tools on passive manipulator arms, some of which can be clamped. In both instances, the tool location is displayed on a screen together with the desired target, generated from pre-operative images and models. This approach keeps the surgeon in control and side-steps safety issues, but fails to take advantage of the full potential of powered robots. A few groups have used standard industrial robots, sometimes modified to move slowly, that hold fixtures appropriately whilst interventions are carried out by hand. Only three groups (including that at Imperial College) have devised special purpose powered systems that allow autonomous cutting actions, all of which have been clinically applied in the last three years. It is still early days in this area of activity and many safety issues have still to be resolved. The second decade of activity should see further developments in the area of micromotion and virtual reality as well as a consolidation of the existing systems.

9.7   Acknowledgements

The author wishes to thank EPSRC for a travel grant which has greatly assisted this work.

9.8 References

1. Kwoh, Y.S., et al, A new computerized tomographic aided robotic stereotaxis system. Robotic Age. June 1985.
2. Kwoh, Y.S., Hou, J., Jonckheere, E.A., and Hayati, S., A robot with improved absolute positioning accuracy for CT guided sterostactic brain surgery. IEEE Trans. Biomedical Eng. Vol 35, No 2. Feb. 1988.
3. Drake, J., Joy, M., Goldenberg, A., and Kreindler, D., Computer and robotic assisted resection of brain tumours. 5th International. Conference on Advanced Robotics, pp.888-892, Pisa, Italy. June 1991.
4. Davies, B.L., Hibberd, R.D., Ng, W.S., Timoney, A.G., and Wickham, J.E.A., A surgeon robot for prostatectomies. pp. 871-875, IBID.
5. Davies, B.L., Hibberd, R.D., Ng, W.S., Timoney, A.G., and Wickham, J.E.A., A robotics assistant for prostate surgery, Proceedings of IEEE EMBS Conference pp. 1052-1054, Paris, France. Oct 1992.
6. Lavallee, S., Brunie, L., Mazier, B., and Cinquin, P., Matching of medical images for computer and robot assisted surgery, Proceedings of IEEE EMBS Conference, Orlando, Florida. Nov. 1991.
7. Sautot, P., Cinquin, P., Lavalee, S., and Troccaz, J., Computer assisted spine surgery: A first step towards clinical application in orthopaedics., Proceedings of IEEE EMBS Conference, pp 1071-1072, Paris, France. Oct. 1992.
8. ADL-1 6D tracking system, Shooting Star Technology, Rosedale, Canada, VOX 1XO.
9. Immersion probe -MD. Immersion Corp. Palo Alto, California, 94309-8669.
10. Mosges, R., et al., A New Imaging Method for Intra operative Therapy Control in Skull-Base Surgery, Neurosurg. Rev. No.11, pp. 245-247, 1988.
11. Davies, B.L., Safety of Medical Robots, Safety Critical Systems. Book Chapman Hall, Part 4, Ch. 15, pp. 193-201. Spring 1993.
12. Davies, B.L., Safety of Medical Robots, Proceedings of 6th International Conference on Advanced Robotics, pp. 311-317, Tokyo. Nov. 1993.
13. Davies, B.L., Safety Critical Problems in Medical Systems, Proceedings of Second Safety Critical Systems Conference, Birmingham, UK, Springer-Verlag. Feb. 1994.
14. Mittelstadt, B., Kazanides, P., Zuhers, J., Cain, P., and Williamson, B., Robotic Surgery: Achieving predictable results in an unpredictable environment, Proceedings of 6th International Conference on Advanced Robotics, pp. 367-372, Tokyo. Nov. 1993.
15. Kazanides, P., Zuhars, J., Mittelstadt, B., and Taylor, R.H., Force Sensing and Control for a Surgical Robot, Proceedings IEEE Conference on Robotics and Automation, pp. 612-617. Nice, France. May 1992.
16. Taylor, R., et al., Taming the Bull: Safety in a Precise Surgical Robot, Procedings of 5th International Conference on Advanced Robotics, pp. 864-870, Pisa, Italy. June 1991.
17. Cutting, C., et al., Computer Aided Planning and Execution of Cranio-facial Surgical Procedures, Proceedings of 14th International Conference of IEE, pp. 1069-1071, EMBS, Paris, France. Oct. 1992.
18. Kienzle, T:C., Stulberg, S.D., Pershkin, M., Quaid, A., and Wu, C.H., An Integrated CAD - Robotics System for Total Knee Replacement, Proceedings IEEE Conference on Systems Man. and Cybernetics, Chicago, USA. Oct. 1992.
19. Hill, J.W., Green, P.S., Jensen, J.F., Gorfu, Y., and Shah, A.S., Telepresence Surgery Demonstration System, Proceedings IEEE Conference on Robotics and Automation, San Diego, California. May 1994.
20. Marbot, P.H., and Hannaford, B., Mini Direct Drive Robot Arm for Biomedical Application, Proceedings 5th International Conference on Advanced Robotics, pp. 859-865. Pisa, Italy. June 1991.
21. Sandeman, D., Marshall, C., and Bret, P., Medical Robotics in Neurosurgery: The Potential of 3D Image Guidance Systems of Intelligent Micro manipulators, Proceedings 1st

International Workshop on Mechatronics in Medicine and Surgery. pp. 153-157, Costa Del Sol, Spain. Oct. 1992.

22. TIDE Project No. 128, A General Purpose Multiple Master Multiple Slave Intelligent Interface for the Rehabilitation Environment.: ECE DG XIII/C3. Brusselles 1993.

23. Davies, B.L., Hibberd, R.D., Coptcoat, M.J., and Wickham, J.E.A., A Surgeon Robot Prostatectomy - A Laboratory Evaluation. J. Medical Eng. and Technology. V13, No. 6, pp. 273-277. Nov. 1989.

24. Davies, B.L., Hibberd, R.D., Timoney, A.G., and Wickham, J.E.A., A Surgeon Robot for Postatectomies, Proceedings 2nd Workshop on Medical and Healthcare Robots, pp. 91-101, Newcastle, UK. Sept. 1989.

25. Ng, W.S., Davies, B.L., Hibberd, R.D., and Timoney, A.G., A firsthand Experience in Trans-urethral Resection of the Prostate, IEEE, EMBS Jl, pp. 120-125. March 1993.

26. Davies, B.L., and Hibberd, R.D., Robotic Surgery at Imperial College, London, Proceedings 6th International Conference on Advanced Robotics, pp. 305-309. Tokyo, Japan. Nov. 1993.

27. Potamianos, P., Davies, B.L., Hibberd, R.D., Manipulator Assisted Renal Treatment, Proceedings of International Conference on Robots for Competitive Industries. pp. 214-227. Brisbane, Australia. July.

28. Bouazza-Marouf, K., Robotic Assisted Repair of Fractured Femur Neck. Proceeding of Sympium on Robotics in Keyhole Surgery, IMECHE. May 1993.

29. Blanshard, J., Brett, P., Griffiths, M., Khodobandehloo, K., and Baldwin, D., A Mechatronic Tool for Microdrilling a Stapedotomy, Proceedings First International Workshop on Mechatronics in Medicine and Surgery. pp. 11-21. Costa del Sol, Spain. Oct. 1992.

30. Buckingham, R.O., Buckingham, R.A., Wood-Collins, M, Brett, P., and Khodabandehloo, K., Kinematic Analysis of Advanced Mechanisms for Intra. Cranial Surgery. IBID. pp. 97-106.

31. Finlay, P.A., Neurobot: A Fully Active System for Assisting in Neurosurgery, Industrial Robot, Vol 20, No. 2, pp. 28-29. April 1993.

32. Watenabe, E., et al., Digitizer (Neuro Navigator): New Equipment for Computer Tomography Guided Surgery, Surg. Neurology. vol. 27, pp. 543-547. 1981.

33. Dohi, T., et al., Robotics in Computer Aided Surgery, Proceedings 6th International Conference on Advanced Robotics, pp. 379-383, Tokyo, Japan. Nov. 1993.

34. Lavallee, S., and Cinquin, P., IGOR: Image Guided Operating Robot, 5th. International Conference on Advanced Robotics. pp. 876-881. Pisa, Italy. June 1991.

35. Lavallee, S., et.al., Geometrical Methods of Multi-Modality Images Registration for Computer Integrated Surgery, Proceedings 1st International Medi Mechatronics Workshop, pp. 127-138. Malaga, Spain. Oct. 1992.

36. Troccaz, J., Lavallee, S., and Hellion, E., PADYC: Passive Arm with Dynamic Constraints, Proceedings 6th International Conference on Advanced Robotics, pp. 361-366. Tokyo, Japan. Nov. 1993.

37. Hayat, S., Vidal, P. and Hache, J.C., A Prototype of a Microtelemanipulator for the Radial Keratomy, IBID, pp. 331-336.

38. Fadda, M., et al., Execution of Resections in Computer Assisted Orthroplasty, IBID. pp. 337-340.

39. Giorgi, C., Casolino, D.S., Luzzara, M. and Ongania, E., A mechanical stereo stactic arm, allowing real and historic anatomical images to guide microsurgical removal of cerebral lesions, IBID. pp. 299-304.

40. Glauser, G., Flury, P., Epitauz, M., Piquet, Y. and Burckhardt, C., Neurosurgical Operation with the Dedicated Robot Minerva, IBID. pp. 347-351.

41. Mosges, R. and Klimek, L., Computer Assisted Paranasal Sinus Surgery, Proceedings 14th International Conference of IEEE EMBS, pp. 1050-1052. Paris, France. Oct. 1992.

Chapter 10

Intelligent autonomous systems for cars

R.H.Tribe

This chapter describes the difficulties of developing an intelligent autonomous car that can drive itself through the current traffic environment. Such systems will not appear for a long time although they may be introduced sooner if the traffic environment is greatly constrained. More viable for the medium term are semi-autonomous systems where the driver maintains responsibility for the overall control of the car. Examples of this type of system are autonomous intelligent cruise control, lane support and collision warning. For each of these examples, the benefits and the likely technologies are described and results shown from demonstrators under development at Lucas.

10.1 Introduction

Ideally, an intelligent autonomous car would have an automatic pilot that can park itself and safely guide the vehicle through dense traffic in towns and at high speeds between towns. At any time the driver would have the ability to switch between automatic and driver control. During periods of manual control the system could act in an advisory capacity warning the driver of hazards and giving information about route guidance or traffic congestion. Such a vehicle would retain the convenience and fun of private transport but would take the drudgery out of driving during automatic operation. Such ambitious goals are now being seriously considered by vehicle manufacturers.

There are immense problems to solve before cars can drive autonomously in today's unconstrained traffic environment. The sensor performance must be very good; there are a large number of different objects and situations that must be detected in a wide range of environmental conditions. System failures are much more critical the driver may not be immediately available in case of failure.

Collision avoidance relies upon predicting your stopping distance in relation to obstacles and that stopping distance can vary by a factor of ten depending on whether you are on ice or dry tarmac. Currently, considerable human judgement is required to infer adhesion correctly. The system must be able to react quickly to incidents. It could be a car crashing ahead, another car swerving into your lane or even a coned off section of the motorway. The reaction to all these incidents cannot be planned in advance.

The greatest difficulty is that the car must be able to anticipate incidents. Most drivers will react differently when seeing an adult by the roadside or a child playing with a ball. In the case of the child, the reaction would probably be to slow down. This decision will be based on an enormous amount of world knowledge that suggests that there is a strong likelihood that the ball could be thown into the road and pursued by the child. The anticipation of that act requires not only a deep understanding of the world but also the difficult sensing task of discriminating between the two scenarios.

Figure 10.1   *Winter testing of autonomous intelligent cruise control on a Saab*

The major issue with autonomous cars is that the driver is not responsible for the vehicle during autonomous operation, and so the system must be extremely reliable. The responsibility will probably reside with the vehicle manufacturer. Some accidents will always be unavoidable, for example when a pedestrian hops out between parked cars immediately into your vehicle path. Product liability will be an enormous barrier to entry in the marketplace, even after most of the technological problems have been resolved. However, their introduction could be limited to highly constrained environments. For instance, motorways are already constrained: all traffic is moving in the same direction, there are no pedestrians, no tight bends and the lanes are well marked. Further constraints, such as infrastructure communications and support would simplify the introduction of autonomous operation.

In the mid term, semi-autonomous systems that maintain driver responsibility for the vehicle will be much more viable and will provide a platform for proving the technology for later fully autonomous systems. This chapter focuses on some of the semi-autonomous systems that are currently being investigated in Europe, the US and Japan. Autonomous intelligent cruise control (AICC), lane support and collision warning are described and results shown from demonstrators under development at Lucas.

10.2 Autonomous intelligent cruise control

Traditional cruise control systems have been in use for many years. They maintain a fixed vehicle speed, as set by the driver, by automatically controlling the throttle, thereby improving comfort in steady traffic conditions. In congested traffic conditions, when speeds vary widely, these systems are no longer effective. The use of cruise control would be significantly increased if the vehicle speed could automatically adapt to the traffic flow.

Figure 10.2 *94GHz radar mounted on Jaguar*

10.2.1 AICC concept

The addition of a radar sensor to the front of the vehicle would provide the necessary range and velocity information for this task. Automatic control of the brakes and throttle would allow the longitudinal controller to maintain a constant time interval behind the vehicle in front. Such systems are commonly referred to as autonomous intelligent cruise control (AICC) and aim to improve driver comfort and convenience. It is important to remember that these systems are quite different from the fully autonomous car. The driver is always responsible for driving and must deal with emergency situations. The system is only capable of fine longitudinal control not emergency braking. Therefore, the driver should have complete control of the system including the right to override, selection of the following interval and selection of targets to follow. The system is further enhanced by allowing operation down to zero speed which would increase comfort during traffic jams. Simulations show that widespread use of AICC would ease traffic flow on motorways by reducing the dangerous *slinky* effect [1]. They also suggest that the resulting synchronisation of traffic would increase throughput in heavy traffic congestion.

10.2.2 Radar sensor

Fundamental to any AICC system is a sensor that can reliably detect obstacles in the traffic environment in a variety of conditions. Microwave radar is a method for detecting the position and velocity of a distant object. A beam of electromagnetic radiation, with a wavelength between 30cm and 1mm, is transmitted and reflected back to the transmitter by the object. Velocity and range can be derived by measuring the doppler frequency shift and the time of flight of the transmission [2]. These techniques have been used widely in aerospace and defence applications for many years. A major advantage of microwave radar is that the performance is not affected by the time of day, and therefore no driver adaptation is necessary for night-time driving. The performance advantages of radar over other sensors are enhanced during poor weather conditions. Systems that rely on visible light are known to suffer significantly in the very conditions for which they are relied upon most. Experience of microwave operation has shown that reliable results can be obtained, even in inclement weather conditions.

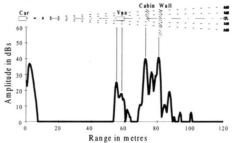

Figure 10.3 *Multiple targets in the radar frequency domain*

The operation of microwave radars fall broadly into two categories: pulse and frequency modulated continuous wave (FMCW). Pulse systems rely on measuring the time of flight of a pulse which is proportional to range. FMCW relies on the linear ramping of the frequency of transmission. If the reflecting object is at a distance R, an echo signal will return after a time $T = 2R/c$, where c is the speed of light. If the echo is then mixed with the transmission frequency, the resulting beat or intermediate frequency f_b will be produced. If there is no doppler shift (stationary target and transmitter) then the beat frequency is proportional to the range. If the rate of change of the carrier frequency is f_o', then

$$f_b = f_o'T = \frac{2R}{c}f_o'$$ (10.1)

Alternative modulation strategies allow the determination of both range and doppler shift for moving scenes [2]. If there are multiple targets then there will be multiple beat frequencies to unravel. Figure 10.3 shows the resulting spectrum for multiple targets after performing a Fourier transform of the radar signal.

Lucas first investigated the possibility of an automotive radar in 1969. Early work resulted in a Ford Zodiac being fitted with a FMCW 24GHz radar to measure range and relative velocity to vehicles in the vehicle's headway [3,4]. Furthermore, this information was used to automatically control the brakes and throttle. Major problems experienced with the system were: large sensor size, multiple objects, poor performance on bends and poor rejection of clutter. Further development in the mid 1970s reduced the size of the sensor by raising the frequency and by using a flat patch antenna. However, the signal processing to overcome some of the other problems was beyond the analogue electronics of the day. False alarm rates should be very low in an AICC system. Performance on a bend was a major problem because the radar would illuminate objects that were safely off the road and miss cars that were being followed. High costs, no angle information and inadequate processing power prevented exploitation of the system [9].

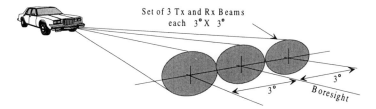

Figure 10.4 *Radar beam geometry*

In 1990 a prototype AICC system was fitted to a Saab 9000 [5]. The microwave radar was a 94GHz FMCW device with a high gain cornucopia antenna. Intermediate frequencies generated by the radar were analysed by a separate real-time digital signal processor. The high carrier frequency led to a further reduction in size, better resolution of targets and improved reflections off shiny targets. Valuable operational experience was gained by testing the system in heavy rain and in extreme conditions in northern Sweden. Many of the original radar problems were solved, and the results from the demonstrator were so encouraging that the system was fitted with some modifications to a Jaguar car, in collaboration with Jaguar Cars Ltd.

On the Jaguar demonstrator the radar is fitted into the front fender and shines through the plastic registration plate, totally disguising the presence of the radar. The electronic control unit and driver interface are similar to the Saab but some modifications were made to the brake and throttle actuators. The throttle is operated by using the standard cruise fitment, a vacuum pump powered device. The auto braking is provided by a modified hydraulic brake booster [5]. Extensive testing and demonstrations of this vehicle to a wide audience have resulted in the evolution of an operating philosophy which should allow AICC systems to be introduced as a stand alone comfort option. Current activities are concentrating on improving the operation of the system in heavy traffic and on bends. The most recent radar is operating at 76GHz (in line with European frequency allocation recommendations) [6,7]. Multiple beams enabled determination of angle, greater field of view and improved tracking of objects around bends. Improved digital signal processing enabled multiple target tracking and clutter rejection.

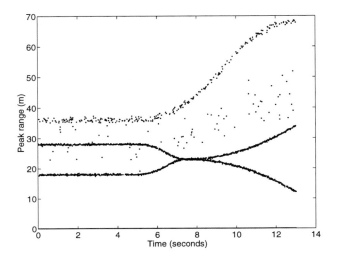

Figure 10.5 *Multiple target tracks merging and splitting*

An extreme example of AICC is called platooning or convoy driving [1]. In this case, the system is designed to have a very quick response to allow cars to travel at speed in tight bunches, with typical separation between cars of only a few metres. The system behaves in a similar manner to the mechanical coupling between carriages in trains. The main benefits are increased capacity of roads and greater fuel economy from slip streaming. Such systems would be highly safety critical due to the likely catastrophe of any failures.

10.3 Lane support

A recent US study shows that 25% of all accidents are caused by unintended road or lane departures [8]. Mounting evidence suggests that many of these accidents are caused by drivers falling asleep, although it is difficult to prove, as many drivers will not admit to falling asleep. Automatic lane following systems have been demonstrated operating at motorway speeds, since 1985 [10]. Such systems certainly prevent lane departure while lane markings are present, but do not prevent drivers falling asleep. In fact, there is a danger that the reduced driver workload could even encourage drowsiness.

The Lucas lane support system [11,12] audibly warns the driver of unintended lane departure. Audio messages are issued as the driver crosses a lane marking without using the indicators and the balance of the stereo sound system is controlled to denote the direction of road departure. The lane markings are detected by processing images from a video camera [11]. The current system achieves lane marking detection at 15 frames/sec. Road edges with no lane markings are also detected although they are less reliable. Preliminary studies show that it is possible for the system to determine driver steering oscillations by tracking the lane marking positions. The frequency of oscillation, seen in Figure 10.6, appears to be about an 1/8Hz for normal alert drivers. It would be useful if this information could be used for determining the state of driver alertness, although further investigations are necessary before this can be proved.

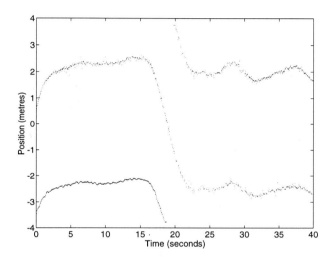

Figure 10.6 *Video measured lane markings during a lane change*

More effective than audible warnings, that could irritate the driver, is the use of haptic feedback in the steering wheel. The system could provide position related assistance/resistance in the steering to give an artificial feel of road camber on either side of the lane. The driver is still expected to steer the car but experiences the sensation of driving along the bottom of a bathtub [13]. Again, the system is deactivated by either the use of the indicators or by the driver exceeding a torque/motion input threshold at the steering wheel. The benefits to the driver would be reduced steering effort, improved steering stability, and safety.

Figure 10.7 *Virtual road camber induced by a power assisted steering system*

The key components of the lane support system are shown in Figure 10.8. A video camera and processing measures the lane markings. Another sensor measures the torque that the driver is applying to the steering wheel. The electronic control unit (ECU) takes this measurement and drives a motor which in turn applies torque through a gearbox to the steering mechanism thereby assisting the driver. The incorporation of a clutch improves the feel (by disengaging the motor) when no assistance is required and also improves the fault tolerance.

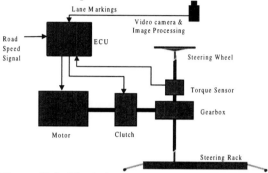

Figure 10.8 *Block diagram of lane support system*

10.4   Collision avoidance

Every year 50,000 people are killed on the roads in the European Community. The figure for the USA is similar. In Britain, out of 240,000 accidents in a typical year, there are 5,000 fatalities and 64,000 serious injuries [14]. Many safety innovations in the areas of braking systems, airbags, body structures, steering and suspension have already had a beneficial effect. However, accident frequency and severity have still remained unacceptably high. Accident studies place most of the blame on drivers [19]. Automatic collision avoidance systems could have many advantages over human drivers: they do not get tired or distracted, they can simultaneously monitor all sides of the vehicle, they generally have faster reaction times and they do not panic. Apart from the human tragedy, these measures would significantly decrease the economic cost of road accidents and reduce the inconvenience associated with the resulting traffic jams.

The main requirement for a collision avoidance system is to be able to predict accurately the likelihood of an imminent collision. If a collision is likely then the system should respond in such a way as to reduce the threat. The response could involve the automatic control of the vehicle or simply an appropriate warning to the driver. Lucas have produced a collision avoidance demonstrator, with Jaguar, that is currently a warning only system.

10.4.1 Collision avoidance sensing

Crucial to the success of any collision avoidance system is the ability to reliably identify and locate different obstacles in the complex and unconstrained traffic scene, where conditions can vary from thick fog at night to bright glare from the sun on a clear day. Collisions can occur at any point on the car, therefore sensor coverage should ideally extend to 360°. It is probable that no one sensor can satisfy these strict demands. Headway detection systems have received most attention because they potentially cover the majority of accident situations [19]. However, they are difficult to implement because of the large maximum range required for driving at speed. Many false alarms are generated by radar systems illuminating objects on bends that are safely off the vehicle's path. False alarm rates should be very low in a collision warning system to avoid driver irritation. If there is automatic braking then they should be eliminated. Calculating collision paths over a long range and on bends requires a detection system with a wide field of view, high angular resolution and the detection of the road geometry.

Many accidents occur due to side impacts or while reversing. Radar beams can cover the rear or blind spots of a vehicle. Fortunately, the speed of travel laterally or in reverse is slow and so the required maximum range is small in comparison with the headway sensors. Parking sensors are already commercially available, some school buses in the USA are already fitted with such devices to warn the driver of children near the wheels [20]. With the strict liability laws in the US such precautions make a lot of sense. The function of the lane support system, discussed earlier, would be considerably enhanced with the addition of blind spot sensors.

Figure 10.9 *Raw image and processed data*

There are a number of possible technologies for collision sensing. The major options are visible or infrared video cameras and ultrasonic, infrared, visible or microwave radar. The passive systems such as video cameras tend to have good spatial resolution and are good for angle determination, object recognition and road detection. However, the calculation of range and relative velocity is complex and the accuracy poor compared with active radar ranging systems. Also, the visible video cameras have worse performance at night or in poor conditions. Ultrasonic is only useful for proximity sensing because its range is limited due to strong atmospheric attenuation. Infrared and visible radar offer good angle information by the use of spinning mirrors and reasonable range information. However, the range deteriorates rapidly in poor conditions, back scatter from foggy patches can create false targets and the use of *class I* lasers creates problems for eye safety.

The current Lucas collision avoidance demonstrator uses a headway video camera and microwave radar. The radar system is similar to the AICC radar and can track 12 vehicles up to 100m within a field of view of 9°. The video based system detects cars and lane markings, as discussed earlier, up to a distance of 60m. The low level image processing acquires images from a miniature CCD video camera situated by the rear view mirror. Edges are extracted using a dual Sobel convolution mask [15] and intelligent thresholding. The edges are then thinned using morphological techniques [16], see Figure 10.9. The thinned edges in the image are then traced by fitting straight lines to them, and turned into line vectors, which are then categorised on the basis of angle. The horizontal lines are used by the car detection algorithm, and some of the non-horizontal by the lane marking detection algorithm. A rear-end car filter locates cars in the image plane by processing the horizontal line patterns. Classification of cars and the calculation of angle are good. Range is estimated from the road to camera geometry by assuming a flat earth. As a result the range data are quite noisy due to car pitching, hilly roads and small variations in the determination of the car position in the image plane.

A very high computational rate is required for the image processing. In the current system a VME rack is situated in the boot of the Jaguar. The rack contains a Sun workstation for development purposes, a Datacube convolution board and two Intel i860 RISC processors. The architecture allows cars to be detected at about 12 frames/second. Ultimately, the algorithms would be implemented in full custom ASICs for production purposes. Ideally, the actual camera would contain the image processing thereby avoiding the need to transfer high bandwidth signals around the car.

10.4.2 Data fusion

Data fusion is a collection of techniques for combining the measurements from more than one sensor to produce an improved unified result. The sensors used can be of the same or different types. This has several benefits:

- The overall estimates of parameters can be more accurate than for individual sensor estimates, as they reinforce each other.

- Any parameter need not be dependent on one sensor only. This has benefits for fault tolerance by allowing redundancy to be introduced into the system. For example, if a system had infra-red and video based sensors, it could survive the failure of either of these and continue to function, although the accuracy and performance in certain conditions would be reduced.

In the Lucas system, data fusion gives major benefits to object detection because the two sensors being used complement each other extremely well. The video produces very good lateral position estimates, but is not able at present to produce good estimates of range. The radar conversely produces very accurate range estimates and hence good relative velocity estimates, but has poor lateral positional accuracy. Thus, by fusing the data from these two sensors the object position can be localised to a better accuracy by considering the intersection of the two areas of positional uncertainty generated by each sensor. The situation can be improved further by considering the additional parameter of velocity [17,18].

Another possible benefit of data fusion is that of object identification, by combining the expected responses of an object in the sensors. For example:

Object	Video parameters	Radar response
Pedestrian	Tall and thin	Poor
Traffic sign	Tall and thin	Good (metal object)

The image processing may confuse a stationary pedestrian and a traffic sign, especially at longer ranges, as they are both tall, thin objects. However, fusing the data would allow an unambiguous decision to be reached immediately as the two objects have radically different radar responses.

Data fusion brings along its own set of issues and problems, that include:

- Dealing with conflicting results from different sensors.
- Dealing with multiple objects in the scene, and maintaining the correct paths for each object.
- Generating sensible models of the system on which to tailor the mathematics of the data fusion process.

10.4.3 Decision strategy

Having detected a hazard the system must respond in the most appropriate way so as to avoid the hazard. Evidence suggests that a small reduction in driver reaction time will dramatically reduce the number of accidents. This can be achieved either by directly intervening with the controls of the car or by warning the driver. Direct intervention will have the most significant effect but causes the greatest liability concern. In the worse case the car could contravene the driver's wishes by emergency braking just as the driver decides to accelerate and overtake. Driver warning is the safer option but there are still drawbacks. In critical situations there may not be enough time to warn the driver or the warning might even distract the driver from the hazard. If there are too many warnings, false alarms, or if the driver is already aware of the hazard then warnings will be irritating.

Usually, warnings are either visual or audible. The visual warnings must be in the driver's field of view at the time of warning. Audible warnings are generally more obtrusive and hence irritating if inappropriate. However, also effective are haptic (tactile) and kinaesthetic (motion) warnings. Haptic feedback in the steering mechanism has already being described for lane support, but the same system can be used to increase the steering reactance if there is an obstacle in the driver's blind spot during a lane change manoeuvre. Kinaesthetic warnings invoke a very quick response by jerking the car or by moving the seat. Even sleeping passengers can be quickly woken by the driver rapidly decelerating the car. If the jerk duration is short then there will be little effect on the car's speed leaving all options still available for the driver.

The detection of hazards can be calculated by considering the trajectories of other objects in relation to your own vehicle, its operating envelope and the state and intentions of the driver. If the driver is inattentive to the situation then action must be taken but if the driver is already taking effective action then no warnings or action should be taken. Therefore, the decision process should include the intentions and states of the driver. Monitoring the driver is a hard problem, it is difficult to discriminate between a genuine lane change and a lane drift due to drowsiness. Progress has been made using neural networks to learn driver behaviour. A backward error propagation network can predict if the driver is going to overtake or brake with a success rate of 90% when trained on only 15% of the total input data [21].

In the situations that require direct control of the vehicle, recent work has shown the use of neural networks for emulating driver brake and throttle control [22,23]. Such a system would be particularly useful for AICC where drivers feel most comfortable when the system mimics their own driving style.

10.5 Conclusion

This chapter has described a wide variety of automotive applications for intelligent autonomous systems. The fully autonomous car is probably not viable in the foreseeable future. Semi-autonomous systems, as discussed above, are technologically feasible, but issues such as driver acceptance, reliability, safety and product liability have yet to be resolved. AICC and lane support systems, not only improve driver comfort, but also reduce the risk of an accident by preventing the car from getting into a critical situation during use. Collision warning will be useful for alerting a distracted or inattentive driver to a hazard, providing the time to impact is not too critical. Collision intervention, although most effective, will probably not appear for a long time due to the sensor unreliability and liability concerns.

When considering the benefits, cost is a major consideration in the automotive market. Much of the technology tends to be expensive because it originates from low volume/high cost military markets. Where volumes have been high, such as for microwave satellite receivers and door openers, then costs have come down and the markets have been successfully exploited. The high volumes associated with automotive production should provide the incentive to design low cost semi-autonomous systems for cars.

10.6 Acknowledgements

The author wishes to thank the directors of Lucas Industries plc for permission to publish this work and also Richard Conlong, Peter Martin, Ian Westwood, Katharine Prynne, Peter Mason, Andy Kirk, Ed Young, Russell Jones, Manzoor Arain and Jaguar Cars Ltd. for their technical contribution to the research presented here.

10.7 References

1. HEDRICK, K.,: Longitudinal control and platooning, collision avoidance systems for intelligent vehicles TOPTEC, SAE, Washington, D.C., Apr., 1993
2. SKOLNIK, M.: Introduction to radar systems, 2nd Edition, McGraw-Hill
3. IVES, A.,BRUNT, WIDDOWSON, and THORNTON, Vehicle headway control, IMechE, Sussex, Sept 1972
4. IVES, A. and JACKSON,: A vehicle headway control system using Q-band primary radar, SAE, Detroit, Feb. 1974
5. MARTIN, P.: Autonomous intelligent cruise control with automatic braking, SAE, Detroit, Feb. 1993
6. Harmonisation of frequency bands for road transport information Systems (RTI), Recommendation T/R 22-04 (Lisbon 1991).
7. European Radiocommunications Committee Decision of 22 October 1992 on the frequency bands to be designated for coordinated introduction of Road Transport Telematic Systems, ERC/DEC/(92)02.
8. The Hanson Report on Automotive Electronics, vol 5,No.6, July/Aug 1992
9. TRIBE, R.H.: Automotive Applications of Microwave Radar, IEE Colloqium on Consumer Applications of Radar and Sonar, London, May 1993
10. DICKMANNS, E. and ZAPP, A.: A Curvature-based Scheme for Improving Road Vehicle Guidance by Computer Vision, SPIE on Mobile Robots, Cambridge, Ma, USA, October 1986
11. TRIBE, R.H., YOUNG, E. and CONLONG, R.: Improved obstacle detection by sensor fusion, IEE Prometheus & Drive Colloquium, London, Oct. 1992
12. TRIBE, R.H., Overview of Collision Warning, Collision Avoidance Systems for Intelligent Vehicles, SAE TOPTEC, April 13, 1993, Washington, D.C.
13. RICHARDSON, M. and BARBER, P.: Advanced Steering: The benefits of a mechatronic gearbox, IEE Prometheus and Drive Colloquium, London, Oct. 1992
14. The department of transport, Road accidents Great Britain 1990 - The casualty report, HMSO 1990
15. BALLARD, D.H. and BROWN, C.M., Computer Vision, Prentice-Hall, 1982
16. MARAGOS, P., Tutorial on advances in morphological image processing and analysis, Optical Engineering, July 1987, Vol. 26 No.7
17. HARRIS, C.J. and WHITE, I., Advances in command, control and communications systems, Peter Peregrinus Ltd., 1987

18. RAO, B.S.Y and DURRANT-WHYTE, H.F., A Decentralised Algorithm for Identification of Tracked Targets, University of Oxford Report No. OUEL 1892/91

19. KNIPLING, R.R.: IVHS Technologies Applied to Collision Avoidance: Perspectives on six Target Crash Types and Countermeasures, Safety & Human Factors Session, 1993 IVHS America Annual Meeting, April 14-17, 1993

20. MENDOLIA, G.: Microwave components for low cost automotive radars, collision avoidance systems for intelligent vehicles TOPTEC, SAE, Washington, D.C., Apr., 1993

21. ARAIN, M.A. and TRIBE, R.H.: Application of Neural Networks for Traffic Scenario Identification, Proceedings of the 4th Prometheus Workshop, Compiègne, October 1990.

22. AN, P.E., HARRIS, C.J., TRIBE, R.H., and CLARKE, N.: Aspects of Neural Networks In Intelligent Collision Avoidance Systems for Prometheus, JFIT Technical Conference, p129 to p125, Keele University, March 1993

23. ARAIN, M.A., TRIBE, R.H., AN, P.E., and HARRIS, C.J.: Action planning for the collision avoidance system using neural networks, Intelligent Vehicles Symposium, Tokyo, July 1993

Chapter 11

Walking machine technology - designing the control system of an advanced six-legged machine

A.Halme and K.Hartikainen

In this chapter design aspects and experiences on walking machines are summarised and critically analysed. Experiences are based on designing the MECANT I - machine, which is a research and development test-bed for work machine applications in an outdoor environment. Considerations are focused to the mechatronics of the machine and especially to the control system. The control system is based on two level hierarchy and distributed control philosophy which follow the canonical layout. Differences between the requirements of engineering design and what nature has realised in animals are considered.

11.1   Introduction

Legs offer some distinct advantages for vehicles operating in difficult terrain like natural ground or environments designed primarily for man, like buildings with stairs. Legs provide more degrees of freedom, which makes the vehicle more flexible in motion and stabilising easier on uneven surfaces. Constructing a successful walking machine, however, is a challenging project and requires special mechanical design knowledge and an extensive on-board computer and software system to provide necessary motion control properties for the machine. Prototype walking machines have been developed this far mainly for research purposes in Japan and USA (some active groups exist also in Russia). In Europe the research activity has been relatively low until recently. There have been no real commercial successes in making walking machine products so far, although a couple of developments in USA and Japan have been marketed as product prototypes (Odetics, Toshiba). MECANT I, shown in Figure 11.1, has been developed in the Helsinki University of Technology in a national technology programme on outdoor robotics. It is a research walking machine which is probably one of the most advanced presently existing and one of the few which carries all its control and power systems on board. It has been designed as an outdoor test vehicle to study work machine applications in natural environments, like forests. The basic design principles, system structures and the main properties of the machine have been described earlier in [1]. The details of the motion control principles have been reported in [2]. The purpose of this chapter is to explain in more detail the hardware and software structure of MECANT and discuss related aspects based on the experiences gained after getting the machine ready and testing it in different conditions. Both structures follow the quite natural distributed philosophy existing in walking machines.

Figure 11.1 *MECANT I*

11.2 MECANT

MECANT is a fully independent hydraulically powered six-legged insect type walking machine. It weighs about 1100 kg and its main geometrical dimensions are illustrated in Figure 11.2. The legs are all identical, rotating pantograph mechanisms with 3 dof in each. The body is constructed from rectangular aluminium tubes forming a rigid light structure. All components, except the hydraulic cylinders actuating the legs, are commercial ones. The vehicle is controlled remotely via the radio link by the operator using two joysticks.

 The motion can be controlled omnidirectionally. The body of the machine is controllable like a free object in the space having all the 6 dof available within the kinematic limits. The leg working volumes overlap both sides of the body making climbing over obstacles possible.

11.2.1 Energy, power transmission and actuation systems

MECANT has a fully self-sufficient energy system. The power is generated by a 38kW 2-cylinder ultralight aeroplane engine with air cooling. The hydraulic system is a traditional one including valve-controlled flow system with central pump, oil reservoir and pressure accumulator, the work pressure being about 300 bar. The high speed gas engine is controlled with a fast analogue rpm-controller to adapt to the system load variations. The energy efficiency of the power system is not very high, but the structure is simple and light.

The actuation system in each leg consists of two cylinders and a motor. The cylinders are tailored with bending stress capability and integrated potentiometric position measurement. The hydraulic motor is a commercial one with potentiometric position measurement. All actuators are controlled with proportional valves.

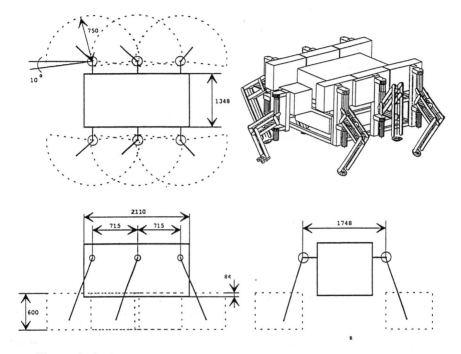

Figure 11.2 *The mechanical design of MECANT I and its main geometrical dimensions*

The machine electrical system is a 24 V system consisting of a 1.5kW generator, battery backup and stabilised DC-power source giving the different voltage levels needed by the electronics through DC/DC converters.

11.3 Control hardware design

The main principle of the control hardware design has been the use of commercial low-cost components. Another important criterion when selecting the control system hardware was effective software development and testing support. The PC technology offers both the features desired and was thus selected. The availability of real time networking hardware and a supporting commercial operating system gave the opportunity for a modular and distributed control system design with effective software development environment. The control system of MECANT is pictured in Figure 11.3.

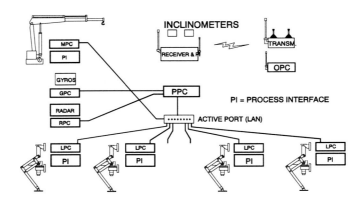

Figure 11.3. *The control system of MECANT I*

11.3.1 Computer system

The computer system of MECANT is built of seven IntelX86 processors (PC-bus boards) connected together with high-speed token-ring network (Arcnet, 1 Mbit/s). The computer system configuration supports hierarchical control of the vehicle motion which is the implementation of the supervisory scheme described in reference [1]. According to that scheme the computers are divided into a pilot computer (Intel486/33MHz) and six leg computers (Intel286/16MHz).

The pilot computer assembly can be seen in Figure 11.4. The pilot computer has a configuration of an ordinary bus PC. The computer acts as a server computer in the network, i.e. the operating system and control software are loaded into the network from the pilot computer. It also includes a flat panel display (DC-voltage operated) and a PC/AT keyboard for system operation and software development purposes. The pilot computer has a hard disk data storage capacity of 120 MBytes.

The leg computer configuration can be seen in Figure 11.5. The operating system and the control software are loaded via the computer network. Both the pilot computer and leg computers have volatile memory storage capacity of 4 MBytes (RAM) each. This is used for on-line data logging purposes.

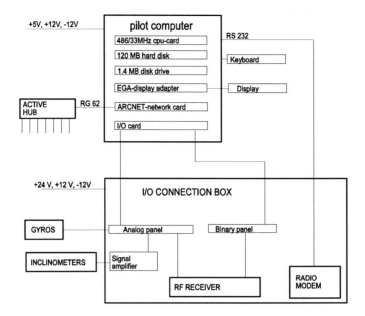

Figure 11.4 *The pilot computer and i/o assembly*

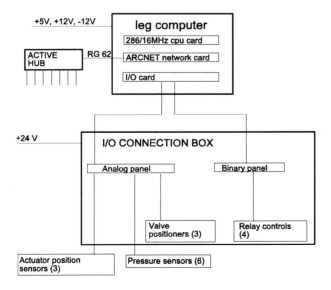

Figure 11.5 *The leg computer and i/o assembly*

11.3.2 I/O system

The i/o system is distributed between the pilot and leg computers depending on the control responsibilities of the computers according to the supervisory control scheme. The pilot i/o system can be seen in Figure 11.4. The pilot computer is responsible for vehicle body motion control, thus pilot computer i/o data include portable operator interface signals and body inclination sensor signals. All leg sensor data are available via computer network at 40 ms intervals. The leg computer i/o system configuration is shown in Figure 11.5. The leg computer is responsible for leg motion control. Separate i/o connection boxes are used to make connection modification and maintenance easy. The connection boxes include the i/o interface hardware: isolated analogue and binary output/input panels. In addition the connection box of the pilot computer includes RF receiver, RF modem and inclinometer signal processing electronics and that of the leg computer servo valve positioner electronics.

Minimum sensor configuration implemented into MECANT includes inclinometers, hydraulic actuator pressure and position sensors. Inclinometers are used to measure body attitude, i.e. roll and pitch angle. They are standard oil damped pendulums with LVDT-position sensors. Filtering of the inclinometer signal is necessary in order to remove the high frequency component created by feet stepping cycles. Leg motion control is executed according to actuator position or pressure difference feedback. Potentiometers are used as actuator position sensors, which are accurate enough for leg servo purposes. Leg ground contacts and collisions can be detected from hydraulic circuits by pressure sensors. They are also used for actuator force control with limited accuracy.

11.4   Control software design

The control software has been developed under a commercial real-time, multi-tasking operating system which supports network operation (QNX). The software development in the network environment is effective, because the network is transparent to the application programmer. The communication is controlled totally by the network hardware. Real time management, task communication and network management C library functions are available. As was the case for hardware, the control software can be divided into pilot software and leg software. Because of the hierarchical nature of the control software, the development of both softwares has been able to be executed quite independently. The block diagram describing the main functions in the pilot and leg software is pictured in Figure 11.6.

In the present software version the overall functionality between the pilot and leg software is divided as follows.

11.4.1 Pilot software

The main function of the pilot software is the execution of *vehicle motion control*, i.e.

body motion planning and control, gait planning and feet trajectory planning according to operator commands, body sensor and leg sensor data. The vehicle motion planning and control is described in detail by [1] and [2]. Body motion planning and control cycle time is presently 200 ms. Gait planning and feet trajectory planning interval is 40ms. Vehicle motion control is executed on-line according to 40ms old data thus motion control is able to react immediately to unexpected events, e.g. leg collisions.

Feet trajectory points produced by the vehicle motion control are sent to the leg computers every 40ms interval by *leg communication* task. Because communication requires receiver confirmation, leg status data can be received by the pilot software during the same communication cycle. Communication is executed independently from other tasks thus maximizing the throughput of the network.

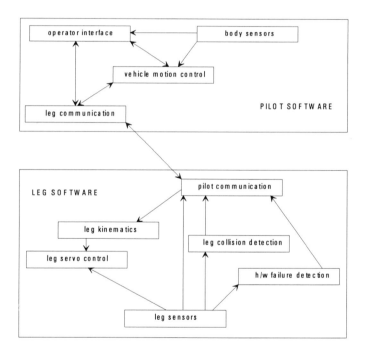

Figure 11.6 *The main functions of the pilot and leg software*

Body sensors and operator inputs are measured with an independent task which is executed every 10ms. This task also includes signal processing. Vehicle status data can be sent via a radio modem link to the operator from an *operator interface*. The execution rate is dependent on the data transmission capacity of radio modems (4800 bit/s).

11.4.2 Leg software

The leg software includes *leg servo control* which controls the actuator motions according to pilot computer supervision. Presently the foot reference trajectory in Cartesian coordinates and the actuator force reference values are sent as supervisory data by the pilot computer. Standard PI-controllers are used as position servos for hydraulic actuators. Actuator force control is also executed by a PI-control algorithm.

The leg software also features *leg collision detection* logic which utilises hydraulic pressure data and leg motion data in detecting the leg collisions with the environment. Leg i/o hardware failures are also monitored to some extent by *h/w failure detection*.

Leg data are collected and sent to the pilot computer every 40ms by the *pilot communication* task which is executed independently. Sending of leg data is synchronised to the pilot computer communication frequency, thus no asynchronous communication between the pilot and legs is presently supported, but the 40ms communication cycle compensates for this deficiency.

11.5   Control requirements of a walking machine

The motion control of a practical walking machine is not a leg motion control problem, but a much more complex one of the motion control of the vehicle body. It is analogous to the end-effector control problem of multiple manipulators sharing a common load. In the case of the walking machine the problem is somewhat easier because of the intrinsically fixed end-effectors and the compliant base of the 'leg manipulators'. Thus the servo and synchronisation errors of some degree can easily be tolerated in the leg motion control. This is why biological organisms can use legs so easily. The 'body motion control' philosophy required in engineering applications differs, however, from that adopted by nature. In most cases motion of the body has to be planned according to the requirements of some specific task. When considering biological systems with statistically stable walking, e.g. insects, postural control is mainly the side effect of feet motion control. The motion control of the legs can be considered to serve the purpose of pure locomotion, i.e. transferring the insect from the present location to a new one. Thus the motion control is more distributed and localised at the leg level in the animal world (higher animals, however, clearly also plan the motion of their body in certain operations). Reasonably accurate body motion control for all six dof of the vehicle body is required in the practical applications of the legged machines which do not possess the capability of grasping support during locomotion. In addition to the demands set by the body motion control, the on-line execution of the motion control at vehicle level without *a priori* knowledge of the environment sets high performance requirements to the distributed control system.

The natural control method evolving from the philosophy of body motion control is the supervisory control scheme. The simplest approach to this strategy is to consider the leg control computers as pure servo controllers. The servo reference values are calculated and sent by the main computing unit e.g. by the pilot computer in the case of MECANT. This control scheme can also be called *servo supervision*.

When designing a real time distributed control system to support the supervisory control scheme of a statically stable legged vehicle some basic knowledge of locomotion processes is required. Most of the performance requirements on the control system can be derived from the estimated maximum velocity of the vehicle. The vehicle velocity naturally affects the trajectory velocities of the feet. If the feet trajectory velocities are known some minimum performance requirements can be established.

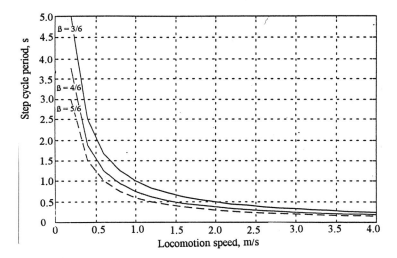

Figure 11.7 *Leg step cycle time as a function of vehicle locomotion speed when using periodic gaits with duty factors 3/6, 4/6 and 5/6. The value of the support stroke used in MECANT is 0.5m*

11.5.1 Leg motion in periodic gaits

Leg step cycle frequency can be derived mathematically if periodic gaits are applied to vehicle locomotion. Different gait types for statically stable legged vehicles are examined quite thoroughly by [3]. For periodic regular gaits the duty factor β describes the normalised ratio of support period to step cycle period and it is

$$\beta = \frac{\tau_s}{\tau}, \tag{11.1}$$

where τ_s is support time period and τ is step cycle time period. When the maximum support stroke R of a vehicle leg is known, the locomotion velocity u can be derived as

$$u = \frac{R}{\tau_s}. \qquad (11.2)$$

Combining equations (11.1) and (11.2) the step cycle time period can be derived as a function of locomotion velocity u and duty factor β

$$\tau(u, \beta) = \frac{R}{\beta u}. \qquad (11.3)$$

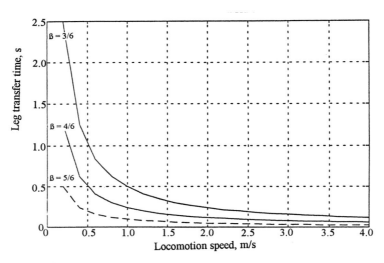

Figure 11.8 *The relationship between the transfer state time period of leg and vehicle locomotion speed. Duty factors and support stroke values equal those of Figure 11.7.*

Fixing the support stroke R and parameterising the duty factor β, Figure 11.7 can be drawn giving the relationship between the locomotion speed and leg step cycle period. A duty factor of n/N, where N is total number of legs, for a six-legged vehicle implies that n legs of six are always supporting the vehicle. Thus for a statically stable legged machine a minimum of three legs should support the vehicle.

Leg step cycle time does not indicate the transfer foot trajectory velocity which is also an important factor when designing the control system. The word 'transfer' is used for the recovering foot which does not support the vehicle and during the recovering phase this foot is said to be in the transfer state as opposed to the support state. The transfer state time period τ_T as a function of the locomotion speed and the duty factor is

$$\tau_T(u, \beta) = (1-\beta)\tau(u, \beta) \tag{11.4}$$

This is shown in Figure11.8.

If the transfer trajectory shape is a half circle so that its radius is equal to half of the support stroke length, the average trajectory velocity of a transfer foot, as a function of locomotion speed is

$$v_T^{mean}(u, \beta) = \frac{d}{\tau_T(u, \beta)} \tag{11.5}$$

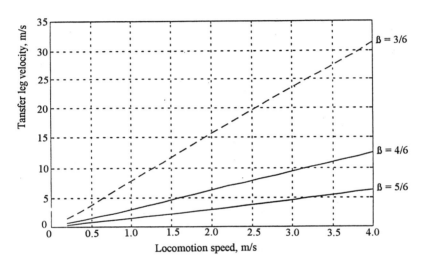

Figure 11.9 *The average velocity of transfer foot as the function of locomotion speed when using periodic regular gait. Duty factors and support stroke values equal those of Figure 11.7*

where *d* is the length of the transfer trajectory and is given by

$$d = \frac{1}{2}\pi R \tag{11.6}$$

The average transfer foot velocity relationship to locomotion velocity with a parameterised duty factor is shown in Figure 11.9.

The performance requirements of the supervisory controller can be examined based on the feet trajectory velocities derived above for periodic regular gaits.

As shown in Figure 11.7, the leg step cycle times are very small even at moderate vehicle speeds, the worst case being the most stable leg configuration (β= 5/6). But when determining the minimum performance requirements for servo supervision the transfer time graph should be studied, Figure 11.8. One can see that already at a locomotion velocity of 1 m/s with β = 5/6 the foot makes the movement in the air within a time period of 100ms. It should be noticed that the value is proportional to the support stroke (in this case $R = 0.5$ m) but the high performance requirement for a distributed control system supporting servo supervision is obvious.

In practice the pure position control of legs cannot be used in irregular soft terrain. At least some support force control of feet is needed.

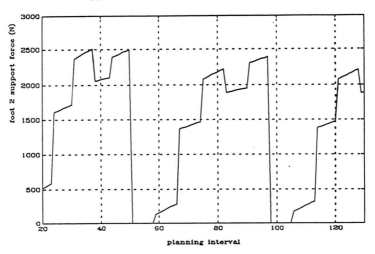

Figure 11.10 *The simulated leg reference support force for MECANT during locomotion. The duty factor of periodic regular gait is 5/6*

Leg force control affects natural body motion trajectory, so it cannot be considered only as a single leg motion control problem but rather a body motion control problem, i.e. servo supervision problem. If closed loop force control is applied at the foot level, the supervisory control loop requirements depend on the vehicle locomotion speed and gait selected. To keep the body motion stable during pure support force control, the foot support force references must be updated frequently. According to the zero momentum rule ([4] and [5]) leg reference support forces change during the locomotion as shown in Figure 11.10.

As is seen from the figure, the reference support force values must be changed five times (corresponding to leg state changes from support state to transfer and vice versa) during the foot support period in order to prevent excessive tilting of the vehicle body. According to Figure 11.10 it can be concluded that with a locomotion velocity of 1m/s the reference support force for the leg should be updated every 100ms. It can also be noticed that the bandwidth requirements increase dramatically for leg force servo thus making the tripod gait the natural choice in high speed locomotion because it does not require force controlled legs. Actually insects always transfer to a tripod gait when their speed increases.

Generally speaking, to enhance the vehicle motion, the body motion control process requires high bandwidth information at the leg level. The performance requirements of a distributed control system in data transfer and acquisition become especially high when reflex type asynchronous control is applied at the vehicle level.

11.6 Conclusions

The control system of a walking machine is perhaps the most challenging part of its design. To utilise all the advanced features of the machine the system must be able to control the motion of the body according to the task in question in variable environmental conditions. A hierarchical and distributed control system architecture is in a way a natural choice, but it also sets quite tough requirements especially for communications in the system. Experiences from MECANT show that even a quite low performance real time LAN can, however, be used when the communication needs are carefully minimised.

Distributed control systems offer some advantages over centralised ones. These are:

- reduced cabling,
- modularity,
- low cost components,
- increased reliability,
- simpler failure diagnostics.

When designing a distributed control system for a walking machine some special control features and requirements of the natural terrain applications should be remembered. These are:

- reactivity at the master and slave levels
- high bandwidth control and communication processes
- fast responses to leg contacts/collisions
- synchronisation
- real time deterministic data transmission.

PC technology offers one possible low-cost hardware realisation of the features mentioned in the control applications of legged work machines.

11.7 References

1. Hartikainen, A., Halme, A., Lehtinen, H., and Koskinen, K., MECANT I: A six legged walking machine for research purposes in outdoor environment. Proceedings of the IEEE International Conference on Robotics and Automation, vol. 1, pp. 157 - 163, 1993.
2. Halme, A., Hartikainen, K., and Kärkkäinen, K., Terrain adaptive motion and free gait of a six-legged walking machine. Preprints of the IFAC International Workshop on Intelligent Autonomous Vehicles, Pergamon Press, pp. 1-7, 1993.

3.   Song, S.M., and Waldron, K.J., Machines that walk; the Adaptive Suspension Vehicle. MIT Press, 1989.

4.   Huang, M.Z. and Waldron, K.J., Relationship between payload and speed in legged locomotion systems. IEEE Transactions on Robotics and Automation, Vol. 6, No. 5, pp.570-577, 1990.

5.   Gorinevsky, D.M. and Yu, A., and Shneider, H.,. Force control in locomotion of legged vehicles over rigid and soft surfaces. The International Journal of Robotics Research, Vol. 9, No. 2, pp. 4-23, 1990.

Chapter 12

Handling of flexible materials in automation

P.M.Taylor

12.1 Introduction

Flexible materials are in widespread use in the home and in industry. A particularly important subset is that of sheet-like limp materials such as fabrics which have low resistance to out-of-plane bending. Yet it is only recently that the automated handling of such materials has received much attention, arising from attempts by researchers to automate the assembly processes. Before surveying the progress made in this area a background will be given of the status of the garment and footwear industries and the reasons behind these efforts at automation.

The apparel and footwear industries rarely get the media, political and academic attention merited by their importance in economic and employment terms. In 1991, for example, the Europe apparel industry, excluding knitting, textiles and footwear, employed an estimated 1.5 million people [1] who produced 3.4 billion items of clothing. With textiles, the apparel industry was the source of over 10% of total industrial output, and in many countries was in the top three industrial sectors in terms of value of production and employment. Similar statistics apply in the USA. It still remains a labour intensive industry and so many sectors are under great threat from countries such as China with very much lower labour costs. Kurt Salmon Associates predict [2] that, by the year 2001, production will have declined to 2.4 billion items with an emphasis on top quality and high fashion items; this is still a substantial industry. It should also be noted that many of the industrialised countries also have significant machinery and IT support industries. In particular, the UK is home to BUSM, a major world supplier of shoe making machinery.

This threat of competition from low labour cost countries has given rise to a number of research programmes which aim to reduce labour content and help ensure consistently high quality production which can adapt rapidly to market needs. Such initiatives have taken place in Japan through the TRASS project, [3], in the USA in various programmes but recently as part of the AMTEX project [4], in Europe under BRITE-EURAM [5], ESPRIT etc. and in the UK under the aegis of the ACME Directorate of the SERC.

12.2 Garment/shoe upper assembly

In general, garments and shoes are progressively constructed from a number of flat pieces of material which will eventually form a 3-D shape to suit the shape and size of the customer. Although ultrasonic bonding and adhesives are used in a few cases, the predominant joining technique is that of sewing, either a lockstitch (2 threads, one above and one below the material(s) and looped over each other in-between) or a chainstitch which is formed from one continuous thread. Above the material there is a needle which introduces the top thread and below, either a bobbin or a looping mechanism. A sewing machine will operate at several thousand r.p.m and even a relatively slow 3000 r.p.m. machine results in a needle reciprocating at 50 Hz. Machine parameters have to be adjusted to give optimum performance for the fabrics being sewn. This is still done manually through experience and trial and error. Clearly this approach is not satisfactory if consistent high quality results are required in a flexible automated system. As a result, the sewing process has been the focus of much world-wide research [6] in 'sewability' which aims to determine the relationships between sewing parameters, fabric properties and seam quality.

On the factory floor, operators handle the material pieces during and between the sewing operations. Typically, over 60% of an operative's time is spent in material manipulation rather than sewing so if automated handling/sewing systems can be introduced (as indeed occurs in ladies hosiery) there should be much saving of non-productive operator time.

However, materials are non-rigid, usually anisotopic, and have mechanical properties which can vary from piece to piece. In addition, previous procedures such as cutting, storage and joining may introduce other uncertainties in the actual shape of the pieces. The eventual customers, people, have a wide variety of shapes and sizes. In the shoe industry many size and width fittings are required for a single style. In the garment industry fashions often change quite quickly giving rise to style and material variations. Manufacturer's margins are low compared with the retailer's so any automation must be relatively low-cost as well as being flexible and tolerant to variations in material properties and earlier processes. Despite these difficulties many impressive advances have been made over the last dozen years or so but, as will be seen, much more remains to be done.

12.3   The automated assembly toolbox

This section describes a toolbox of techniques now available to the designer of handling and assembly systems.

12.3.1 Movement of pieces of material

If material panels are to be moved from one place to another there must be some holding/fixing/supporting means provided to connect the panel(s) to a moving medium. The choice of technique depends on the accuracy of movement required, the location of the points on the material for which such accuracy is required, the properties of the material and the required speed of operation.

12.3.1.1 Pick/carry/place

This addresses how to pick up a single, separated panel of material from a table (say) and place it at another location. The panel may be grasped at a number of points using pins, pinching mechanisms, vacuum suction or adhesives. All other points on the panel are unconstrained apart from internal material forces and thus may move during transportation. The advantages and disadvantages of these techniques are as follows:

i) Pins: They give a good, mechanically secure contact but may damage the structure of delicate materials or the appearance of others, e.g. the surface of leather. Pin insertion depth and geometry may require adjustment between material types. Material release is fast and generally has good repeatability and reliability, but occasionally materials may stick to a pin.

ii) Pinching mechanisms: These give a good mechanically secure contact for materials with low bending stiffness if pinched from above, i.e. with the pinch movement in the horizontal plane. Figure 12.1 shows the operation of the Clupicker [7]. Pinching (or clamping) from the side (pinch movement in the vertical plane) may be possible and is equally good but access to the edge of the panel is required. Material release is fast, effective and has good repeatability. Occasionally a fibre may become trapped in the gripping fingers or tend to stick within one gripper thereby distorting the final placement.

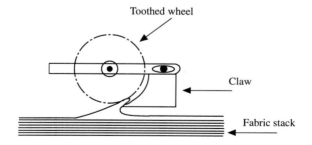

Figure 12.1 *Gripping by pinching*

iii) Vacuum suction: This technique is best with materials having low air porosity and smooth surfaces so that vacuum cups can mate closely to the material surface.   Many gripping points may be required to support the panel adequately. For example, Jarvis et al. quote a requirement of 400 nozzles to hold dry (no resin), fibre reinforced, plastic composite panels of up to 1m x 0.7m [8]. Variations in panel sizes and shapes can cause unwanted gripping points to become exposed thereby sucking in air and reducing the effectiveness of suction at the wanted points. This can be avoided by having reconfigurable vacuum nozzles or by having programmable valves on each suction point. These techniques can considerably decrease the cost-effectiveness of the approach.

iv)   Electrostatics: Flat plate electrostatic grippers [9] can be used to pick up a surprisingly wide range of materials from light fabrics, through leathers and carbon fibre mats to copper sheet. The cohesive forces are low (a function of the molecular structure and polarisability of the material and its relationship with the dielectric used to coat the gripper surface) [10] and so a large ratio of surface area to weight is required. Thus heavy and hairy materials (such as wool) or ridged materials (such as corduroy) are difficult to pick up. It is recommended in these cases that materials are transported in a vertical plane since the electrostatic shear forces are much greater than the cohesive forces. The release of the material can be a problem if the material is very lightweight and tends to cling to the plate. A secondary removal mechanism is required in these cases. However, since the flat plate gripper picks up the material and carries it in a purely flat form with no disturbances it can be released in perfect condition. Very large panels of fabric can be picked up using the roller version [9] seen in figure 12.2. The energised roller is placed along one edge of the fabric and rolled along until all of the fabric panel is wrapped around the roller. Release is easy as the trailing edge does not adhere to the gripper and so may be simply unrolled onto any convenient surface. This technique is limited to materials with low bending stiffness, such as fabrics, which can be easily rolled up.

Figure 12.2 *Electrostatic roller gripper*

v)   Adhesives: Some early work was carried out using adhesive tape to grip fabric. This tape was dispensed from a roll and then discarded, see Figure

12.3. More recently, permanently tacky adhesives have been used [11-12] in laboratory situations. The adhesive surface can be reused but attracts lint and debris which then reduces its adhesive properties. The surface can be cleaned with water to remove these unwanted particles but, in practice, some fibres and dirt particles become embedded in the surface and it gradually loses its effectiveness and must be renewed. The technique is best used with relatively lint-free and clean materials such as leather. A secondary removal mechanism is essential for a flat surface device, Figure 12.4 shows a version with a pneumatically activated ejection collar. Roller versions are possible for fabrics and have the advantage of making ply removal easy.

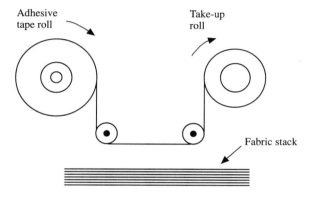

Figure 12.3 *Gripping using a roll of adhesive tape*

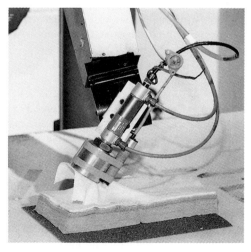

Figure 12.4 *Adhesive gripper*

vi)   Freezing: If a damp surface is put in contact with a very cold surface the
        formation of ice will cause the surfaces to adhere. This concept has been used
        by Schulz et al. [13] but suffers from the requirement of having to have a
        heating/cooling cycle which may be too long for a practicable system.

12.3.1.2 Conveying

Perhaps the simplest form of transportation is a conveyor which can be
stopped/started, typically under computer control, and onto which a piece of material
is placed. However, the apparent simplicity is deceptive. Conveyor surfaces are often
slightly bowed from side to side and are subject to vibrations. A piece of material,
whilst sitting on the moving conveyor surface may well be moving slightly on it. The
actual movement will be a function of the vibratory amplitudes, the shape of the
surface and the frictional properties (which will vary according to direction) between
the material and conveyor surface. Such movements are usually small but even a
1mm displacement for fabric or a 0.1mm displacement for leather might have a
significant effect on the following process and so it may be necessary to realign or
sense the new position of the panel. Sometimes an overhead conveyor is used with a
sliding surface or another conveyor beneath. This allows greater normal forces to be
applied at the conveyor/material interface(s) and therefore reduces unwanted
movements.

12.3.1.3 Vibrating surfaces

The movement of material when vibrated can be used deliberately in the equivalent of
a linear vibratory feeder [14]. A piece of material placed on an inclined plane which
is vibrated in the vertical direction will tend to move down the slope given an
amplitude of vibration and inclination angle appropriate to the weight of the material
and its coefficients of friction with the surface of the plane. A number of different
forcing strategies are possible using multiple excitation points, again analogous to the
more traditional mechanical feeding systems [15].

12.3.1.4 Rolling surfaces

Rollers may be regarded as short conveyors. If a small force is applied through the
roller to the material, very high pressures can be exerted in the material immediately
under the roller axis. This gives a very tight grip. Again, the situation is not so simple
as it appears. If a piece of leather is transported back and forth through a pair of
rollers arranged like a mangle it is found experimentally [16] that the leather will
gradually precess in a consistent but seemingly arbitrary direction. Although the
movements are small they could be unacceptable in tasks such as precision stitching.
Further study shows that the direction of movement is related to the grain direction of
the leather and it is postulated [16] that the rolling action causes distortion of the
internal structure of the leather which combines with the surface friction properties to
cause the precession. A similar effect is seen to greater effect when we walk on mats

placed on a thick pile carpet with the result that the mats tend to move along the direction of the pile.

12.3.1.5 Pallets/Jigs

A seemingly good method of moving material panels is to place them in pallets, restraining them with clamps whilst the pallets are moved. Unfortunately pieces of fabric and leather come in many shapes and sizes even for a single design. This leads either to a multitude of different pallets or adjustable pallets which can cope with a limited range of sizes or shapes. Either way, this restricts the use of clamping pallets to special cases where their excellent holding properties during transportation are essential.

12.3.1.6 Mechanical progression

Local movements of limp material can be effected by a dog feed mechanism, found typically in the bed of a sewing machine close to the needle point. This mechanism is used to advance intermittently the material under the needle. This mechanism can be made small enough to operate in the close confines of the sewing point and can be mechanically linked to the sewing mechanism to ensure correct synchronisation with needle movement. Slippage may be a problem in automated systems.

12.3.1.7 Sliding

In the simplest case, the material is placed on an inclined plane with low coefficients of friction and allowed to slide under gravity, perhaps assisted by some small intentional or unintentional vibrations. A more controlled method uses a hand to slide the material over a surface. One embodiment of this uses a robot with a foam block and pad as its end-effector. This is placed onto the top of the material panel which is then slid across the low frictional surface, see Figure 12.5. Clearly the pad must have a higher coefficient of friction with the material than the material with its supporting surface. The use of a compliant block accommodates irregularities in the surface or robot movement.

If a porous surface is used as the top of a pressurised air box, a piece of material placed on the surface will float above it. If the surface is inclined the panel will float downhill.

A variation on this is to incorporate a number of small directional air jets into the surface. The panel will now both float and travel across the surface in the direction of the directed air jets.

Figure 12.5 *Sliding fabric over a smooth surface*

12.3.2 Automated assembly : destacking

Stacks of material are usually the starting point in the automated assembly of garments so the removal of single plies of material has been a natural starting point for automated handling research. This task can be relatively simple for cleanly cut stiff materials or can be extremely difficult given a stack cut with a blunt blade which results in fibres entrained or fused between the edges of adjacent layers. The problem can be compounded if the stacks are transported long distances (it is not unusual to centralise cutting at one factory and assembly at another). By the time the stacks reach the assembly point the edges are unlikely to be perfectly vertical. In 1980 only one device was available: the 'Clupicker' [7] a combination of toothed wheel and claw used to pinch and grip the topmost ply as seen in Figure 12.1 but with the addition of a small, but crucially important, restraint to prevent the under-plies being removed too. Once adjusted it works well, particularly on woven shirt material, with consistent material properties. Good results for knitted materials have been gained with the air-jet finger device of Figure 12.6 developed by the University of Hull and Autotex Ltd. [17-18] and pin techniques developed initially at Durham [19]. Subsequent research has seen the emergence of electrostatic rollers and plates [10], and permanently tacky adhesive rollers, plus various combinations of the basic methods. None is universally applicable: given a range of materials from fine lace and silk to heavy denim and corduroy this is not too surprising.

A feature of all successful methods is the combination of a method of gripping only the topmost ply with a technique to restrain or break the links with the remaining plies. Sensors can be used to determine faulty picking and if corrective/recovery action is available their use can considerably raise the reliability of the separation process. In the example of the air jet/finger gripper an improvement from 90 to 99.5% reliability was reported [17]. Infra-red crossfire sensors can successfully detect the condition of zero or one or more than one plies between gripping jaws [17]. Alternatively the thickness of the gripped fabric can be measured, for example by detecting the distance between gripping jaws [20].

Figure 12.6 *A development of the airjet/finger ply separation system*

12.3.3 Orientation/Alignment

Uncertainties and variations in the various picking and handling processes result in material panels not being quite where they ought to be in translation and rotation. Fabric panels may also be distorted out of shape. If subsequent processes cannot cope with these errors then some sort of corrective action must be taken.

12.3.3.1 Robotic correction with vision sensing

If a single piece of material is to be picked up from a surface it may be possible to get a visual image with good contrast between material and surface. Top lighting is one possibility using a dark or light supporting surface as appropriate. This method becomes less viable if a wide range of colours and patterns are used or if the surface may get dirty or if lighting cannot be well controlled. If material panels are large with respect to the accuracy required, a single overhead camera may not have the resolution needed. However, given that there are usually just a few key features such as edges and corners that have to be found, a satisfactory alternative may be to use multiple cheap solid state cameras which can be selectively connected to an image processor. *A priori* knowledge is used to direct each camera at areas of interest such as the collar points or gusset corners as seen in Figure 12.7 [21].

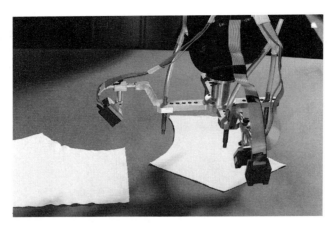

Figure 12.7 *Multiple low cost gripper mounted cameras for feature detection*

Better contrast may be obtained with back lighting. The problem of keeping the back-lit supporting surface sufficiently clean may be obviated by moving the material across a back-lit slit and under a linear array camera. This method is used in the Autoscan sewing system [16] which is capable of discriminating between over three thousand different shapes and sizes of shoe materials and determining the position and orientation of the scanned shape.

12.3.3.2 Edge alignment during conveying
An air jet technique can be used to push the edge of conveyed fabric up to a predetermined mechanical constraint as exemplified by the Zyppy device [22]. The amount of air pressure used should be related to the thickness, weight, frictional characteristics of the material, and the amount of edge correction needed. Errors in edge position just prior to the alignment device may be detected visually with a linear array and the air jet pressure adjusted automatically. In simpler cases a few photodiodes may suffice. Two alternatives to the Zyppy device are the 'star wheel' of the Felix System™ [23] which mechanically moves fabric sideways, and steered overhead conveyors which direct the material with a velocity component normal to the direction of movement of the main conveyor underneath.

12.3.3.3 Sliding against edge constraints
If material is slid across a surface by vibration or air flotation and hits a constraint its orientation will change. This has been used to good effect [14] to reorientate panels of knitted cotton using a constraining bar against the straight edge of the panel. A vibrating table can be seen just under the robot gripper in Figure 12.8. If the bending stiffness of the material is too low or alignment errors are too high then the material panel can buckle when it hits the constraint.

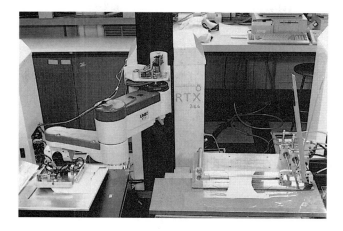

Figure 12.8. *Vibrating alignment table with robotic transfer to assembly jig*

12.3.4 Controlled distortion

The distortability of fabric can be exploited to simplify subsequent handling and processing in 2-D and 3-D.

12.3.4.1 2-D distortion of fabric

It is considerably simpler to sew in straight lines rather than curves and so it can be advantageous to deliberately straighten a curved edge before, say, adding a trim to it. Figure 12. 9 shows a laboratory prototype device which very successfully straightened the leg edges of briefs prior to elastication/binding [24].

Figure 12.9 *Straightening leg seamlines prior to sewing*

12.3.4.2 3-D distortion

Garments and shoes are designed to fit human bodies and must therefore be 3-D when worn. They are generally put on/taken off by stretching over feet, legs or arms and as such are 3-D in nature. There are of course exceptions when a 2-D object such as a nappy is assembled on the body and fastened to form a secure 3-D shape. Ignoring these cases there will be some stage during the assembly process at which a garment/shoe upper forms a 3-D shape after starting with 2-D panels. This complicates the handling and may require completely new techniques which do not apply to 2-D or rigid 3-D objects.

An example is turning inside out. Garments are usually constructed inside-out so that seams and fabric edges are not visible when the garment is worn, so at some stage the garment must be everted. Some early work was done on this by Vitols on turning a jacket inside out and at Hull [25] on turning a concealed gusset construction inside out as depicted in Figure 12.10. Their technique was later used in Hull's CIMTEX automated production line for the folding of cuffs and waistbands at the output of a cuff/waistband making machine. The final module in this assembly also required the flattened sleeve openings to be opened up to be placed on a former for sewing to the cuffs.

Figure 12.10 *Turning a gusset construction inside-out*

12.4 Joining cells

12.4.1 Two dimensional sewing

Examples are taken to illustrate the choice of different techniques.

12.4.1.1 Sewing with pallets

The right hand side of Figure 12.8 shows fabric panels loaded onto a pallet for joining along one edge. Three clamps restrain the three panels (two gusset and one back panel) close to the sewing edge and the linear movement of the pallet is co-ordinated with the speed of the sewing head. The pallet dimensions must be large enough to accommodate the largest size of back panel. The design of the clamps is such that all sizes can be held. In this example the cycle is: load the pallet (load piece1, clamp piece1, load piece2, clamp piece2, load piece3, clamp piece3), sew, unload the pallet, and return the pallet. The total cycle time is long but in this case balances the rest of the assembly line. The main reason for the use of a pallet is the need to hold securely the edge of the three pieces of fabric during the sewing operation.

12.4.1.2 Sewing with roller drives

This is exemplified by the Autoscan sewing system [26] of Figure 12.11, designed to apply decorative stitching to single pieces of shoe upper material. Rollers which move the shoe upper material back and forth are mounted in a cradle which gives the side-to-side movement. This x-y movement is co-ordinated with the sewing needle control to achieve the desired stitch pattern. Prior to stitching the shoe upper piece is fed in any orientation, scanned and its position and orientation determined. The corresponding stitch pattern is selected and then translated and rotated in software in order to match the piece. The main benefits of this method are the lack of pallets (thereby giving no tooling costs and times, no storage requirements and no pallet loading/unloading tasks) and precise, repeatable control of stitching which can have any complexity.

12.4.1.3 Sewing with conveyors

Part of the Hull CIMTEX assembly line has the function of sewing the side seams from the cuff up the sleeve, to the arm pit and then down the side of the garment to the waist. The straightening is not exact so adjustments are necessary just prior to the sewing point. The straightening mechanism is an overhead conveyor which can be rotated about the vertical axis at one end thereby driving the fabric beneath in the desired direction. The edge of the fabric is detected by photocells which provide the signals to close the feedback loop. Excellent results were obtained with overlock seams being produced with only about 1mm of trimmed material.

12.4.2 Three dimensional joining

Two contrasting techniques are selected here. The first arose from the Japanese TRASS project described more fully below. One of the cells was designed to attach pre-assembled sleeves to a jacket body. The approach taken was to put the jacket body onto a 3-D torso-like clamping system and then to clamp the sleeves over the armholes. A large robot was then used to hold and move a new lightweight sewing machine along the desired 3-D contour to sew the pieces together.

Figure 12.11a *The autoscan decorative stitching machine*

Figure 12.11b *Autoscan detail of the rollers, carriage and sewing head*

The Hull CIMTEX project required made-up cuffs and a waistband to be attached to the sleeves and waistband respectively of the leisurewear top. In this case a stationary cuff attaching machine is used. A made-up cuff is loaded onto a nose cone in front of the sewing machine. A multi-point gripper then grasps and opens out the end of a sleeve (connected to the main body) as presented at the end of a conveyor. The sleeve is then drawn over and slightly back over the cuff to eliminate fabric curl. The two parts are then clamped and drawn up to the needle point and guides for sewing together. One guide adjusts its position in response to edge detection photocells to give reliable and straight feeding as the cuff is rotated. The waistband and main body can be regarded as a very large cuff and sleeve so the same basic technique can be used.

The Hull briefs assembly system made use of a folding jig to hold the brief during the side seam, waistband, side seam sequence [27]. Figure 12.12 shows the addition of the waistband after the first side seam has been joined.

Figure 12.12 *3-D sewing: adding a waistband*

The TRASS 3-D system is potentially very flexible and can sew any 3-D contour. It is also complex, costly and slow and requires much preparation of the garment pieces. The Hull cuff attachment system is essentially built up from a number of the elements from the handling toolbox. Likewise, the 3-D briefs assembly module is designed for the specific task in hand and can therefore be made much simpler and cheaper than a very general purpose device.

12.5 Integrated systems

Several integrated assembly systems have been produced as the end result of these worldwide research efforts.

12.5.1 High-tech robotic systems

This approach uses many high performance robots and sophisticated vision systems to move parts and join them in a way which typically mimics the way a human might perform the task.

The Japanese TRASS project [3,28], a nine-year research programme supported by MITI and industry ended in 1991 with a demonstration of four basic systems for high speed laser cutting, flexible sewing, high technology assembly and 3-D flexible pressing. These were applied to the assembly of a ladies blazer made from woven fabrics. A schematic of the assembly system is shown in Figure 12.13. Extensive use was made of general purpose robots and vision systems. Monitoring sensors were applied to sewing machines which could have critical parameters set from an external computer. Thread cutters were improved and thread joiners and bobbin changing mechanisms developed. The 3-D sewing system described above completed the garment.

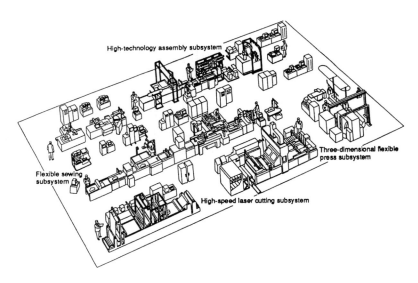

Figure 12.13 *TRASS automated sewing system: test plant layout [28]*

12.5.2 Low-tech robotic systems

The Hull brief's assembly system, shown in Figure 12.14, was designed to be capable of making either men's or women's styles [29]. It was originally envisaged that robots would be used to guide the fabric panels through the sewing machines but it was found that all sewing could be constrained to be in straight lines apart from the 'top taping' operation on the front of the men's brief [30].

The final system used pallets for accurate sewing, and low cost UMI-RTX robots for transfer and manipulation of parts between each sewing cell. Simple photodiodes were used to detect presence or absence of fabric panels and vibratory tables to realign and reposition the panels at critical stages (as, for example, seen in Figure 12.8), thereby avoiding sophisticated vision systems and the associated lighting requirements. The realignment stages meant that inaccuracies caused by the robot could also be tolerated thereby allowing the use of the cheap UMI-RTX devices.

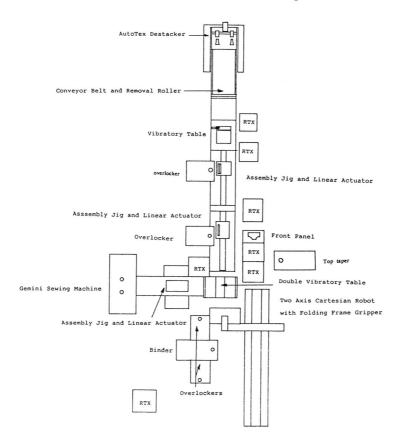

Figure 12.14 *Schematic layout of the Hull briefs assembly system*

12.5.3 Systems with a minimum of robots

The Hull CIMTEX leisurewear top assembly line took the 'keep it simple' concepts even further. The garment assembly sequence is shown in Figure 12.15 with the corresponding system layout in Figure 12.16. Much preparation time was spent in refining the overall system concept and a key part was the decision to carry out the 2-D assembly tasks, folding and side seaming using conveyors as the transportation medium. A typical cell has one conveyor for the left side and one for the right with the gap between used to accommodate the excess fabric in the larger sizes. Simple edge alignment techniques are used to ensure good fabric control even with the slightly misshapen fabric pieces (of a medium weight fleecy cotton or cotton/man-made fibre mix). The outputs of one stage are completely compatible with the inputs of the next thereby minimising handling of the large fabric pieces. Only two UMI-RTX robots are used in the whole process, to transfer fabric panels from an overhead transportation system onto the conveyor at the start of the process. No array cameras are used at all.

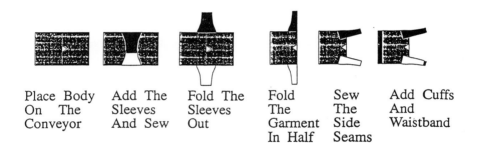

| Place Body On The Conveyor | Add The Sleeves And Sew | Fold The Sleeves Out | Fold The Garment In Half | Sew The Side Seams | Add Cuffs And Waistband |

Figure 12.15 *The CIMTEX leisuretop garment assembly sequence*

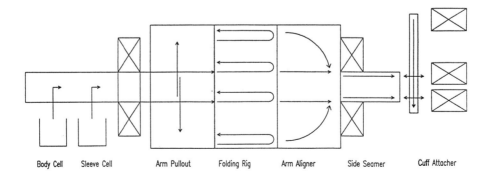

Body Cell   Sleeve Cell   Arm Pullout   Folding Rig   Arm Aligner   Side Seamer   Cuff Attacher

Figure 12.16 *The CIMTEX at Hull automated assembly cell layout*

12.6 Evaluation and exploitation

Many handling techniques have been invented but in practice they usually work satisfactorily for just a limited range of materials and in some cases may require manual setting up. This has severely restricted their application to situations where there are large variations in material properties. There remains a lack of basic science which relates these mechanical parameters to the material behaviour during automated handling. Humans acquire the ability to handle limp materials from childhood and have sophisticated hand/eye/hearing co-ordination which allows them to perform complex handling tasks. Given that some garment assembly operations require weeks or months of operator training before peak performance is attained, it is not surprising that automated systems which try to emulate human handling are rarely capable of achieving even the same order of magnitude of performance metrics.

Many early automation projects were very much in the human emulation mode and resulted in very expensive, slow systems which were impossible to commercialise. Later projects have shown that a focus on the task requirements can lead to simpler solutions which are then much cheaper and more reliable. Great attention must be paid to detail to ensure that, as far as possible, all likely contingencies are catered for. This is especially true for the integrated systems where reliable total system operation requires extremely high reliabilities in all subsystems. Finally it must be accepted that limp materials made from natural materials will have variable properties, that current handling processes and sewing operations introduce their own uncertainties and any automation must cope with these.

The efforts at automation have progressed through industrial views that 'it is all impossible' through over expectation of the capabilities of sensory robotic systems to disillusionment with the results of over-ambitious projects, compounded by the effects of the recession and the perceived benefits of the alternative of overseas sourcing. The state of the sewing machinery industry, one natural route for exploitation, has not helped. In Europe there have been many alliances, take-overs and mergers and even some Japanese manufacturers have experienced difficult trading conditions. It is unfortunate that this cycle is also reflected in support of long-term research from public funding.

Fortunately there are signs that industrial perceptions are maturing and becoming more realistic. The basic motivation for automated handling remains, and progress is being made through smaller scale, well-focused projects some of which are using automated handling as a supplement to, rather than a replacement of, labour.

12.7 Conclusions

Up to 1980 very little work had been carried out on automated handling of flexible materials. There are many techniques in the automation toolbox but the scientific knowledge is lacking to apply them correctly and to improve upon them. This is akin to the early stages of many technological developments such as steam engines, electronic amplifiers and aeroplanes with the engineering outstripping the science which then catches up and allows further improvements. It is to be expected, therefore, that much more progress is possible in the long term. Meanwhile, in the short term, small scale exploitation of existing research is being targeted at specific industrial problems.

12.8 References

1. CIRCEA, Clothing Industry Research Centres European Association, PRACTICE: Project for Research in Automated and Computerised Technologies for the Industries of Clothes and Knitwear in Europe, 1991.
2. KURT SALMON ASSOCIATES, Scenario: the textile and clothing industry in the EC until 2001, K.S.A., Altrincham, Cheshire, 1993.
3. IGUCHI, K., Automated Sewing System, JIAM '90, Tokyo, Journal of Textile News, pp.90-96, 1990.
4. BLACK, S.S., New Day Dawns in Apparel, Bobbin, pp.2-4.
5. BRITE-EURAM Programme, Synopses of current projects 1990-1991, Commission of the European Communities, 1991.
6. HARLOCK, S.C., LLOYD, D.W., and STYLIOS, G., Sensory Robotics: Identifying Sewing Problems, in Taylor P.M.(ed) Sensory Robots for the Handling of Limp Materials, NATO ASI Series, pp.85-96, 1988.
7. DLABOHA, I., Cluett's Clupicker enhances robot's ability to handle limp fabric, Apparel World, Knitting Times, 1991.
8. JARVIS, S.D.H., et al, Design of a Handling Device for Composite Ply Lay-Up Automation, Proceedings IEEE-91 ICAR, Pisa, pp.790-795, 1991.
10. MONKMAN, G.J., TAYLOR, P.M., and FARNWORTH, G.J., Principles of electroadhesion in clothing robotics, International Journal of Clothing Sci. Technol., 1(3), pp.14-20, 1989
11. MONKMAN, G.J., and SHIMMIN, C., Use of Permanently Pressure-Sensitive Chemical Adhesive in Robot Gripping Devices, International Journal of Clothing Sci. Technol. 3(2), pp.6-11, 1991.
12. PATTERSON, A., Integration of the Cutting and Sewing Rooms through Automated Stripping, Proceedings 2nd International Clothing Conference, Bradford, 1992.
13. SCHULZ, G , Grippers for Flexible Textiles, Proceedings IEEE-91 ICAR, Pisa, pp.759-764, 1991.
14. GUNNER, M.B.G., Automated 2-D Fabric Handling, PhD thesis, The University of Hull, 1992.
15. GUTMAN, I., Industrial Uses of Mechanical Vibrations, Business Books Ltd., London, 1968.
16. SMITH, D.L., The Manipulation of leather workpieces for the assembly of shoe uppers, PhD thesis, The University of Hull, 1991.
17. KEMP, D.R., TAYLOR, P.M., and TAYLOR, G.E., The Design of an Integrated, Robotic Garment Assembly Cell, Proceedings 4th IASTED Symp. on Robotics and Automation, Amsterdam, 1984.
18. KEMP, D.R., TAYLOR, G.E., TAYLOR, and PUGH P.M., A., A Sensory Gripper for Handling Textiles, in Pham D.T. (ed), Robot Grippers, IFS Publications Ltd., pp.155-164, 1986.
19. SARHADI, M., NICHOLSON, P.R., and SIMMONS, J.E., Advances in Gripper Technology for Apparel Manufacturing, Proceedings Conference on UK Research in Advanced Manufacture, I.Mech.E., London, 1986.
20. GILBERT, J.M., TAYLOR, P.M., MONKMAN, G.J., and GUNNER M.B., Sensing in Garment Assembly, Nato ASI, Side, Turkey, 1992.

21.   TAYLOR, P.M. and BOWDEN P., The Use of Multiple Low Cost Vision Sensors in Fabric Pick and Place Tasks, Proceedings IFAC Symposium on Low Cost Automation, Valencia, 1986.

22.   PROFEEL , Zyppy, Profeel Sales Brochure, 1988.

23.   SCHIPPS , JIAM '90 Exhibition Catalogue, 1990.

24.   PALMER, G.S., WILKINSON, A.J., and TAYLOR, P.M., The Design of an Integrated, Robotic Garment Assembly Cell, Proceedings 6th National Conference on Production Research, Glasgow, 1990.

25.   TAYLOR, P.M., and KOUDIS, S.G., The robotic assembly of garments with concealed seams, Proceedings IEEE Conference Robotics and Automation, Philadelphia, 1988.

26.   SMITH, D.L., TAYLOR, P.M., TAYLOR, G.E., JOLIFFE, I., and REEDMAN, D.C., Decorative Stitching of Randomly Fed Shoe Parts, Proceedings IEEE-91 ICAR, Pisa, pp.765-768, 1991.

27.   PALMER, G.S., WILKINSON, A.J., and TAYLOR, P.M., The Manipulation of 3-D Fabric Shapes, Proceedings Control '91, Edinburgh, 1991.

28.   OGAWA, S., R&D in Automated Sewing Systems, JIAM Seminar, 1991.

29.   TAYLOR, P.M., WILKINSON, A.J., GIBSON, I., PALMER, G.S., and GUNNER, M.B., The Robotic Assembly of Underwear, Proceedings IEE Control Conference, Edinburgh, 1991.

30.   GIBSON, I., TAYLOR, P.M., and WILKINSON, A.J., Robotics applied to the Top Taping of Men's Y-front Briefs, Proceedings International Clothing Conference., Bradford, 1990.

Chapter 13

Robotics in food manufacturing

K. Khodabandehloo

Over the past decade, interest in the use of robotic technology has grown in the food sector. There is a strong indication that this sector and, in particular, industries dealing with meat, fish and poultry products would use robotics if such technology were to meet the specific needs of the food production environment.

Meat, fish and poultry processing present important challenges both from the engineering and business view points. Perhaps in years to come it will be common for food products to be labelled 'food untouched by human hands'. Automating the complete production cycle, from raw materials to finished and packaged goods has been achieved only in a few sectors of the food industry. Use of robotics technology provides a means for automatic inspection, handling, packaging, cutting and general processing of products in circumstances where flexibility and adaptability are considered important. Automating the 'skills' associated with tasks currently performed by people requires a detailed knowledge of the processes and, in many cases, considerable improvements to robotics must be achieved. Where the conditions are harsh or indeed dangerous for people to work in, automation is clearly the solution. The level of sophistication of the automation technology would however depend on the task and the nature of the environment. It is also conceivable that some products cannot be manufactured without the use of robotics.

Many applications exist for robotics from ready meal assembly through to the packaging of cooked products in highly demanding conditions for this industry. The use of robotics may be more obvious in the meat, fish and poultry sectors. This will be the focus of this chapter which is a short review of the relevant aspects of a recent book edited by the author.

13.1 Introduction

It is an essential requirement that the introduction of robotics in meat, fish and poultry processing meets with the production demand, consistency and hygiene standards. In addition under no circumstances can the industry afford a compromise in quality. The introduction of robotic technology can be made to improve quality in today's applications and also in the future.

The use of industrial robots in the more conventional production lines such as those seen in the motor car industry has been well established. Indeed the capabilities of the systems currently available are not being fully utilised even in the established manufacturing application. A more significant improvement in capability through the uses of sensors and advanced software will encourage the automation of the more difficult tasks throughout a number of industrial sectors.

In food processing, the robot arm, its controller and the ancillary devices must be made to withstand the environment and the conditions for cleaning. Cleaning often involves the use of high pressure hot water and various chemical agents. The design of the system should meet with all the required standards and legislation for food processing machinery. This is an essential requirement which is now being seriously considered by robot manufacturers. Flexible automation systems for food processing can now be purchased from some specific robot suppliers. The main limitation, however, is the immediate use of such robots in tasks currently performed by a 'skilled' labour force. With the range of tasks and the variability in shape, size and properties of the products, novel end-effectors and sensors need to be introduced. Through the integration of these technologies and the implementation of intelligent software, the more demanding or highly skilled tasks may be automated. Skilled robots will need to be developed but much research is needed before they can be used in industry. *A robot can be said to have skill if it demonstrates the necessary sensory perception and dexterity equal to or better than that which can be produced by a human.* This gives a principle definition for a 'skilled' robot.

Clearly there are many food industry tasks that do not require sophisticated automation and the current technology can be made to perform such tasks fairly readily. For the more varied products, the choice of the end-effector (grippers, cutting tools, etc.) and the capabilities of sensory technology with the corresponding software and strategies for manipulator control, play a major role. In almost every case integration can lead to viable solutions, however, applications in food processing have yet to be exploited. The arguments for using robots by these industries are the same as those for any other application; cost savings, reduction in overheads, quality, repeatability in production, yield, speed, safety and reliability benefits are amongst those that feature highly. Perhaps the strongest reasons for using robotics in the food sectors are related to the high labour turnover and difficulties in recruiting skilled staff because of the harsh work conditions, particularly in butchery and slaughter lines. Although considerable research and development has been in progress in the USA, Australia, Europe and New Zealand, the exploitation of robotic systems in this sector is yet to be achieved by the equipment suppliers.

Although it is hard to estimate the number of systems utilising robots or robotic elements likely to be installed in these sectors, the opportunities for exploitation of robotic technology can be considerably higher than what has been achieved in all the other industrial sectors so far. The factors that prevent a more immediate exploitation include:

a)   The range of tasks that industry expects to automate demands a degree of skill from automated systems, currently beyond the capability of industrial robots.

b)   The cost of systems that could be implemented are generally higher than what industry is able to afford. Systems with comparable capability to skilled staff performing butchery or packaging should cost no more than about £40,000

($70,000) per robot replacing one person. The level of engineering and the corresponding lead times together with the high value of the essential elements of a system make it difficult for some applications to be automated. It should be noted that several low cost devices for processing meat have been developed already and it is envisaged that commercial units for integration into systems can be made available in the foreseeable future.

c)   The sensory functions inherent in human systems, which enable the interpretation of complex shapes and the control of manipulation, cutting and handling could not be duplicated easily in a robotic system. The integration of existing sensors with robotic devices can nevertheless provide solutions in specific cases. Deboning with the use of force sensing is an example.

d)   The hardware used in the design of robots does not cater for the wide range of hostile operating conditions. This is a major consideration, and although solutions are available, equipment that can withstand the food processing environment, particularly in the meat industry, is difficult to find.

13.2 Current robotic devices in use in the food sector

The meat, fish and poultry industry has been able to use many specific automation solutions for processing its products. Due to the high throughput many companies have operated a continuous flow line with automated transfer systems in the form of overhead rails or conveyor lines. In general there is very little 'flexible' automation of robotics and most of the work is performed manually.

Over the recent years, considerable effort has been placed on the development of specific purpose tools to assist with difficult tasks, such as deboning. Examples of this may be seen in the meat processing sector with tools such as the pneumatic rib loosener from SFK Meat Processing AS. The device in this case performs the task of separating rib-bones by the use of a flexible but high strength cord. This makes the task of deboning considerably easier and more efficient. It is possible that a robot system capable of handling the tool could perform this task. This was indeed the subject of study in the project entitled 'The Deboning of Bacon Backs' at the Centre for Robotics and Automated Systems, Imperial College.

Automatic systems for deboning other products such as poultry can also be seen in many plants world-wide. Such systems are usually dedicated to one specific task.

Filleting machines capable of processing 40 breasts per minute may be purchased readily. It is claimed that the meat recovery is comparable to manual filleting but the speed is higher. Clearly this class of machines can offer cleaner, smoother and more hygienic production capability. The next natural step in machine development would be in providing additional capabilities through the introduction of sensors and actuators that change the attitude of each cutter in the machine to compensate for the variation in the size and shape of the poultry carcasses. The same principle is valid for fish filleting too. The next stage in the development however, could go beyond the adaptation of such dedicated systems. Robotic techniques may be employed to perform similar tasks giving the advantage of adaptability as well as flexibility in use.

In contrast to poultry products that come in variable shape and size carcasses, production of sausages may seem more straightforward. The challenge here has been the production of variable size and shape end products and machinery for this purpose is available from several sources.

Once the products are produced in accordance with the requirement, further handling of sausages is required, particularly for packaging. Here the challenge for robotics is not simply to handle a variety of flexible products, but to manipulate and package units at a rate comparable with dedicated handling systems.

The ultimate target for robotics in these applications would be to achieve high volume production, but also to adapt to variations in the tasks, products and the environment. This goes much further than the current application of robots which tend to be restricted to low volumes and batch production of components of known geometry. The most recent application of robotics which has attempted to move in this direction is based on the use of the adept robots. The system has been used for packaging burgers and has uses in handling variable shape or size products of a similar nature including chops and steak portions.

One of the underlying reasons for using robots has been the removal of people from a wet and hostile environment. The slaughterhouse applications are good illustrations of this kind of environment. Dedicated hardware has been widely used in the production of poultry and fish. For carcasses of larger size like pigs, the technology is also becoming available. In particular devices for carcass opening, back splitting and removal of 'leaf lard' have been developed for pig carcasses.

These represent robotic systems in the slaughterhouse, which incidentally perform some basic sensing tasks in order to compensate for the variations in carcass size. The grasping or holding methods have been designed to allow alignment and correct orientation of the carcasses. With such arrangement, the accuracy of the process or the cut in each case is at least as good, if not better than the manual process.

A recent Brite-Euram project has been exploring the use of sensory guided robots in the evisceration of pig carcasses. The project aims to develop and demonstrate the sensing, controlling and handling of automation technology for separation of flexible food materials with evisceration of pig carcasses as the model process. The main objective is to develop technologies for automation of handling and processing of flexible materials in manufacturing processes with small tolerances, high quality and better productivity. The project intends to produce a fully automated system which will measure the pig in three dimensions, process the information, and transmit the relevant data to the mechanical system which is to perform the evisceration process of the pig carcasses. New methods for handling and manipulating flexible products such as pork are envisaged. A positive outcome of the project may also be applied for automation in other parts of the slaughtering process. Furthermore, the results could be applied for automatic production of other flexible heterogeneous products.

An area of concern for automation is often related to the cleanability of the machines. On one hand, it is desirable to minimise the contact between people and the food to achieve a more hygienic end product, and on the other hand the use of automation must be free of contamination or cross-contamination possibilities. Using facilities that are self-cleaning may be desirable. More generally if robotic tools could be employed to perform all the cleaning tasks there could be tremendous benefits. The task of cleaning in a meat production line is considered one of the least attractive for employment. For this task there is at least one product which can be described as a multi-degree of freedom cleaning device, or a robotic cleaner, available on the market.

Robotics may be a new technology in the meat, poultry and fish processing industries and some applications could be considered for exploitation, but there is no limit to the scope of robot use.

A Brite-Euram workshop held in Bristol gave an account of recent European Developments.

13.3 Success in automated production of primal cuts

There is very little literature on the automation of primal cutting. The most successful system seems to have been the semi-automated flat belt conveyor system. Most of the research activity in Europe, Australia and New Zealand has gone into automating slaughtering operations rather than cutting or boning.

More recent advances at the Danish Meat Research Institute have led to the development of a production machine now available commercially from KJ Maskinfabriken. The machine has a vision system which analyses the carcass side features and defines the path for the cuts required. In this example, all the cuts are restricted to 2-D. The carcass is manipulated in the plane of the conveyors on which it is travelling by adjusting the speed of adjacent conveyor belts and the position of a moving conveyor section horizontally, perpendicular to the direction of belt travel. In this way the carcass travelling towards a large rotary blade is positioned according to the information from the vision system to align it relative to the cutter for the cut to take place.

Three dimensional cutting of carcasses resulting in production of primal cuts would be the next obvious step. Robotics in this case would have a significant role to play. Another Brite-Euram project entitled 'Robotic Butchery of Carcasses for Meat Production' aims to develop a demonstrator system that performs a number of cuts on a pork side. The aim is to perform anatomical cuts in a hot butchery environment using a robotic system that can achieve a variety of cuts on carcasses of different size and shape. The focus of the project is to achieve a demonstration of a robotic system for the butchery of pork carcasses into primals.

The programme of work has been examining the products and the processes of butchery aiming to define automation solutions based on developments in the following fields:

* Artificial vision for the recognition of carcasses for the purposes of cutting.

* Sensors and cutting devices for robot guidance. Automatic handling and fastening for automatic manipulation of carcasses for cutting of primals.

*   Software for control and real-time optimisation of the cutting process with the use of an integrated robot system.

*   Modelling to support decisions and strategies for automatic handling manipulation and cut-up of pork.

Within the next decade it is anticipated that many new applications of robot technology will come of age.

13.4 Fish processing application

In any robotic application, the transfer and handling of components from one point in the production sequence to another is a requirement. Often in the food industry, such components or products are of varying shape or size and randomly placed on conveyors or feeding units. In fish processing, it is common to see manual feeding of machines with specific functions, such as heading or filleting. Robotization of the handling task requires manipulation of each fish from a random position and orientation into a desired location. In addition to the complications caused by shape and size variation, there are problems for robotics related to the flexibility (or non-rigidity) of fish, the non-uniformity in external or internal texture, and anatomical differences from one fish to the next, and one family of fish to another. Conventional multi-degrees of freedom, serial or anthropomorphic robots do not meet the exact requirements. There is a need for more novel approaches where such manipulators may be integrated with vision devices, and with the addition of control software duplicating the human expertise necessary for the task of fish handling (see Figure 13.1), an integrated system may be developed.

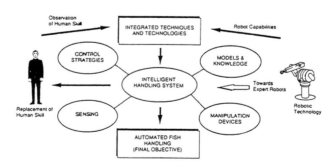

Figure 13.1 *Intelligent handling of fish: system integration provides solutions*

Without this integration of 'intelligence', computer vision and robotic devices, it is unlikely that a robotic system can handle such products, whilst taking account of the slippery, non-rigid and variable nature of the fish in the same way as people do. An added advantage of a robotic handling approach in this way, rather than specific solutions, will be that the skills associated with the manual handling operation at each stage of production could be automated. In contrast, dedicated solutions will be costly and achieve automation of only a specific part of the production line.

The task of fish manipulation will need specific attention with respect to the following:

a)   *Cleanability:* Many of the devices available to date cannot be easily cleaned, and by provision of covers and special air-conditioning for the robot and its control systems this problem may be tackled.

b)   *Slippery nature of fish:* This creates complications for the gripper, which will need to achieve a stable grasp without damage under all handling conditions.

c)   *Speed:* The ability of the manipulator to cope with the large number of fish is the key, and handling speeds approaching 2 fish/second are common. To achieve this, multiple handling stations may be necessary as no robot developed to date can achieve this, even when the tolerances in positioning the fish are low.

d)   *Positioning tolerance:* The placement of fish in specific positions and orientation onto a fixture, a carrying unit or a packaging tray varies according to the steps in production. Typically, positioning tolerance of ± 1 mm is the tightest specified in most handling tasks. Control of fish orientation, however, is the more difficult as fish move due to their non-rigid structure, and the control of their absolute orientations would depend on the velocities of motion during and after grasping. In some cases use of multi-grasp or multi-arm robots will be necessary (see Figure 13.2). This is a new area of development which is specifically relevant to the handling of flexible products.

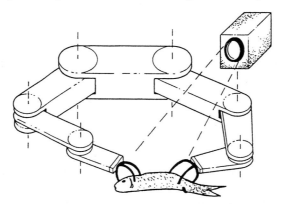

Figure 13.2   *Multi-arm robots for handling fish. Object to be manipulated is: (a)highly flexible, (b) of unknown geometry, (c) damaged*

e)   *Yield*: The process of handling fish manipulation may affect yield if the required position, and to some extent orientation tolerances, are not achieved for each fish. This is particularly the case where the handling and positioning operation affects the cutting, deboning or other operations in the production sequence. Manipulation of the fish by the robot must be more precise than the manual process if savings of up to 2% in yield are to be made, say at the stage where the robot places the fish on a fixture for a heading machine. In general, the task of handling will require a manipulator capable of meeting the above basic needs of the process. In some cases, the robot may be only required to cut fish say into slices, in which case the positioning and handling of fish becomes a matter of concern, and the robotic task will be confined to that of cutter-tool manipulation. As an instance in slicing smoked salmon, the robot is required to manipulate the cutting tool through specific trajectories defined by a vision system. This system should take account of variations in shape and size, and set the trajectories in such a way as to minimise waste and produce slices of about 1.5 mm thickness. It should be highlighted that in all circumstances, the systems envisaged for fish processing must be able to withstand the environment, whether for handling or cutting.

13.5 Poultry processing application

Worldwide, poultry production was estimated to be 36,000,000 tonnes in 1988, since when the market has been growing by about 5% annually. Processing plants, planning to expand to meet this demand, cannot do so due to labour shortages. Automation may be the most appropriate way ahead. The unpleasant, repetitive and hostile working environment of this industry further supports the need for automation.

Off-the-shelf solutions already exist to many of the automation problems that the chicken sector of the poultry industry faces, but the technology tends to be specific to a given process such as cutting. Turkey has seasonal market variation and is a lower volume product than chicken. Turkey processing therefore is usually done manually. Different birds within a turkey flock may vary a great deal in size, further complicating the problems of automation.

The strengths of robotic technology lie in specific areas and the application of these strengths will yield benefits. Visits to several main poultry plants in the United Kingdom revealed that over 30% of the workforce is handling the product. Along the length of the process line, after evisceration, people perform relatively mundane tasks involving limited decision-making. A wide range of products are handled including whole birds, portions (including boneless portions) and quarters. These products may be moved between the lines, between conveyors, from bin to conveyor, bin to overhead line, loaded onto machinery, packaged and so on.

The packaging of portions tends to be highly labour-intensive in most factories. Packaging may mean simply putting thighs or quarters into a bag until it has the right weight and then sealing the bag. Alternatively, it may be the more complicated task of arranging portions neatly in a tray, which is then sealed and priced. Although the most labour-intensive area of the plant is packaging, this by no means represents the only potential for robotisation.

The packaging process involves the manipulation of a poultry portion off or on the bone from an unknown position and orientation into a known location within a packet. The task may also require some degree of transformation by straightening or turning the portion as well as translating it from one location to another. A robotic system that recognises a randomly orientated portion and places it into a tray would need to consist of a robot and gripper, a robot controller, a vision system and a central computer for cell control. A schematic diagram illustrating the cell operation is shown in Figure 13.3.

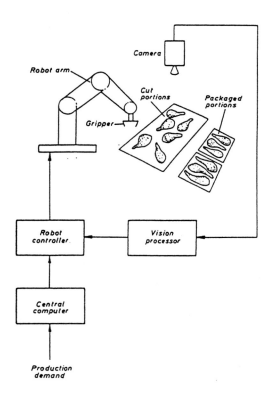

Figure 13.3 *Schematic of the proposed cell for poultry packaging*

An important aspect of automation of this task is the development of grippers that could be directly applied to handling poultry portions. The dextrous hands are unnecessarily complex and the contour adapting vacuum gripper would cause skin damage. The pneumatic rubber muscle developed at Bristol University, however, could be adapted for use in this application, giving a degree of compliance as well as a safe and hygienic power supply required for use in a hose-down, food environment (Figure 13.4).

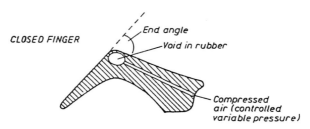

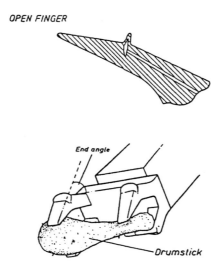

Figure 13.4   *The pneumatic rubber muscle finger and gripper developed at Bristol University*

The speed of response of such a gripper may be a limiting factor. It is necessary for any gripper used for this application to be simple and capable of hygienic use. The main features of robotic grippers or hands for poultry handling include:

*   Ability to handle non-rigid products without damage
*   Hygienic and easily cleaned
*   Handle up to 1.0Kg
*   Pick up moving objects
*   Handle different portions

* Operate at high speed
* Light weight
* Inexpensive so that it can be replaced regularly
* Rugged and reliable
* Deliver portion with correct final orientation
* Easily attached or detached
* Capable of accommodating a variety of touch sensors

13.6 Robot system integration

A detailed evaluation of the market has revealed that the most immediate application for a robotic system of this kind is for fixed-weight-fixed-price (FWFP) packaging of fresh portions of poultry and more specifically chicken. FWFP packaging is a process by which chicken portions of a particular type are packaged in a specific number to meet a predetermined but fixed weight. This involves the assembly and packaging of chicken portions of different weight, say three chicken breasts at a time, into trays which will be wrapped and then given the same price. FWFP thus involves the weighing of each portion and selection of a group of portions, say three, that gives the required weight for pricing in this way. It should be noted that the packet weight must not be below the weight at which it is priced.

The main reason for choosing robotics for FWFP is that sorting of portions by weight using sort-weighing machines still relies on the judgement of the human packer to determine the weight of each pack. For people to assemble a given tray with a fixed and predefined number of portions, obtaining the exact weight can be extremely difficult and time-consuming. The advantage of a robotic system is that each portion can be mechanically weighed and the weight information transmitted electronically to the computer controlling the robotic cell (Figure 13.5), to optimise the process of FWFP packaging.

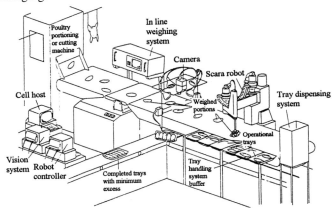

Figure 13.5 *Conceptual system of a commercial fixed weight, fixed price robotised cell*

The scheme for assembling each portion should be defined in such a way as to maximise productivity and minimise excess weight per packet. This in itself will result in considerable savings in addition to labour savings. Since excess weight reduction is one of the main advantages of the system, a simulation study was performed to define the assembly and packaging strategy for the robot. This simulation suggested that a payback of less than 18 months may be possible due to the significant reduction of give away weight.

13.7 Conclusions

The conventional machinery in the meat, fish and poultry processing industry has given greater productivity in certain dedicated applications. The pressures on this industry to achieve higher throughput, better quality and lower cost is greater than ever. Staff shortages for skilled tasks such as butchery, and the problems in the industry to attract staff due to the harsh working conditions are amongst the many problems of the industry. Robotics can provide solutions, and as described there are many levels at which current technology can be adapted to assist with current manual procedures and to remove people from dangerous working conditions. Several different tasks are being considered for robotisation. Handling, packaging, cutting, inspection and deboning may be possible. These tasks represent the most labour intensive in the food industry, and although the examples may be concerned with one type of product it is often the case that the robotic solutions apply to other tasks or processes involving a variety of variable shape and size components or carcasses. The use of the techniques presented here are limited only by the imagination, and the level of investment that can be afforded in each application. The development of generic robotic systems is limited by our understanding of the processes by which human skills may be artificially duplicated.

13.8   References

1.   BRITE-EURAM WORKSHOP on Handling Automation and Inspection in Manufacturing, proceedings 3 and 4 December 1992, Bristol, LK.
2.   BUCKINGHAM, R. O., BRETT, P. N. and KHODABANDEHLOO, K. (1991) Analysis for two arm robots for applications in manufacturing industry, Proc I. Mech. E. part B, vol. 205, pp43-50.
3.   ROBOTICS IN MEAT, Fish and Poultry Processing, Edited by K. Khodabandehioo. Blackie Academic and Professional, 1993, Chapman and Hall.

Chapter 14

Robotic milking

R. C. Hall, M. J. Street and D. S. Spencer

The Silsoe Automatic Milking System consists of a novel high speed pneumatic robot and associated hardware, software systems and dairy parlour equipment. The system will allow voluntary attendance by cows over 24 hours with minimal human supervision. The control software for the robot and end-effector has to guide the teat cup onto an unpredictable, moving target and cope with different cows, udders and teat sizes. The pneumatic actuators are compliant and not likely to damage the cow. Mechanical contact sensors pressed onto the cow by air springs monitor the position of the cow when in the milking stall. This information is combined with signals from position sensors in the end effector to guide the robot. The present robot has been used on several extended trials including voluntary milking. This chapter addresses system design, control problems associated with the pneumatic milking robot and discusses some milking trial results.

14.1   Introduction

Silsoe Research Institute has been working on automatic milking systems (AMS) since 1987 and now has an operational system on a farm. Early research sought to solve the problem of how to attach teat cups to a cow's udder.   The unique applications and animal welfare considerations led to a specially developed robot using compressed air as the drive medium. The considerations included restricting the amount of robotic hardware under the cow for her safety, compliancy in all actuators, simple end-effector design for robust operation and the avoidance of hydraulics and electrics for contamination and safety reasons.   The first robot (MK I) verified the concepts and pointed the way to a new and improved design.

The control software for the robot guides the teat cup onto an unpredictable moving target, copes with different cows, udders, teat sizes and occasional hard kicks. Economic and commercial issues require that the system should be predominantly autonomous, unattended, and be operational for at least 18 hours in a day.

14.2   The MK I system

The stall was simple with rudimentary length and width adjustments to ensure correct positioning of the cow. The MK I robot [4] was based on a spherical co-ordinate design (see Figure 14.1) mainly to obtain a low profile suitable for working under a cow. Axis one provided a rotation to carry the end-effector under the cow, and gave the fore and

Figure 14.1 *MK I robot*

aft components of the end-effector movement. Axis two rotated about a pivot point to give the vertical component, and axis three extended the arm to give side to side movements. The cow was milked by attaching the fixed teat cup to each teat in turn. Location of the teats was calculated using a stored relative teat position, combining it with the cow position feedback and using a sensor in the end-effector for fine adjustment. A number of sensing methods were considered and two were tested:

- Electrical conductivity: A simple design (see Figure 14.2) which sensed the change in conductivity when a teat touched one of the sensing plates. It had the effect of irritating the cow, causing her to flinch and leading the robot to chase the teat.

- Light beam matrix: A matrix of six light beams mounted across the top of a box carrying the teat cup detected the presence of the teat without making contact, but jets of milk from the teat occasionally confused the sensor.

Figure 14.2 *Contact sensing end-effector*

Trials were carried out to test the system using the sensors and were successful in demonstrating that robotic attachment could be done. The robot had limitations; axis 1 carried all the moving mass and, despite removal of excess weight its acceleration was inadequate. Axis three had long pneumatic pipes and a long (500 mm) cylinder which suffered from stick-skip characteristics. Based on this experience it was decided to design a new robot, of lighter construction with all axes operating on much lower masses.

14.3   MK II system

The new robot (see Figure 14.3) was integrated in to a custom designed milking area with a redesigned and improved stall and a complete milking system in order to allow full milking trials to take place. The stall has a 200 mm step at the front, a side exit has also been incorporated and a teat cup magazine placed opposite the robot. Air driven mechanical paddles, with potentiometers attached, sense the position and the movements of the cow. The main advantage of the new stall is that it encourages the cow to adopt a posture suitable for robotic milking, with good separation between legs and udder (see Figure 14.4) [2].

Figure 14.3 *MK II robot*

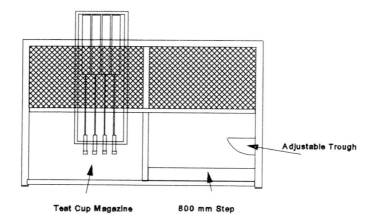

Adjustable Trough

Teat Cup Magazine   800 mm Step

Figure 14.4 *Milking stall*

The MK II robot was designed to pick up and attach each teat cup in turn. Special attention was given to maintaining short pneumatic time constants and avoiding sliding joints that could introduce excessive friction. The robot was constructed from aluminium tubing and the arm carrying the end-effector was a single tube. The arm support is a trapezoidal rocking frame and T link, used to project the suspension point to infinity and give a substantially horizontal arm motion. The maximum force available at the end of the robot arm is about 75 N. Movement in the y direction is sufficient to allow retraction from under the cow hence reducing the need for the robot to move away from the stall after each milking. The robot end-effector is open ended (see Figure 14.5) so it is able to slot onto a teat cup which is retained by a small gripper. The light beam matrix is skewed (see Figure 14.6) to leave an open end for the teat cup entry. A pneumatic motor is incorporated in the robot arm to rotate the gripper actuator and sensor through 180 degrees so that inverted teat cups are picked up from the magazine.

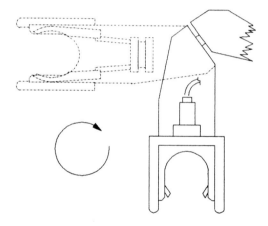

Figure 14.5 *MK II robot end effector*

Three main computers are involved in the process. One is used for closed loop control of the three axes, another for control of the robotic process and the management of the system is controlled by the Automatic Milking Management System (AMMS). A database is incorporated into the AMMS which contains all required information for milking a cow. There are five resources with which the AMMS exchanges data:

- Cow identity: The cow's registered number is read from a transponder around the cow's neck.
- Stall controller: Cow dimensions, feed ration, trough position.

- Robot controller: Teat positions, attachment errors, attachment process start, attachment process complete.
- Milking controller : Yields, flow rates.
- User: System reports and analysis, initial cow information

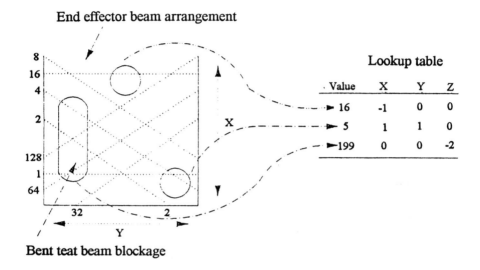

Figure 14.6 *End-effector and look up table relationship*

Cow traffic management is an essential part of robotic milking. The current system will prevent a cow from gaining access to the parlour within four hours of its previous visit or if she had been milked the maximum number of times for that day. The AMMS also holds cows waiting for milking if the parlour is already in use and records the performance figures for the robot and stall.

14.4 Attaching teat cups

When a new cow is introduced to the system relevant dimensions must be recorded. Initially the cow is given access a few times to the parlour to accustom her to the new surroundings. When it is felt that she is ready (normally a maximum of two days is enough for settled behaviour) the operator will allow her access to the stall, adjust the trough position and allocate her feed. The trough is used to shorten or lengthen the stall according to cow size. When the cow is stationary the robot is driven to each teat in turn using a joystick, and a button is pressed to store each teat position.

During automatic milking, the robot picks up a teat cup and moves to each teat position in the x-y plane at approximately 80 mm below the teat tip. It then raises the teat cup on the z- axis until it detects the teat in the end-effector. When the teat is in the end-effector the pattern of blocked beams allows the robot controller to centre the teat and to continue lifting the teat cup until entry of the teat into the cup is sensed. Teat entry is detected by monitoring milking vacuum, which is turned on when the teat cup is collected; blockage of the teat cup by the teats causes an increase in vacuum level. The attachment process may be terminated by :

- Timeout: Robot has been searching for the teat for 30 seconds.
- Above programmed height: The robot has moved to a height that is obviously above the sensible teat height.
- User over-ride: The operator has aborted the attachment of the current teat.
- Vacuum increase : The teat cup has been attached to the teat.

To assist with attachment a modulation routine is used that causes the robot to make a small oscillation in any of its axes, when the end-effector detects a teat. The oscillation is currently used to move the arm 10 mm up in the z-axis and 8 mm in the x-axis. This reduces the attachment time and improves the success rate.

14.5   Robot tracking

Robot demand calculations are initiated by a 20 ms interval interrupt. When tracking the cow, the stall sensors are read and processed with the other information to generate the demanded x, y and z positions. Devices that have control of the robot position (i.e. joystick and end-effector) are treated as separate variables which, as part of each axis calculation, are totalled with the stall sensor positions to give the new robot position. The axes positions are limit checked and passed to a rate restriction routine. This routine has a maximum predefined rate of movement set by software. If the robot is required to move rapidly the rate limit is high, the opposite is true for slow movement. The main reason for implementing this control is to avoid harsh and erratic movements due to transients that occur when switching between tracking cow movements and static teat cup collecting i.e. to obtain bumpless transfer.

The y axis calculations are corrected (see Figure 14.7) to compensate for the arc movement on the x axis due to the pivot point mounting of the arm. This correction allows the end-effector to move in a straight line. A table defines the corrections required at intervals of 64 mm. Interpolation is used to calculate intermediate values. The axes positions are finally converted to robot position values and transferred to the axes controller.

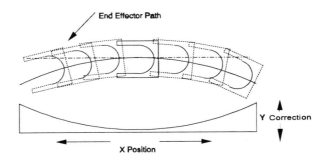

Figure 14.7 *Corrections to y axis for arc of robot arm.*

14.6   Comment on operation

The system described has been used on several trials covering various aspects including cow response to feed, milking frequency and diurnal pattern behaviour as well as engineering trials.

The robotic operation of transferring a teat cup from the magazine to the teat takes an average of 20 seconds with a best time of 17 seconds. The teat cup attachment process takes about 80 to 100 seconds which is acceptable in a voluntary attendance system where attachment time is not critical. Mounting the teat cups on the opposite side of the stall to the robot results in the attachment process taking place with minimal risk of collision. The separation of the robot and teat cups means that one robot could be used to serve several stalls.

The compliant nature of the robot provides significant benefits. Cows kicking and treading on the arm have occasionally caused minor damage in a year of operation. The robot performance has exceeded expectations. On occasions it has been kicked but still carried on to make a successful attachment. The robot's tracking system can follow most cow movements. When the robot is not able to out-manoeuvre the cow it will always catch up when she settles.

The freedom that the cow is given within the stall does occasionally lead to failures.   Some cows do not stand on the step when entering the stall causing the stall sensors to provide incorrect teat positions, which can lead to failures.

The most recent trial (unpublished) included a week of supervised milkings at fixed intervals and a week of voluntary milkings. The success rate was 89%; defined as the number of successful fully automatic attachments out of the total approaches made. The voluntary week of the trial indicated that cows were willing to be milked without supervision. All cows voluntarily attended the stall at least twice a day.

14.7 System problems

The local sensing system comprises an eight beam infra-red matrix that crosses the top of the end-effector about 15 mm above the teat cup. When a teat enters the matrix the photocells produce an 8 bit binary number that is referenced to a look-up movement table (see Figure 14.7). The end effector is sampled every 100 ms and corrections applied to the robot arm until the teat and teat cup are aligned.

Two general problems occur :

- Transient beam blockages : Occasionally milk from the udder flows before the teat cup is attached. Sometimes, dependent on quantity, it can interfere intermittently with the beams. A simple average routine is used to eliminate noise due to drops of milk.
- Beam blocking: Occasional problems have occurred with muck accumulating over the sensors. This failure is detectable as all beams should be clear under normal conditions and a warning can be generated if this is not the case.

Occasional problems occur with the pre-programmed teat positions. There appears to be a relationship between some cow postures and the system's ability to re-locate the cows teat. This problem is causing us to develop a vision and laser stripe system. The proposal is to use a simple vision system to locate the teats initially rather than relying on the pre-programmed positions. The tracking and end-effector sensors will continue to be used to complete attachment. We do not intend to use vision as a continuous sensing system, because of the high probability of the images of the teats being obscured by the robot as it approaches for attachment.

14.8 Future developments

Over the next three years improvements will take place in the software to allow it to cope with the unexpected cow behaviour (e.g. kicking). Artificial intelligence will be required to replace the knowledge of the herdsperson and safety critical software will also be needed.

The development programme is now aimed at improving the performance of the system and introducing expertise to deal with normal and unexpected problems. The design is fundamentally safe and would not damage cows, but additional techniques are required to ensure that alarms are raised and cows released properly in the event of problems. Other future work will include automatic performance monitoring and self testing to ensure that everything is operating correctly.

14.9 Conclusion

A prototype robotic milking system has been designed, built and used extensively over a period of two years. Due to the use of pneumatics it is inherently safe and robust and has survived numerous kicks and general abuse. The fundamental concepts have been proven and further developments will make performance acceptable to commercial users. The addition of expertise will allow the system to operate substantially unattended.

14.10 Acknowledgements

This work was funded in part by the UK Ministry of Agriculture, Fisheries and Food.

14.11 References

1. MARCHANT, J.A., STREET, M.J., GURNEY, P and BENSON, J.A.: 1992. Pneumatics for robot control. - Conference proceedings, 3rd Symposium 'Automation in Dairying', IMAG, Wageningen .

2. MOTTRAM, T.T., 1992. Passive methods of modifying cow posture for automated and robotic milking systems. Journal of Agricultural Engineering Research, 52.

3. SPENCER, D.S. and STREET, M.J. 1992. The design of the management system for the Silsoe automatic milking system. In: A.H. Ipema, A.C. Lippus, J.H.M. Metz and W. Rossing (Eds.): Proceedings of the International Symposium on Prospects for Automatic Milking, EAAP publication No. 65, Pudoc Scientific Publishers, Wageningen, The Netherlands, pp 309-314.

4. STREET, M.J., HALL, R.C. and WILKIN, A.L. 1990. A pneumatic milking robot, structure, performance and first results. In: Robotereinsatz in der Landwirtschaft am Beispiel des Melkens; Tagung Braunschweig-Volkenrode. VDI/MEG Kolloquim Landtechnik, Dusseldorf, Germany, pp 188-201.

5. STREET, M.J., HALL, R.C., SPENCER, D.S. and WILKIN, A.L. 1992. Design features of the Silsoe automatic milking system. In: A.H. Ipema, A.C. Lippus, J.H.M. Metz and W. Rossing (Eds.): Proceedings of the International Symposium on Prospects for Automatic Milking, EAAP publication No. 65, Pudoc Scientific Publishers, Wageningen, The Netherlands, pp 40-48.

6. STREET, M.J., SPENCER, D.S. and HALL, R.C. 1993 The Silsoe automatic milking system. Measurement and Control, Volume 26, September 1993, pp. 197-201.

7. WINTER, A., TEVERSON, R.M. and HILLERTON, J.E. 1992 The effect of increased milking frequency and automated milking systems on the behaviour of the dairy cow. In: A.H. Ipema, A.C. Lippus, J.H.M. Metz and W. Rossing (Eds.): Proceedings of the International Symposium on Prospects for Automatic Milking, EAAP publication No. 65, Pudoc Scientific Publishers, Wageningen, The Netherlands, pp 261-269.

8. DIJKHUIZEN, A.A., HUIRNE, R.B.M. and HARDAKER, J.B. 1992. Scope of management and management decisions. In: A.H. Ipema, A.C. Lippus, J.H.M. Metz and W. Rossing (Eds.): Proceedings of the International Symposium on Prospects for Automatic Milking, EAAP publication No. 65, Pudoc Scientific Publishers, Wageningen, The Netherlands, pp 299-308.

9. PARSONS, D.J., 1988. An initial economic assessment of fully automatic milking of dairy cows. J. Agric. Eng. Res. 40, pp 199-214 .

10. WITTENBERG, G., 1993 A Robot to Milk Cows. Industrial Robot Vol. 20 No. 5, pp. 22-25.

11. HALL, R.C. and STREET, M.J. 1993 'Teat Cup Attachment using a Compliant Pneumatic Milking Robot'. IEE Computing and Control Division, Colloquium on "Dealing with non-rigid robots and products', Digest No: 1993/071.

12. FROST, A.R., MOTTRAM, T.T . STREET, M.J., HALL, R.C. SPENCER, D.S and ALLEN, C.J. 1993 Password 'A field trial of a teatcup attachment robot for an automatic milking system'. J. Agric. Eng. Res., 55: pp 325-334.

13. ROSSING, W., IPEMA, A.H. and VELTMAN, P.F. 1985. 'The feasibility of milking in a feeding box'. IMAG, Wageningen, The Netherlands. Research Report 85-2.

Chapter 15

Intelligent sensing as a means to error free semiconductor wafer handling

J.E.Vaughan, G.J.Awcock, R.Thomas and P.Edwards

This chapter describes an on-going collaborative research project between the University of Brighton and Applied Materials Ltd., a manufacturer of semiconductor process equipment. A robot manipulator places semiconductor wafers with extreme reliability within this equipment. Advances in semiconductor technology have prompted demands for even higher levels of reliability in the manipulator, but have also led to greater diversity in the physical properties of the wafers, posing great problems for the manipulators' sensing systems. A comprehensive survey of suitable sensing techniques has found vision to have the potential to overcome these problems, and a systemic approach to the implementation of machine vision is presented.

15.1 Introduction

The ion implanter is a piece of equipment used to introduce dopant into a semiconductor substrate in order to produce the required electrical characteristics. Atoms or molecules of dopant are ionised and then electrostatically accelerated to sufficient velocity to penetrate the target surface and achieve implantation. The wafers are then passed through the beam with their crystal structures aligned in a critically controlled direction. This process takes place under conditions of vacuum and extreme cleanliness.

During the machine load cycle, a robot manipulator orients and attaches a batch of wafers to a large wheel, which will rotate at high speed to expose the wafers to the beam. Any wafers that are not securely attached may leave the wheel under the influence of centrifugal force.

It is imperative that no wafers are mishandled, since the effects of a wafer breakage within the implanter are severe. Not only is the wafer concerned lost, but the resultant particles may contaminate the rest of the batch and necessitate shut-down and thorough cleaning of the machine.

In order to prevent the mishandling of wafers and minimise the generation of particles, the sensing system must be able to detect all events likely to lead to a mishandle, including failure of the sensors themselves. It is therefore insufficient to measure the position of the manipulator alone; the precise location of all the wafers in the system must be known.

The currently employed sensing techniques, which require expert set-up, allow the implanter to operate within its specified reliability. However, recent trends in semiconductor technology have led to increases in the scale of integration and the number of stages in the fabrication process [1], and therefore an increase in the intrinsic value of the wafers. The increase in the scale of integration is facilitated by a reduction of device feature size, which has in turn caused the particulate contamination criteria to become more stringent.

Although wafer breakages occur very infrequently, the above factors provide sufficient motivation for research into the application of intelligent sensing to further reduce the frequency of wafer mishandles.

There is also a rising demand for equipment capable of implanting transparent wafers, which have proved difficult to detect with traditional sensors.

15.1.1 Practical considerations

The hostile environment of the ion implanter poses many challenges to the sensor designer. The time taken to process sensory data, and therefore successfully place each wafer, must be minimised if current levels of throughput are to be maintained. In order to preserve the quality of the vacuum, materials that *out-gas* and equipment containing *trapped volumes* should not be located within the process chamber. The stringent particulate contamination criteria mean that abrasion within the chamber must be eliminated.

Further constraints are imposed on the location of electronic equipment by the poor heat transfer capabilities of a vacuum and the difficulties associated with the transfer of signals through the chamber walls.

The environment of the implanter does not remain consistent with time. For example, the process chamber walls distort slightly during evacuation. Certain implantation processes cause heating of the wheel, causing slight expansion. Changes in the surface properties of the heatsinks also occur due to *sputtering*. To cope with these changes, any sensing should therefore be adaptive.

15.2   The nature of intelligent sensing

Intelligent sensing has been the subject of a great deal of research, and many techniques for processing sensory data are in existence [2]. However, the majority of these techniques were developed for industrial inspection applications, and intelligent sensing systems for the control of manipulators have received relatively little attention [3].

Although traditional sensors, such as proximity sensors, can perform adequately when correctly setup [4], they are intolerant to even small changes in their environment. Intelligent sensing systems can apply inherent knowledge of the objects under investigation in order to adapt to such changes.

15.2.1 Smart sensors

Recent years have seen research into *smart sensors* [5]. These devices contain in-built processing, calibration, self-test, diagnosis and communication capabilities. There is increasing pressure from the process control industry for the adoption of a standard communication protocol, such as that offered by Fieldbus [6], which allows a network of sensors and actuators to be connected by just two wires.

15.2.2 Levels of adaptation

Since this application is concerned with the development of dedicated sensing systems to be embedded in a product, an economically viable solution is sought. It is therefore important that the requirements are precisely defined early on in the design stage, so that the usual engineering functionality versus cost trade-off can be applied to derive an optimal system configuration. This requires careful consideration of the degree of adaptation that the sensing system must make in order to tolerate environmental changes. In a wider context, the processes of installation and calibration can also be viewed as forms of product adaptation. In total, four distinct stages of adaptation have been identified.

15.2.2.1 Product design level
The ion implanter is a specialised piece of equipment built to individual customer requirements. Rather than develop a sensing system for one particular configuration, a generic sensing strategy, applicable to all product variants, is being formulated. This involves a modular approach whereby sensing modules generate data that are easily integrated into a consistent framework. It is increasingly recognised that the investment in thorough design results in a reduction in the costs associated with the manufacture, testing, calibration and routine maintenance of a product [7].

15.2.2.2 Unit level
During the installation and routine maintenance of each machine, machine specific adjustments may be required. Since this occurs at a customer's site (anywhere in the world), the time and manpower required at this stage should be minimised. This can be achieved by replacing manual hardware adjustments with the modification of software parameters, and by the incorporation of diagnostics to assist the engineers.

15.2.2.3 Execution level
During normal operation, a small, but critical, degree of adaptive behaviour is required to accommodate changes in system parameters, such as those caused by fluctuations in temperature and pressure, and by machine wear. This adaptation should be performed without the need for human intervention. The time and computational

resources at this stage are limited by the specified performance and complexity of the machine.

15.2.2.4 Exception level

It may not be possible to position every single wafer successfully without extending the overall process time. Provided no wafers were actually dropped, it would be desirable to employ a slightly more time consuming error recovery routine every time the normal method failed, requiring operator intervention if appropriate. The overall cycle time would not be significantly altered, as long as the frequency of recoverable wafer mis-handles is negligible.

All of these levels of adaptation have been considered during the formulation of an overall wafer handling sensing strategy, and are discussed further in appropriate sections.

15.3   Sensing technologies

Robot sensors can perform two broad classes of sensing, proprioceptive, in which the internal status of the manipulator system is being measured, and exterioceptive, in which the external environment is being sensed [8]. Since proprioceptors are widely available and understood, only exterioceptors are discussed here.

15.3.1   Contact sensing

Contact sensors interact with their environment in order to make their measurements. In many cases, this can provide a simple and effective means of sensing.

15.3.1.1 The microswitch

The microswitch provides a simple binary output which can be used to indicate the presence or absence of an object. The control algorithm for microswitch based systems is very simple. In practice, the inertia of moving parts makes the control of applied forces and accurate positioning difficult, and microswitches tend to be used to monitor relatively static states, such as door and valve positions.

15.3.1.2 Force sensors

Strain gauge sensors are often incorporated into robot joints to determine the torque applied at that point. On its own, joint force sensing can only be used to detect error conditions before damage occurs, but when used in conjunction with compliant joints (usually integrated with the sensor in a force pedestal), tasks such as the assembly of mating parts can be achieved without the need for high accuracy positioning [9]. This technique, known as remote centre compliance (RCC) is now widely accepted [10] and is available commercially from several manufacturers. It is not suited, however, to the task of wafer detection, since the friction involved would generate particles. Nevertheless, force sensing could be applied to detect movement other than at joints, for example to detect deformation of the manipulator system components.

15.3.1.3 Tactile array sensors

Tactile sensing differs from force sensing in that an image (consisting of *taxels*), is generated, in order for high level interpretation, such as shape recognition and location to take place [11]. Tactile sensing is particularly suitable to applications such as flexible manufacturing [12], requiring highly sensorised dextrous manipulators. The application of semiconductor wafer sensing, however, is more concerned with determining the location of known objects than with the recognition of unknown objects.

15.3.2 Non-contact sensing

In order to meet the particulate contamination objectives, physical contact with wafers should be kept to an absolute minimum, and for this reason non-contact sensing is preferable.

15.3.2.1 Acoustic sensing

Acoustic sensing, in particular ultrasonic range-finding, is widely accepted as an effective method of obtaining accurate estimations of distance irrespective of the properties of the target [13]. For this reason, automatic guided vehicles (AGVs) commonly use ultrasonic sensing for obstacle avoidance [3]. However, the lack of acoustic transmission in the low pressure environment of the ion implanter makes this technique unsuitable for wafer detection.

15.3.2.2 Proximity sensors

Proximity sensors rely on either the capacitance or inductance of the target object. The capacitive sensor is a versatile device, which can sense all materials, liquid and solid. However the response varies for different materials, so the target material must be known for calibration [14]. This makes them unsuitable for the detection of semiconductor wafers that may have been subject to a wide range of fabrication processes, causing variations in the permittivity of the material. Due to the nature of their operation, inductive sensors cannot detect electrically non-conductive materials, such as semiconductor substrate.

15.3.2.3 Photo-electric sensors

The photo-electric sensor provides a binary indication of the presence or absence of an object without contact. The three most common configurations are, reflective, through-beam and retro-reflective. The through-beam configuration requires careful alignment of transmitter and receiver.

A variation on the through-beam principle uses a retro-reflector to return light to the photodetector. Since the area of the reflector is larger than the beam of light, alignment is not critical, and the optical properties of the retro-reflector are such that light will always be reflected at the angle of incidence. One major problem with the retro-reflector configurations is that reflections from shiny objects may cause spurious triggering.

The reflective configuration can be extended to form a position sensor with the use of a position sensitive diode. However, the optical characteristics of wafers vary with oxide thickness, and a single wavelength light source may be totally absorbed.

15.3.2.4 Optical time-of-flight sensors

The principle employed in optical time-of-flight sensors is optical radar. However, since the distance between the sensor and the target is small, and the speed of light is large (3×10^8 m/s), it is more practical to modulate the light source and measure the phase shift due to distance [15]. Unlike the previous two sensing techniques, this method provides a measure of distance that does not require knowledge of the physical properties of the target material, although not all targets will reflect sufficiently.

15.3.2.5 Vision sensors

Of the human senses, vision is the richest source of information; parameters such as absolute and relative position, scale, depth orientation, velocity and range can be estimated without the need for physical contact. Vision sensing can be made suitably adaptive to meet the demanding requirements of this application. Smart vision sensors, with integrated processing and communication capabilities, have been the subject of recent research [16], and are now even commercially available [17]. These factors together indicate vision to be the most suitable technology for semiconductor wafer sensing.

15.4 A pragmatic approach to machine vision

Because humans are good at vision, the task of implementing vision artificially is often underestimated. Even without any financial constraints, it is still not possible to artificially implement vision comparable to that of humans. Fortunately, this is not required or even desirable for industrial sensors, since the sensing problem can invariably be greatly reduced in complexity by applying *a priori* knowledge. An area of research, commonly referred to as *industrial machine vision,* is concerned with the development of visual systems for use in constrained industrial environments.

15.4.1 A generic model of a machine vision system

As with any system, a machine vision system can be decomposed into a number of functional modules. This approach has two main advantages. Firstly, each of these modules can be verified independently, simplifying system design and validation. Secondly, the data are refined from stage to stage, which reduces the computational load and memory requirements.

A generic model of a machine vision sensing system has been proposed [18], and is presented pictorially in Figure 15.1.

This model is very simplistic, since the complexity of the various modules in the model is dependent on the specific requirements of the application, and the boundaries between modules are often hard to define precisely. However, thorough analysis of each module's requirements during the design stage allows precise specification of the whole sensing system.

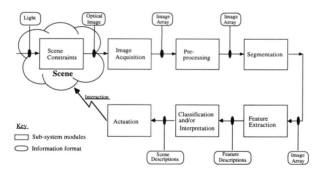

Figure 15.1 *A generic model of a machine vision system*

15.4.1.1 Scene constraints

Greatest advantage can be taken of any *a priori* knowledge in the *constraint* of the scene that forms the image to be processed. Knowledge of the objects to be sensed, such as their physical appearance and size, likely position and orientation may be thought of as *existing* constraints. The use of carefully designed lighting, to highlight salient features and suppress unwanted ones, can be considered as *imposed* constraints. In general, it is almost always easier to improve the lighting than the image processing [19].

An example of the imposition of scene constraints is shown in Figure 15.2. This shows two images of a silicon wafer, (a) with back-lighting, and (b) the edges detected in this image. Due to the opaque nature of the wafer, back-lighting produces a very clear silhouette of the wafer, from which its boundary can be deduced.

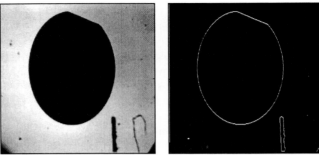

Figure 15.2 *Images of a silicon wafer (a) with back-lighting (b) after edge detection*

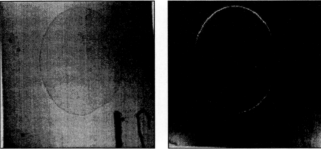

Figure 15.3 *Images of a glass wafer (a) in ambient light (b) with back-lighting*

The use of suitable lighting techniques also allows transparent wafers to be reliably detected, as can be seen in Figure 15.3. Figure 15.3 (a) shows a transparent wafer in ambient light. Not surprisingly, it is barely visible, even to the human eye. The image is also cluttered by a reflection of the image sensor. Figure 15.3 (b) shows the effect of directing light obliquely at the wafer edges. Although the wafer remains transparent, the boundary is now clearly defined, allowing its position to be inferred. The reflection of the image sensor is now much less prominent, since the ambient light is now 'swamped' by the applied lighting.

15.4.1.2 Image acquisition

An 'image' may be acquired using a one-dimensional *line* sensor, or a two-dimensional *area* sensor. It is increasingly recognised that the most suitable image for an automated vision system is not necessarily an image that would be chosen for human interpretation. Humans dynamically interpret the context of a scene, whereas an automated vision system is directed to use certain cues to extract features of interest from an image, and therefore requires far less contextual information. This difference can be considered analogous to the difference between an interpreted and a compiled computer language. A human voyeur can readily interpret a previously unseen scene, whereas an automaton requires the inherent knowledge of the scene and assumptions made in its interpretation to be presented in a pre-digested format.

A little careful thought about the essential characteristics can yield a solution which is effective, elegant and achievable with inexpensive computing equipment. Far from requiring an image of an entire scene, many sensing solutions can be found by simply replacing the single photodetector of a through beam sensor with a line or array of photosites, creating an adaptive sensor requiring little mechanical set-up.

Early automated visually guided robots employed fixed overhead cameras to acquire an image. A high spatial resolution was required to capture all the necessary detail, and extra processing had to be performed to convert from sensor to world co-ordinates. Mounting the image sensor on the manipulator, a configuration commonly referred to as *eye-in-hand* [20], overcomes these limitations, and allows optimisation of the sensing system.

Unfortunately, freedom of sensor placement is severely limited within the ion implanter. In order to reduce communication bandwidth and increase noise immunity, the use of smart sensors is advocated. However, since these sensors are still largely under development, they are still not available in the form of a single integrated device. For reasons of out-gassing and particulate contamination, discussed previously, the placement of printed circuit board based equipment within the vacuum chamber is undesirable.

This leaves two main options, placement of the entire sensor system outside the chamber and acquiring an image through a window, or employing a coherent fibre-optic bundle to convey the image to a remotely situated sensor. For many sensing problems, the former is preferable, provided that the optical distortion introduced poses no great problem.

15.4.1.3 Pre-processing

Although the imposition of scene constraints at the image acquisition stage greatly simplifies the processing required, some pre-processing is often required to minimise the complexity of the segmentation and feature extraction stages.

In particular, pre-processing performs three basic functions:

- Increasing the contrast between homogenous regions in the image to aid segmentation.
- Compensation for any distortion in the image caused by the optics.
- Removal of noise from the image.

If the illumination and acquisition of the image have been carefully considered, then the first two of these will be fairly trivial. Algorithms for contrast stretching, image sharpening and inverse perspective transformation are widely accepted and documented.

Random noise may be efficiently removed from images, without much degradation, using techniques such as median filtering and frame averaging.

15.4.1.4 Segmentation

The process of feature extraction is greatly simplified if the image is segmented into meaningful regions. Segmentation is vital if a machine vision system is to be robust to noise and the presence of unexpected objects in the scene. There are two basic approaches to segmentation: region growing and edge based segmentation.

Region growing techniques search the image and attempt to group together pixels that are in some way similar. Region growing is commonly used to segment complex images, such as natural scenes, but is rarely used in machine vision applications, due to its computational intensity.

Edge based segmentation techniques offer a more efficient means of segmenting images of constrained industrial scenes, which are characterised by more clearly defined regions. Edges in grey-scale images can be detected by applying a first or second order discrete difference filter. The position, strength and orientation of the individual edge points can then be used to generate a complete description of the boundaries between regions.

The task of edge detection is greatly simplified if the number of grey levels in the image is reduced to two. Figure 15.4 shows the distribution of grey-scale values in an image that has been acquired from the scene of a back-lit object. The histogram is essentially bimodal, and a simple threshold can be applied to reduce the image to two grey levels.

A method for automatically selecting a suitable threshold from the distribution of grey scale values has been developed by Otsu [21]. The use of a dynamically determined threshold, as opposed to a manually entered constant, allows the system to adapt to different levels of lighting, affording it extra robustness.

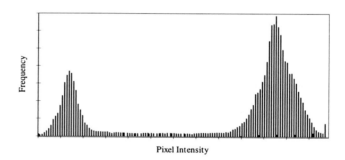

Figure 15.4 *A bi-modal image intensity histogram*

Binary images require less storage space, and can be processed with much simpler algorithms. However, complete segmentation cannot be achieved by thresholding alone. In a simple case, an image may comprise a single object and a background, but in reality, this is rarely the case. If the sensing system is to be resilient to noise and tolerant to the presence of unexpected objects in the image, then the image must be further segmented into connected components before any feature extraction is attempted.

The image intensity histogram itself conveys information useful during the set-up of the sensing system. For example, the lens aperture could be adjusted (either manually or automatically) until the histogram consists of two sufficiently distinct peaks.

15.4.1.5 Feature extraction

It is the task of the feature extraction module to produce suitable descriptors from the regions identified by the segmentation module. Quite often, this may require two passes, one to extract sufficient features from all the regions in the image to enable regions corresponding to objects of interest to be selected, and a second to extract all the features required by the interpretation or classification module from these regions.

Unless calibrated measurements are required, the descriptors should, as far as possible, be insensitive to the scale, position and orientation of the object, although these are often useful parameters in themselves. For example, a dimensionless feature is the *circularity shape factor*. Given by the ratio of the square of the perimeter to area, this shape factor falls to a minimum for circular objects.

15.4.1.6 Interpretation

The interpretation stage uses the features that have been previously extracted from objects in the image to produce a meaningful description of the environment. In the control of simple robotic systems, the output from a single sensor would usually be used to derive a single control variable. The high reliability required in this application has necessitated a different approach. This is discussed in detail in Section 15.5.

15.4.2 The development of a wafer orientation sensor

It is required that the circular wafers are oriented correctly before implantation, using a small notch or flatted portion as a datum. Suitable sensing techniques have been developed to detect the orientation of wafers so that they can be placed in the correct orientation.

Images of the wafers were acquired in appropriate lighting conditions and thresholded at a suitable value. Segmentation of the binary image is then completed using chain coding [22]. The image under inspection is searched by raster scan until the transition from light to dark is detected, indicating the edge of an object. At this point an outlining routine is initiated. This routine traces around the boundary of the object, calculating some parameters on the way. As each outline point is detected, it is recorded in an output image to prevent the same object being processed more than once. After the outlining routine has finished, the raster scan continues until another object is found.

Tracing the outline generates parameters such as the area, perimeter and centroid of each object. This method has the advantage that white objects on a black background return a positive area, while the area of black objects on a white background is negative, allowing 'holes' and objects to be easily distinguished. The area and circularity shape factor together provide enough information for the wafer to be reliably distinguished from any other objects in the image, and selected for further processing.

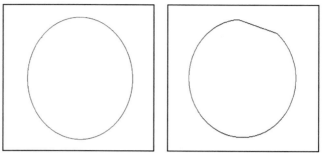

Figure 15.5 *Wafer edge maps (a) notched wafer (b) flatted wafer*

Figure 15.5 shows the edge maps of two common types of wafer. The wafers are sufficiently different to warrant subtly different processing. Rather than develop a technique sufficiently adaptive to cope with both wafer variants, the key to adaptivity lies in the ability of the system to automatically select the most appropriate processing technique.

15.4.2.1 Detection of notched wafers
Insensitivity to the position of the wafer within the image can be achieved by subtracting the co-ordinates of the centroid from the previously detected edge point co-ordinates, thus transforming absolute position into relative position.

To ensure that notches at any position in the wafer edge can be detected, the relative edge co-ordinates can be used to produce a list of radii, which has the effect of reducing the data to be manipulated from two dimensions to one. Figure 15.6 shows the radial profile of a notched wafer. Due to the aspect ratios of the camera and digitiser, the wafer produces an elliptical image. The elliptical nature of the wafer object produces a sinusoidal variation in radius. The notch is not easily detectable, since the digitisation errors give rise to fluctuations of a similar amplitude.

The sinusoidal variation in radius can be removed by multiplying one set of edge co-ordinates by an aspect ratio correction factor. This could be a constant value, but if it were automatically derived, then the system would be adaptive to minor changes in the camera viewing angle

In practice, this is achieved by deriving the equation of the ellipse from the set of outline points, using the method of least mean squares. The aspect ratio is then given by the ratio of major to minor axis. Figure 15.7 shows the radial profile after this has been performed. A moving average has also been subtracted from the image, to further increase the signal to noise ratio.

The interpretation stage searches the radial profile for a trough corresponding to the notch. Some simple confidence metrics are also calculated to verify that the peak does, in fact, correspond to a notch and is not induced by noise.

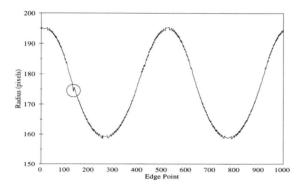

Figure 15.6 *Radial profile of a notched wafer*

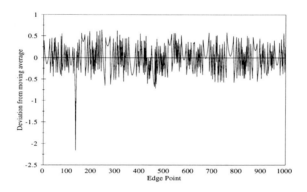

Figure 15.7 *Aspect compensated radius*

15.4.2.2 Detection of flatted wafers

The problem of detecting flatted wafers is complicated by the fact that the centre of area of the wafer does not coincide with the origin of the ellipse, so the centre of area must be derived by other means.

The Hough transform provides an effective means of detecting simple shapes, such as lines and ellipses from incomplete boundary information. In most real-time applications, votes for candidate object parameters are accumulated in a *parameter space* from the strength, position, and orientation of each edge point in the image. A large number of variations on this theme have been devised [24].

A particularly elegant method for detecting the centroid of ellipses is the bisecting chord method devised by Davies [24]. This method uses two one-dimensional accumulator arrays instead of a single two-dimensional array method, and this has been shown to dramatically increase the computational efficiency.

A chord is constructed between each pair of edge points with the same X co-ordinate, and the mid-point of these chords is selected as candidate Y co-ordinate of the centroid. Votes for these candidates are accumulated in parameter space, which is then searched for peaks. Figure 15.8 shows the accumulation of centroid Y co-ordinates for the flatted wafer in Figure 15.5 (b). A similar process is then performed to determine the X co-ordinate of the centroid.

Since the image was segmented and only edge points corresponding to the wafer were considered, there is only one distinct peak in parameter space.

The position of the flat, and therefore the orientation of the wafer can then be determined from the radial profile, which is calculated from the edge point co-ordinates and the centroid position. The position of the trough relates to the edge point in the centre of the flat. A typical radial profile is presented in Figure 15.9.

The voting process causes the median of the candidate centre points to be selected, which is an integer approximation to the centroid location. This makes this technique unsuitable for the processing of notched wafers with the currently employed spatial resolution, since minor errors in the calculated centroid position will vastly reduce the signal-to-noise ratio of the radial profile.

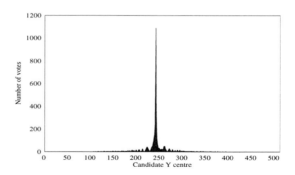

Figure 15.8 *Centre co-ordinate accumulator array*

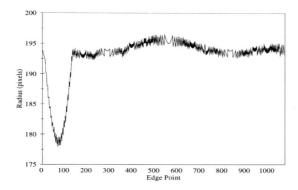

Figure 15.9 *Radial profile of a flatted wafer*

15.4.2.3 Review of orientation sensing experiments
Through the adoption of a rigorous systemic design procedure, solutions to the wafer orientation have been developed which are elegant, robust and computationally efficient. The prototype sensing systems were able to detect the orientation of both notched and flatted wafers at speeds approaching video frame rate using standard hardware.

15.5 Integration of sensory data

Most current robotic sensor systems employ a single sensor to extract one piece of information from their environment for each required control signal. These sensors are limited in their ability to resolve ambiguities, identify spurious information, and detect errors or failure [25].

The generation of an accurate description of the environment requires the integration of the data from a number of sensors. Multi-sensor systems can derive accurate measurements based on the data from all the correctly functioning sensors and disregard erroneous data. Integrating redundant information can reduce overall uncertainty and thus increase accuracy [26]. However, this does not imply the need for a large number of sensors, since more than one physical quantity can often be inferred from one sensor.

An area of work known as *sensor fusion* is concerned with the integration of sensory data to overcome these limitations.

15.5.1 A summary of sensor fusion techniques

Tanner [27] presents a comprehensive taxonomy of multi-sensor fusion techniques, and proposes several classes of sensor fusion.

15.5.1.1 Uniquely determined dependent data fusion

With the uniquely determined class, there is precisely enough data to determine the fused value. Tanner uses the derivation of wind-chill temperature from air temperature and wind speed as an example of this.

15.5.1.2 Over determined dependent sensor fusion

Over determined dependent sensor fusion occurs where there is more data available than is actually required to determine the fused value. When all sensors are functioning correctly, the data from one or more can be disregarded. The inherent redundancy of the system can be exploited to detect erroneous sensory data.

Durrant-Whyte [25] describes the use of Bayesian state estimators [28] to group sensors that provide a consensus. The Kalman filter [29] and extended Kalman filter [30] can be used to provide a less computationally intensive way of achieving Bayesian state estimation, and has been adopted by many researchers for use in real-time systems [31-36].

Evidential reasoning techniques, such as Dempster-Shafer reasoning, have also been applied to multi-sensor fusion [37]. However, this technique is only suitable for combining data from similar sensory sources, and is therefore not suitable for applications where data from different sensory sources are fused to provide an accurate description of the environment [30].

15.5.1.3 Under determined dependent sensor fusion

Under determined dependent sensor fusion takes place where there are not enough sensory data available to determine the desired value. This is really a process of estimation or extrapolation, and, in general, assumptions must be made to replace sensory input. An example of under determined sensor fusion given by Tanner is the computation of an object's velocity from distance measurements made by a range sensor. Since the range sensor only provides a measure of distance in one plane, the calculated velocity will only be correct if the object moves directly towards or away from the sensor.

15.5.1.4 Sequential sensor fusion

Sequential sensor fusion describes a class where the data from some sensors may be used to control others. Durrant-Whyte [25] considers the sensors in a multi-sensor system as members of a team. Each individual sensor can make local decisions based on its own observations, but must co-operate with others to achieve a common objective. A blackboard data structure [38] allows the communication of partial and complete results between team members.

Crowley [39] extends the blackboard to form a Composite Local Model. This model begins with the *a priori* knowledge of the environment, and is updated as information from sensors supports or contradicts it.

15.5.1.5 Behaviour based control

Behaviour based control is discussed separately here, because it can be applied to perform any of the previously discussed classes of sensor fusion.

As the resolution and sophistication of sensors increases, the processing of sensory data clearly becomes more complex. The relationship between high level sensory inputs, such as vision sensors, and the required motor commands becomes increasingly difficult to specify in mathematical terms. This sort of problem is often addressed with an artificial neural network (ANN), which can be trained to give the correct response to input stimuli.

Researchers have being investigating the concept of adaptive behaviour based sensory-motor controllers [40]. The complexity of ANNs can be shown to increase at an exponential rate with respect to problem complexity, making the manual design of ANNs unfeasible for sophisticated behaviours. Instead, artificial genetic evolution techniques have been found to produce near optimal network configurations that are tolerant to noise [41]. The placement of sensors and selection of their parameters has also been performed under genetic control with success.

Brooks [42] argues that the low-level insect-like behaviours must be simulated first, and then the results can be built upon, eventually aiming towards much more complex behaviours employing high resolution sensory data.

ANNs offer a robust solution, since they are inherently redundant, and can therefore continue to function to some degree in the event of a partial failure. However, a major disadvantage of this technique is that the mathematical transforms applied are not easily quantified, making in-service verification difficult.

15.5.2 Selection of suitable techniques for further work

The environment of the ion implanter is more constrained than that of systems such as an automatic guided vehicle, which must navigate in an uncertain environment. In the same way that industrial machine vision is a sub-set of image processing that is constrained and optimised for individual applications, the term *machine perception* is used here to make a similar distinction from sensor fusion as a whole.

Machine perception should allow *a priori* knowledge to be integrated in such a way that allows product adaptation to be easily performed. This makes Crowley's composite local model very attractive as a platform for further work.

Also of interest are behaviour based control techniques. The simulation necessary during the machine learning could be completed off-line, using the available CAD descriptions of the wafer handling system.

15.6 Conclusions

A challenging sensing and control problem has been briefly presented. Thorough analysis of the requirements and the relative merits of a number of sensing techniques has found machine vision to have the adaptivity necessary to overcome the difficulties encountered by traditional sensors.

The application of *a priori* knowledge and adoption of the generic model of machine vision has allowed the development of sensors that are near optimal in terms of robustness and processing performance. Once a similar technique has been established to allow the integration of visual sensory data with data from traditional sensors, a complete sensing and control system can be developed and evaluated.

15.7 Acknowledgement

This project is supported by the Science and Engineering Research Council under CASE award 92567547.

15.8 References

1. O'REILLY, H., Robotics and Wafer Transfer, European Semiconductor, Vol.14, No. 1, Angel Publishing 1992.
2. VERNON, D., Machine Vision: Automated Visual Inspection and Robot Vision, Prentice Hall, 1991.
3. WALLACE, A., Vision: Meeting the Future Needs of Industry, BMVA Seminar, Birmingham, 1992.
4. CLERGEOT, H., PLACKO, D. and DETRICHE, J. M., Electrical Proximity Sensors, NATO Workshop on Sensors and Sensory Systems for Advanced Robots, Springer-Verlag, 1988.
5. HUDSON, C., Research in Distributed Sensor Systems, MMSG Seminar, Warwick, December 1993.
6. WOOD, G., Fieldbus Standards, MMSG Seminar, Warwick, December 1993.
7. BOWSKILL, J., DOWNIE, J., and KATZ, T., A Facility for Concurrent Design and Production in an Engineering Facility, CEEDA '91, Bournemouth, UK, March 1991.
8. NITZAN, D., Assessment of Robotic Sensors, 1st RoViSec, Stratford, England, April 1981.
9. ESPIAU, B., An Overview of Local Environment Sensing in Robotics Applications, Sensors and Sensory Systems for Advanced Robots, Dario, P. (ed.), 1989.
10. KANADE, T., and SOMMER, T. M., An Optical Proximity Sensor for Measuring Surface Position and Orientation for Robot Manipulation, Proc., Society of Photo-optical Instrumentation Engineers, Vol. 449 pp 667-674, 1984.
11. PAUL, R. B., Compliance and Control, Proc. Joint Automatic Control Conference, Purdue University, Ind. , 1976.

12. PUGH, A., An Overview of Visual and Tactile Sensor Technology, NATO Conference on Highly Redundant Sensing in Robotic Systems, Tou, J. T. (ed.), 1990.

13. NICHOLLS, H. R., and LEE, M. H., A Survey of Robot Tactile Sensing Technology, International Journal of Robotics, 8(3), June 1989.

14. SATO, N., HEGINBOTHAM, and PUGH, A., A method for Three-dimensional Part Identification by Tactile Transducer, Robot Sensors, Vol. 2: Tactile and Non-vision, Pugh, A (ed.), IFS (publications), 1986

15. POPERAY, S.C., Ultrasonic Distance Measuring and Imaging Systems for Industrial Robots, ., Robot Sensors: Volume 2 Tactile and Non-vision, Pugh, A. (ed.), IFS Pubs, 1986.

16. AWCOCK, G., and THOMAS, R., Systems Considerations in Industrial Computer Vision, Proc. 6th International Conference on Systems Engineering, Coventry, 1988.

17. VELLECOT, O., The Imputer™ - A Programmable Smart Vision Module, MMSG Seminar, Brunel University, April 1993.

18. AWCOCK, G., PhD Thesis, Chapter 7, Brighton Polytechnic, 1992.

19. HALL, E., Computer Image Processing and Recognition, Academic Press, 1979.

20. LOUGHLIN, C., Eye-in-hand Robot Vision Scores Over Fixed Camera, Sensor Review, No. 3, 1983.

21. OTSU, N., A Threshold Selection Method for Grey-Level histograms, IEEE Transactions on Systems, Man and Cybernetics, Vol. 9 Part 1, 1979.

22. WILL, J. M., Chain-code, Robotics Age, March/April 1981.

23 .LEAVERS, V. F., Which Hough Transform, CVGIP: Image Understanding, Vol.58, No.2, September 1993.

24. DAVIES, E. R., Machine Vision: Theory, Algorithms, Practicalities, Academic Press, 1990.

25. DURRANT-WHYTE, H., Integration, Co-ordination and Control of Multi-Sensor Robot Systems, Kluwer Academic Publishers, 1987.

26. LUO, R.C., Multi-sensor Integration and Fusion in Intelligent Systems. IEEE Transactions on Systems, Man and Cybernetics, Vol. 19, No. 5, October 1989.

27. TANNER, R.,A Taxonomy of Multi-sensor Fusion, Journal of Manufacturing Systems, Vol. 11, No. 5, 1992.

28. BERGER, J.O., Statistical Decisions, Springer-Verlag, 1985.

29. KALMAN, R., A New Approach to Linear Filtering and Prediction Problems, ASME Transactions Series D. J. Basic Engineering, Vol. 82, 1960.

30. HULLS, C. W., Multi-sensor Fusion for Robots Controlled by Relative Position Sensing, University of Waterloo, Canada, 1993.

31. ATTOLICO, G., CAPONETTI, L., CHIARADIA, M.T., DISTANE, A, and STELLA, A., Dense Depth Map for Multiple Views, Sensor Fusion III: 3-D Perception and Recognition, Proceedings SPIE 1383, 1991.

32. BAY, J.S., A fully Autonomous Active Sensor-based Exploration Concept for Shape Sensing Robots, IEEE Transactions on Systems, Man And Cybernetics, Vol. 21, No. 4, 1991.

33. BECKERMAN, M., A Bayes Maximum Entropy Method for Multi-sensor Data Fusion, Proceedings of the IEEE 1992 Conference on Robotics and Automation, Vol. 2, 1992.

34. CLARK, J.., and YUILLE, A., L., Data Fusion for Sensory Information Processing Systems, Kluwer Academic Publishers, 1990.

35. CROWLEY, J.L., and RAMPARANY, F., Mathematical Tools for Modelling Uncertainty in Perception, 1987 Workshop on Spatial Reasoning and Multi-Sensor Fusion, Morgan Kaufmann, 1987.

36. HALL, D.L., Mathematical Techniques in Multi-sensor Data Fusion, Artech House, 1992.

37. PEARL, J., Reasoning with Belief Functions: An Analysis of Compatibility, International Journal of Approximate Reasoning, Vol. 4, Part 5-6, November 1990.

38. NII, P.H., Blackboard Systems: The Blackboard Model of Problem Solving and Evolution of Blackboard Architecture, AI Magazine, Vol. 7, Part 3, 1986.

39. CROWLEY, J.L., Principles and Techniques for Sensor Data Fusion, Signal Processing, Vol. 32, Elsevier Science Publishers, 1993.

40.   CLIFF, D., HUSBANDS, P. and HARVEY, I., Analysis of Sensory-Motor Controllers, Technical Report CSRP 264, University of Sussex, December 1992.
41.   CLIFF, D., HUSBANDS, P. and HARVEY, I., Evolving Visually Guided Robots, Technical Report CSRP 220, University of Sussex, July 1992.
42.   BROOKS, R.A., Achieving Artificial Intelligence Through Building Robots, A. I. Memo 864, MIT A. I. Lab, May 1986.

Chapter 16

The concept of robot society and its utilisation in future robotics

A.Halme, P.Jakubik, T.Schönberg and M.Vainio

Societies are formed as collaborating structures to execute tasks which are not possible or are difficult for individuals alone. There are many types of biological societies, but societies formed by machines or robots exist only in laboratories at the moment. The concept offers, however, interesting possibilities especially in applications where a long-term fully autonomous operation is needed and/or the work to be done can be executed in a parallel way by a group of individuals. This chapter introduces the basic control and communication structures for robot societies by using a model society. Also a mini-scaled mobile robot which is under construction, and will be duplicated for testing as a physical society demonstrator, is introduced. Simulation results illustrating the behaviour of the model society are given. Some potential applications are introduced and discussed.

16.1 Introduction

Societies are formed as collaborating structures to execute tasks which are not possible or are difficult for individuals alone. In some cases societies can also be seen as alternatives to large size more complex individuals. There are many types of biological society formed by animals and by humans. Societies formed by machines or robots are still rare. The related research activity is, however, quite lively today, if we consider such topics as different kinds of multi-robot environments, autonomous agents and artificial life (see e.g. [1] - [4]).

There are at least two basic reasons why the concept of a robot society is of potential interest as an engineering solution. The first reason is fault tolerance. A machine society has a high redundancy, because in a society concept the functions of a faulty individual can always be easily replaced by other members without any redefinition of tasks or overall control system. The second reason is the member to member communication structure which makes it possible to increase or decrease the number of the members in a society easily, without any redefinition of the communication structure. Applications where a long-term fully autonomous operation is needed may utilise the society concept by avoiding the ultra reliable design of the machines. Another interesting basic feature of robot societies is related to the task division in the overall work. Because every member can independently execute its tasks (this does not mean that the members cannot do coordinated work!), a society is especially effective in work which allows vectorising in parallel tasks. Material transportation, environment mapping, collecting of objects, cleaning etc. are examples of easily vectorised works.

The concept of robot society has potential in many future applications of robotics. These are related to tasks other than to traditional factory production, for example maintenance and inspection, environment cleaning and protection, societal and medical applications. However, in the future ecofactory, in which materials are circulated effectively, the disassembly lines may offer interesting possibilities to apply the concepts. One of the most natural application areas will be in micro-machines or micro-robots, because they are used as groups rather than individually due to the limited capacity of one individual. A couple of possible scenarios are considered in more detail later in the chapter.

This chapter introduces the basic control and communication concepts for a robot society based on the research done in our laboratory. The concepts are illustrated by a model society which is formed by mobile robots searching for and collecting stones in a limited but initially unknown area. The society has two types of members: work units which collect the stones and supporting units which carry energy resources and act as work coordinators. The work units have a limited energy resource which is refilled from the support units (called also energy units). Thus the society can also distribute energy like it distributes information. Although the model society looks quite a simple one at first glance, it includes many interesting features and can be used to illustrate applications of various types. To illustrate the properties of this model society we have made a computer simulator for it, and to have a more concrete idea of how it works, we are also presently constructing a physical society which consists of mini-scaled mobile robots. Both are briefly described and some basic behavioural properties of the model society are illustrated.

16.2   The concept of robot society

Basically every society is a group of individuals, which we call members, with an information and a control structure. The members are independent in the sense that they know completely how to control their behaviour and how to respond to the information coming from the environment through the sensory system or from the society through the communication system. All members of a society need not be similar. Members having the same properties can form clusters or classes. The information structure defines how information is spread within the members and how an individual member communicates with the other members of the society. The control structure defines the way the society affects its members. Because all working power is produced by the members, the control structure manages the task execution of the society by utilising the control system implemented through the members. Since both structures in human societies are very complex, we are not trying to analyse or imitate them when modelling robot societies. In addition, robot societies are artificial ones and we can define both structures exactly. Firstly, we assume that these structures are constant and the society cannot redefine them as happens quite often in human societies. More suitable model societies can be found by considering the societies formed by lower level animals like social insects, e.g. ants or bees [5-6]. It should be noted, however, that the society structure which we are looking for is more than just a group or herd of animals with no clear information or control structure.

The ultimate practical goal of the robot society concept is to construct a kind of 'distributed robot' or 'group robot' which can execute tasks which are defined by the user or 'society controller', as in the case of a conventional individual robot. This means that the behaviour of the society must be controllable outside and that the society must have this information connection to the controller. However, it is important from a practical point of view that this connection is not made to every member of the society, but rather to the information system of it. This is because a society may include a large number of members which are located in places where a communication system is difficult to build. Basically the communication in a society is carried out on a member-to-member basis.

Although the concept of a society does not limit the size of the members it is quite natural to think that members are small in size. In a very small scale, i.e. the scale of micro-machinery, the society concept is a very natural one to be utilised. One can see the same in biology, too. Many small animals are represented in groups (not all forming societies in the sense we are talking about!) and so they have in a sense divided one body into many pieces. In spite of the small size of an individual and consequently its limited capacity, a large number of individuals can execute amazing jobs.

One of the very attractive properties of a society, is that its functioning is not dependent on one individual. The functions of an individual can always be replaced completely by others and this does not need any reconfiguration of the society. This is because each member is fully self sufficient in its function and the work division within the society is not dependent on a named individual. Thus the redundancy of the system is high. The other interesting property which might be utilised in some applications is an information structure which is based on member-to-member communication. This has the advantage that the communication system is very flexible and the capacity is not limited by the number of members. Also new members can easily be added to the society at any time.

Biological societies like ants, or bees, have been formed through evolutionary process over millions of years. It is therefore sensible to study the properties of these societies to obtain some reference models. The basic tasks done by both societies are to collect food, construct a nest and protect their own society. Analogues to these tasks can be found in many technical applications, for example monitoring, cleaning, separation, environment mapping, etc.

16.3 A model society

The robot society concept is illustrated here with a model society which is formed by mobile robots searching for and collecting stones (or in general some recognised objects) in a limited but initially unknown area. The stones are detected, picked up and carried to a base station. The base station serves also as the interface with the user. The operator can send commands through it and information (like map information) collected by the members can be communicated to the user. The robots are assumed to be able to avoid obstacles and each other in a simple reactive way, and they can navigate by a simple dead-reckoning type of navigation system. Communication between the members can take place only locally by using the broadcasting principle so that only members within a certain maximum distance can hear what a talking member is saying. The talking robot reserves the space within the communication distance and forces the members inside it to listen. A set of simple messages can be sent and received to communicate the following types of information: co-ordinates of location, stones found, help needed, detection of nearby units, change of operation mode. The society has two types of members: working units which collect the stones and energy units which carry energy from the base station and act as work co-ordinators. The workers have a limited energy resource which is refilled from the energy units. Two members can also equalise their energy resources when meeting. Thus the society can also distribute energy like it distributes information.

The above society can serve as a simple model for the following tasks:

- mapping an unknown environment
- separating valuable or dangerous material
- finding and destroying specified objects in the environment

Depending on the detailed definition of the communication and control structures the society behaves in many different ways. Some properties are illustrated by a simulation model in the next section. Because a physical realisation always brings some additional details we are also presently constructing a mini-robot society to study and demonstrate the model society in the real world. The work unit robot is explained in more detail later.

16.4 Simulated behaviour

16.4.1 Computer simulation

A computer simulator has been constructed to illustrate the basic properties of the simplified model society [14]. Many researchers who are working with robot societies uses simulations, because analytical approaches are extremely complicated especially if the communication structure is included. Comparable simulations have been previously introduced e.g. by Mataric [7] and Hara [8].

The simulator models a 2-D environment which includes obstacles, stones, one base station and energy units, the symbols of which are shown in Figure. 16.1. The environment is a 1000 by 1000 area illustrated in Figure 16.2. The algorithms of the working units execute the following tasks: random searching, moving towards a given destination and obstacle avoidance. The working units can communicate with each other, with the base and with the energy units. Communication is used to ask the position of another member and the energy level of a working unit. Obstacle avoidance is done in a reactive way by using five sensors (like ultrasonics) that cover 180 degrees in front of the robot. The range of the sensors are 40units for point type obstacles (stones or other members) and 25units for line type obstacles (walls of larger obstacles). The robots move with a step size of 20units. When the energy level of a working unit drops below a specified minimum level it can navigate to the nearest energy unit for refilling assuming that this can happen before its energy is completely finished. If a working unit loses its energy completely it dies. The energy units cruise in the environment by avoiding obstacles and other units. They can hear the working units within the communication distance and move to any member which has used its energy. The simulator has been realised in a 486 PC using an MS-Windows programming environment and ACTOR object oriented programming language.

Figure 16.1 *The parts of the environment as they look in the simulator are from the left, working unit, stone, base-station, energy unit, line-obstacle (walls of bigger obstacles or the border of the arena). The line on the working unit and energy unit shows the direction*

The behaviour of the society and its efficiency when executing the collection task can be studied by running the simulator with different parameters. As noted before, even in this quite simple case analytical approaches are very complex and simulation is the only practical way to find out the behavioural details of the society. The behaviour is stochastic in nature, and thus every simulation is a realisation of a stochastic process. Several simulations with the same parameters have to be run normally to obtain a mean behaviour. In the following some characteristic features of the model society are illustrated. All simulations lasted 5000 time steps but if all 225 stones are collected before that or all working units are dead the simulation is stopped. The data shown in Figures 16.3-16.9 illustrate the energy distribution, the effect of communication distance and the number of members. The data is collected at every point by averaging results from several simulation runs (only 2-5 runs are used as several hours are required to complete one run).

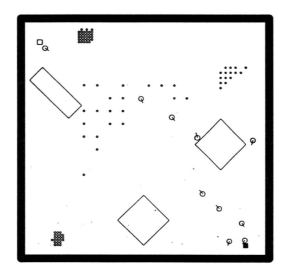

Figure 16.2 *The simulation area during the simulation.*

16.4.1.1 Energy distribution

The energy threshold is the energy level which causes a working unit to stop working and to start moving towards the nearest energy unit for refilling. The energy threshold can be anything between 0-1000 and after recharging the energy level will be 1000. One interesting feature which can be found for example in ant societies is energy equalising. This means that two members, when they meet, can equalise their energy storage. It seems that this feature improves the energy distribution within the society, as is illustrated by simulation in Figure 16.3 and Figure 16.4. In these simulations the working units communicate all over the area. The interesting result is that the energy equalising property clearly helps to distribute the energy more efficiently which results in a smaller number of dead members especially when a low energy threshold value (<300) is used. The energy per collected stone, however, increases. The number of stones collected during the simulation is about the same in both cases. To illustrate the effect in a clear way the energy unit does not move in these simulations (saving dead units is not possible).

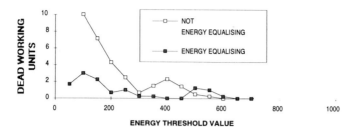

Figure 16.3 *With energy equalising the working units will not run out of energy as easily as they do while not using equalisation. The difference is seen best at energy threshold values smaller than 300. The energy unit is not moving in the simulations*

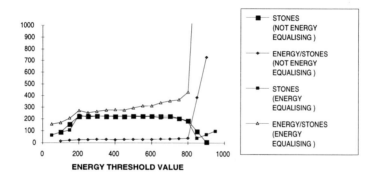

Figure 16.4 *The energy used for collecting stones is greater when using energy equalising than without it but the number of collected stones remains about the same. The energy unit is not moving in the simulations*

16.4.1.2 Effect of communication distance
The effect of the communication distance is illustrated in the Figures 16.5-16.7. The energy unit is now moving. An interesting observation is that the energy equalising property loses its meaning when the communication distance increases. When a small distance (<200) is used almost all working units without energy equalising properties die during the simulation (i.e. within the maximum 5000 time step). The energy threshold used is 200 which causes the agents to run out of energy relatively quickly.

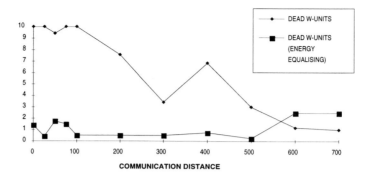

Figure 16.5 *At bigger communication distances there is not a big difference between using energy equalising or not. For communication distances less than 200 almost all working units without energy equalising die during the simulation but working units with it do not die*

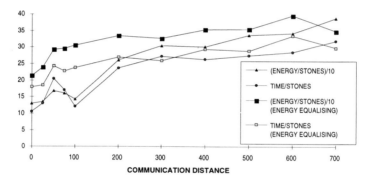

Figure 16.6 *At bigger communication distances there is not a big difference between using energy equalising or not*

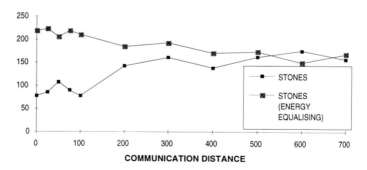

Figure 16.7 *At small communication distances when using energy equalising most of the time is spent collecting stones since energy is always available. If energy equalising is not used time is spent for searching for energy and robots die*

16.4.1.3 Number of members

The number of members also has some interesting effects on how the society can carry out work. As illustrated in Figures 16.8 and 16.9, the time and energy needed to collect a stone is strongly dependent on the number of working units. There is also a minimum number of agents needed for completion of the task. The communication distance is again assumed to cover the whole area. Several energy units are used and they are moving. The number of energy units does not have much influence, probably, because the area used in these simulations is easily covered by one energy unit. The energy threshold used is 200.

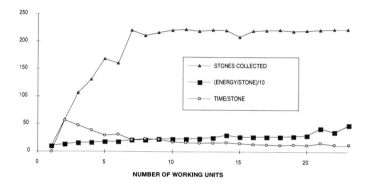

Figure 16.8 *When using more members the time needed for collecting a stone decreases but the energy used per stone increases*

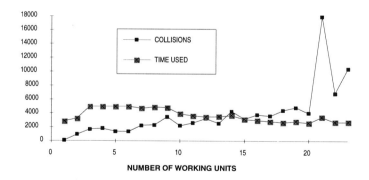

Figure 16.9 *Using more members the time needed to complete the task decreases but the number of collisions increases. The peak at 21 is believed to correspond to a deadlock situation in a couple of simulations*

16.4.2 A physical realisation

HUTMAN (Helsinki University of Technology's Mobile Autonomous Navigator) is a mini-scaled mobile robot which was designed and built to provide a test-bed and a prototype for the member of the model society (working unit). It is still under development and duplication has not been done yet. Figure 16.10 shows a simplified picture of the robot with the basic functional processes included.

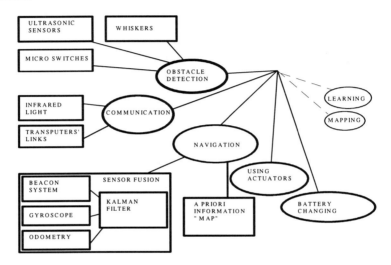

Figure 16.10 *Basic processes in a robot society member*

16.4.2.1 Construction

Due to the need to use several mobile robots in an office environment the size of the robot is quite small: the diameter of this turtle-type robot is 15cm and the height is 14cm. The total weight is less than 1.5kg. It has two driven wheels and four balancing swivel castors. The robot uses two narrow beam ultrasonic sensors for obstacle and stone detection and it has two whiskers to give tactile information from the environment. Its navigation system is dynamic, using a modified Kalman filter to fuse the information from separate systems. To give HUTMAN the ability to fulfil the stone collecting task, a simple one degree of freedom gripper was designed for it. The robot, shown in Figure 16.11, is completely autonomous, it carries batteries and the processing takes place only in processors located on-board. A more detailed description of HUTMAN can be found in [9]. The following sections summarise briefly the hardware and software used.

Figure 16.11 *HUTMAN*

16.4.2.2 Hardware
In an autonomous mobile robot the amount of data to be processed is large even if only relatively simple functions are included. The solution chosen was to use transputers and their support for parallel processing. The general hardware architecture is shown in Figure 16.12.

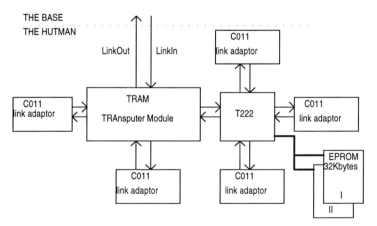

Figure 16.12 *General hardware architecture*

Only two transputers, a Transputer Module (16 or 32 bits) and a T222 transputer (16 bits), are used. The network is connected to the outside world via link adaptors, which operate as interfaces between the processors and the sensors/actuators. All communication in the network is done through the transputers' standard links. Examples of transputer based hardware solutions can be found in [10-11].

16.4.2.3 Communication

HUTMAN communicates using locally restricted randomly accessed broadcasting principles. The size of this local broadcasting space is adjustable and the actual communication is done with infra-red light.

16.4.2.4 Navigation

The navigation system was designed to make the robot as autonomous as possible. The solution is a 'hybrid' system. It is a combination of two different methods, a dead reckoning system with a miniature gyroscope and a beacon detection system (based on sound pulses and infra-red light detection). The results from these two systems are combined in a Kalman filter, where the 'true' position values from the beacon detection, will make the final estimates much more reliable than those achieved by pure dead reckoning (for general principles, see e.g. [12]).

16.4.2.5 Software

The control structure of the robot and thus the foundation of the software is basically behaviour based reactive control. The behaviour which the HUTMAN performs depends on three individual factors: the energy status, the communication, and the operational mode. Figure 16.13 illustrates these concepts. The triangle describes the robot and the three main factors are shown as intersecting circles. The shadowed area where the three circles intersect describes the actual working state of the robot.

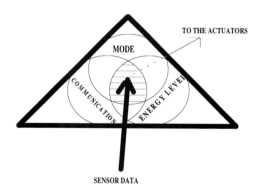

Figure 16.13 *The control structure of the HUTMAN*

The actual code is physically divided into two separate blocks. The basic programs, which are essential for any kind of operation at all, are loaded to the EPROM. The more sophisticated programs, which are not so vital, such as map building or travelled curve smoothing, are stored to the module's RAM.

16.5   Possible applications

It is a fact that today there exist no real applications for robot societies (at least the authors do not know of any). Of course, when one is looking at a modern car production line where the car bodies are welded simultaneously by many robots, the idea of a welding robot society is visually quite close. However, the robots lack the basic communication and control structures which form a society. They are not able to replace each other without a considerable effort in reprogramming, and they cannot help each other in task execution. Even real multi-robot applications, where the robots are controlled by a central controller, are rare today. However, when looking again at the assembly line of the same car factory, where humans are doing most of the assembly tasks, it is quite common today to use so-called team control which means that the people in the team do the assembly work in a parallel way. This means that each one is able to execute several assembly tasks and mutual communication controls the whole work process. Robotising this work while keeping the same feature would lead to a society of mobile assembly robots which are capable of executing individual and cooperative tasks.

16.5.1 Disassembly society

Because of the tradition of systematic design of parts and the related assembly tasks the above development in production lines is, however, not probable. However, the car factory (like consumer electronics and domestic appliance factories) will change in the future to so-called ecofactories where materials and parts of the products are circulated. Among other things this means that the factory must include a disassembly line where the used products are disassembled completely so that all usable materials and parts can be separated. This work is much more disordered and complicated to do with a traditional type of robotics. The robot society concept offers a potential possibility in this area as in other similar disassembly work.

16.5.2 Environment mapping

Looking at applications of robots outside factories there are cases where the environmental map can be used to find out if there is something valuable or dangerous or to simply monitor its condition. The problem is that the environment itself and the location of the target are not known *a priori*. This search problem together with the task of learning the nature of the environment form the primary goal. It can be shown that a multi-robot society with a proper information structure and an optimal control policy is a very effective tool for this work. An environment mapping society can be considered as a special case of the model society considered previously. The information concerning the location and type or measured value of the findings can be brought back or communicated to the base station. The energy problem is in many practical applications one of the most challenging, and the use of mobile energy carriers and the energy equalisation concept offer interesting possibilities to solve it.

16.5.3 Dumping area monitoring and cleaning society

A possible practical scenario for the model society could be a dumping area society. Dump sites pose big problems as environmental risks because of the harmful materials included. On the other hand they also include valuable materials, e.g metals, which could be reused. Typically everything is mixed and the basic task is to find and separate the reusable materials from the other waste matter. The robot society could include specially designed dumping area diggers which have sensors to detect dangerous and valuable materials, can collect them and/or raise alarms. The energy carriers serve energy to the society as well as acting as work leaders who are also connected to the operator station. The society is able to work continuously 24 hours per day and can produce a remarkable power output in spite of the slow operational speed and low power output of individual machines. The work does not require high accuracy and is parallel in its nature. Technical solutions to the problems of mechanics, communication and control systems are not easy, but could be achieved.

16.5.4 Swarming of mobile robots

A special form of a society is a swarm. In a swarm the members are moving and doing work in very close contact to each other so that the whole group can be considered as one large object. Nature uses swarms mainly for two purposes: to move a group safely from one place to another and to search for food effectively in a restricted area. The motion of a swarm can be easily controlled by just one leader. Swarming of autonomous mobile vehicles is an interesting application which may possibly be applied in harvesting or cleaning.

16.5.5 Monitoring and cleaning inside of processes

The next consideration is a future scenario which uses technology not available today, but might become available when micro-machinery technology takes further steps in the near future. Along the same idea some medical applications have been suggested in the large national programme on micro-machine technology launched in Japan two years ago.

In the process industry, e.g. paper industry, the use of natural water is becoming closely regulated for environmental reasons. Plants have to close their water circuits, which causes problems in maintaining the water quality inside the process. Problems caused include contamination coming from unfavourable bacterial growth. Cleaning can be done by chemicals, but because of the vast amounts of water it is very expensive and ineffective. What is needed is a kind of 'precise weapon' which destroys the bacterial flocks (indeed the bacterial or algae growth often happens only in some places) where they are met. A robot society which could do this job could include small 'bacterium robots' moving along the process streams as illustrated in Figure 16.14. They are equipped with biosensors to detect biological growth and two types of actuators. With the magnetic actuators they can fix to the process walls or to other individuals to form a more powerful flock, and with the enzyme actuator they can destroy the contaminates. The communication and control systems and the energy supplying system are nontrivial problems, but might be solved.

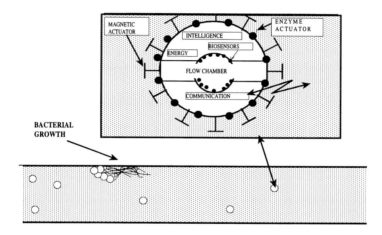

Figure 16.14 *The bacterium robot*

16.6   Conclusion and future research

A society formed by robots is a new concept which gives us a possibility of controlling the work of a large number of mobile robots so that a common goal for the work can be realised. The control structure which utilises the member-to-member communication is fully distributed and allows changes in the number of active members without any reconfiguration of the system. A noticeable difference between the society concept presented here and many multi-robot systems presented in the literature is that the central controller does not communicate directly with the members (as it does in most multi-robot concepts), but the commands controlling the work of the society are injected into the information system of the society through its members.

The behaviour of a robot society is stochastic in nature and this has to be taken into account when applications are sought and when considering the problem of how to program and control a society to execute the work defined.

This is one of the most important problem areas currently being studied. Many approaches, from classical state theory to utilisation of neural nets, are possible [13], but little is known at the moment. This area forms a very fruitful topic for future research.

The information and control structure of a robot society, even a simple one, can make the whole system difficult to analyse by analytical methods. As illustrated, a simulator is an efficient tool for analysing its behaviour experimentally. Physical demonstrators are, however, also important because some features in the society might be very dependent on the technical solutions used.

16.7   References

1.   MATARIC, M.M., Minimizing Complexity in Controlling a Mobile Robot Population, Proc. of the 1992 IEEE International Conference on Robotics and Automation, Nice, France, pp 830-835, 1992.
2.   NORELIS, F.R., Toward a Robot Architecture Integrating Cooperation between Mobile Robots: Application to Indoor Environment, The International Journal of Robotics Research, Vol. 12, No 1, 1993.
3.   UEYAMA, T.,and FUKUDA, T., Structural Organization of Cellular Robot Based on Genetic Information, Prints from the ECAL'93, Brussels, Belgium, pp 1060 - 1069, 1993.
4.   GOSS, S., DENEUBOURG, J.L, BECKERS, R., and HENROTTE, J.L., Recipes for Collective Movement, Prints from the ECAL'93, Brussels, Belgium, pp 400 - 410, 1993.
5.   SUDD, J.H., and FRANKS, N.R., The Behavioural Ecology of Ants, Chapman and Hall Newyork, 206p, 1987.
6.   SEELAY, T.D., Honeybee Ecology A Study of Adaption in Social Life, Princeton University Press New Jersey. 192p, 1985.
7.   MATARIC, M. and MARJANOVIC, M.J., Synthesizing Complex Behaviours by Composing Simple Primitives, Prints from the ECAL'93, Brussels, Belgium, pp.698-707, 1993.
8.   HARA, F. and ICHIKAWA, S., Effects of Population Size in Multi-Robots Cooperation Behaviours, Proceedings of International Symposium on Distributed Autonomous Robotic Systems, Wako, Saitama, Japan, pp.3-9, 1992.
9.   VAINIO, M., Design of an autonomous small-scale mobile robot, HUTMAN, for robot society studies. Master's thesis, Helsinki University of Technology, p83, 1993.
10.   BARSHAN, B. and DURRANT-WHYTE, H., An inertial navigation system for a mobile robot, Preprints of the 1st IFAC International Workshop in Intelligent Autonomous Vehicles, Southampton, UK, pp.54 - 60, 1993.
11.   Hu, H., Probert, P.J., Rao, B.S.Y., A Transputer Architecture for Sensor-Based Autonomous Mobile Robots, IEEE/RSJ International Workshop on Intelligent Robots and Systems '89, Sep. 4-6,Tsukuba, pp.297- 303, 1989.
12.   Murata, S. and Hirose, T., Onboard Locating System of Autonomous Vehicle, IEEE/RSJ International Workshop on Intelligent Robots and Systems '89, Sep. 4-6,Tsukuba, pp.228-234, 1989.
13.   NAGATA,S., SEKIGUCHI,M., and ASAKAWA, K., Mobile Robot Control by a Structured Hierarchial Neural Network, IEEE Control Systems Magazine, pp.69- 76, 1990.
14.   Schönberg, T., Simulator Studies of the Robot Society Concept, Licentiate thesis, Helsinki University of Technology., p.65, 1993.

Chapter 17

Miniature and microrobotics: technologies and applications

P. Dario, R. Valleggi, M.C. Carrozza,
M.C. Montesi and M. Cocco

17.1 Introduction

There is a growing interest worldwide in the concept and possible applications of miniature, micro and even nano scale devices, including robots [1] [2] [3]. This interest is motivated by the widespread perception that advances in microfabrication technology can lead, together with the progresses of robotics research, to the development of autonomous or tele-operated machines capable of carrying out useful tasks at a miniature, micro and perhaps nano scale, thus substantially extending the range of application of current machines and possibly leading to a technological breakthrough whose effects could be comparable with those determined by the micro-electronic revolution [4] [5].

As is often the case for new areas of application, and despite the increasing popularity of the field, the very same notions of 'micromachine' and of 'microrobot' have not yet been defined clearly. In fact, such terms as 'microsystems', 'microelectromechanical systems', 'micromechatronics', 'micromechanisms', 'micromachines' and 'microrobots' are used almost like synonyms to indicate a wide range of devices whose functioning is related to the 'fuzzy' concept of operation at 'small' scale. The first serious attempt to provide some guidelines for research in this area has been made in Japan: during the preparatory phase for the 'Large Scale Project on Micromachine Technology', the Japanese MITI has promoted the activity of panels of experts, who were asked to provide definitions, identify application domains, and highlight theoretical knowledge and technologies which are necessary to implement real devices. The results of this analysis are presented in papers and reports which represent perhaps the most accurate and comprehensive discussion so far available for approaching the field of micromachines and microrobots [6] [7].

More recently, some of the authors of this chapter have presented a critical analysis of the field of microrobotics, aimed at providing information primarily to researchers in the field of robotics, who are generally unfamiliar with the microfabrication technologies which are required to implement real microrobots. Some of the considerations presented in those papers [8] are summarized in the following. Whereas the term 'microsystems' is appropriate to define the whole range of 'microdevices', an important conceptual distinction should be made between the terms 'micromachine' and 'microrobot'. The definitions proposed for the two terms are the following: a micromachine is a device capable of generating or modulating mechanical work, without necessarily possessing any on-board control. Thus, a micromotor, a microvalve or even a fluidic microamplifier can be regarded as micromachines. A micromachine can also comprise a number of different micromotors and micromechanical components to produce complex motions. Basically, however, a micromachine is a 'passive' device and it is usable only as a component of a more complex system.

The definition of 'microrobot' refers to the features attributed to a robot in the macro world, that is, some form of programmable behaviour (as in the industrial robot used in factory automation), or some degree of 'adaptivity to unpredictable circumstances' (as in the class of so called 'advanced' robots for unstructured environments), or remote controllability (as in tele-operated robots). This definition comprises the features that T. Hayashi has attributed to the devices he named 'micro-mechanisms' [9]. Conceptually, the only difference between a microrobot and an ordinary (macro)robot resides in the scale of the application domain. This change of scale, however, has a number of dramatic implications on the operation of the microrobot, whose consequences can be clarified by an additional sub-division of the microrobot into three distinct classes [10][11]: the "miniature" robot, the "microrobot" and the nanorobot. This subdivision is proposed to help identifying more precisely the practical capabilities of current technology (in particular micromachining) in the field of micro-operation; on the other hand it helps to identify the intrinsic engineering limitations of present technologies in the micro world.

17.2 The 'miniature' robot

This class of robots has a size of the order of a few cubic centimetres. The miniature robot operates in a workspace and generates forces comparable to those applicable by human operators during fine manipulation. A miniature robot can be fabricated by assembling conventional miniature components, including micromachines. A miniature robot can be either tele-operated, or can incorporate some degree of intelligence in order to operate autonomously. There are a few examples of real devices which can be considered as belonging to the class of the miniature robot.

The first example is the 'gnat' robot developed by Flynn *et al.,* at MIT, AI Lab [11], which corresponds almost perfectly to the proposed notion of 'miniature' robot. The really significant aspect characterising this kind of miniature robot is the idea of scaling down an entire robotic system to a small scale, still saving the ability to carry out useful 'human scale' tasks (such as picking up small objects, carrying loads, exploring and analysing new territories, etc).

A second representative type of miniature robots are those navigating inside pipes for inspection and repairing such as the 'midget mobile robot' developed by Aoshima and Yabuta [12], the 'giant magnetostrictive alloy (GMA)-actuated' mobile robot prosed by Fukuda *et al,* [13], and the 'MEDIWORM' described by Ikuta [14].

17.3 The 'microrobot'

This class of robots has a size of the order of a few cubic micrometres. Therefore, the structure of a microrobot can be conceived as a sort of 'modified chip' fabricated by means of silicon micromachining techniques, and containing micromotors, sensors and processing circuitry. The workspace of the microrobot is visible only under a microscope. At this scale of dimensions a few intriguing modifications occur in the behaviour of ordinary mechanical machines: in fact, the various forces which are important for the functioning of mechanical actuators, such as surface tension, pressure related forces, electrostatic and magnetic forces, change their hierarchical order [15]. The designer of a microrobot must take into account these modifications and scale his or her preconceptions of the macro world into the micro domain.

A class of applications in which a microrobot could be particularly useful is the manipulation and the assembly of nanomachines and nanorobots. As will be explained in the next paragraph, nanomachines and nanorobots have sizes comparable to those of biological cells. A microrobot could also manipulate cells (scale of a few hundred nanometres) because an appropriate scale factor is retained between the manipulator and the object to be manipulated. Although microgrippers suitable for manipulating cells have been already fabricated [16], non-contact manipulation techniques, such as those based on electric fields [17], can also be exploited for handling such tiny objects.

17.4 The 'nanorobot'

This class of robots is characterized by operating at a scale similar to the biological cell, that is in the order of a few hundred nanometres. In contrast to the microrobot, the 'nanorobot' would not be capable, in general, of manipulating biological cells using a 'conventional' mechanical approach. Rather, a nanorobot could interact with cells or similar objects using electrochemical means, just like its biological counterparts.

At the scale of integration of a nanorobot, the efficiency and simplicity of biological organisms makes them ideal models to imitate. Cells are an excellent example of a nanorobot; in fact they possess a sort of mechanical structure which confines the internal components, energy processing capabilities, and the ability to carry out tasks by interacting with other cells through biochemical exchanges of signals and matter [18].

Although some efforts are being made towards the development of 'nanofabrication' technologies and tools for obtaining nanostructures (including gears, bearings, harmonic drives) by handling individual molecules and even atoms [19], solid state technology is not suitable, at present, for nanoscale fabrication. Traditional and non-traditional (for example Langmiur-Blodgett monomolecular layer deposition) polymer chemistry techniques can provide a solution to the problem of assembling materials and structures having the functions required by a nanomachine [20]. The recently established field of 'intelligent materials' [21] describes very well the blend of material technology, and embedded sensing, actuation and control capabilities which are necessary to develop nanorobots.

If a real nanorobot could be fabricated, its range of application would be similar to that of micro-organisms; conceptually, there would be only minor differences between a nanorobot and a modified (for example by genetic engineering processing) bio-organism. Both for a nanorobot and for a living micro-organism, the possibility of executing useful tasks requires that their activity is confined in a limited space.

17.5   Possible configurations and applications of microrobots

According to the proposed definition, a microrobot (in the following, unless differently specified, we shall use the term 'microrobot' to indicate collectively miniature, micro and nanorobots) will consist of the following main components: a physical structure, a controller, a power source, sensors and actuators. These components can be combined in different ways to design different types of microrobots. Microrobots can be either fixed or mobile: in both cases, a useful microrobot should possess the ability to 'manipulate' objects.

A number of possible arrangements of the components of a microrobot have been proposed by T. Hayashi [9], with reference to the mobility of the microrobot. This classification emphasises the critical importance of the following components of a microrobot: the control, the power source, the actuators necessary for positioning the microrobot, and the actuators necessary for operation (i.e. manipulation and locomotion). A scheme of this classification and related microrobot configurations is depicted in Figure 17.1.

Referring to the classification proposed above, a range of possible different applications has been identified for the microrobot. Such applications are reported in Table 17.1.

If we consider in some detail the robots sketched in Figure 17.1, we can observe that the microrobots illustrated on the left side of the figure are those potentially easiest to fabricate and to control, and perhaps most widely applicable. In fact, microrobots of types (a), (b) and (c) are basically tele-operated, and this largely facilitates the connections between the energy supply and the controller, on one end, and the operating part of the robot, on the other. In the case of robot (a), all the elements of the robot are physically connected. Just like in an ordinary industrial robot, only the 'actuators for operation' and the gripper have to be miniature, whereas the remaining structure has not. An example of this type of (micro)robot is the scanning tunnelling microscope (STM), which in fact could be regarded as a real micromachine [22]. As mentioned previously, additional feasible applications of this type of robot could be in the assembly of micromechanisms; in this framework the microrobot could be a key component of a future 'microfactory for automated microassembly'. Type (b) robots are also quite feasible and are particularly suitable for the manipulation of biological objects. The connection between the driving source and the micromanipulator could be made by flexible hydraulic, pneumatic or electric lines. Robot type (c) is also connected by cable to the control unit and to the power source, but it is mobile. This feature adds considerable operational power to the robot, but also poses a number of challenging technical problems to its practical realisation.

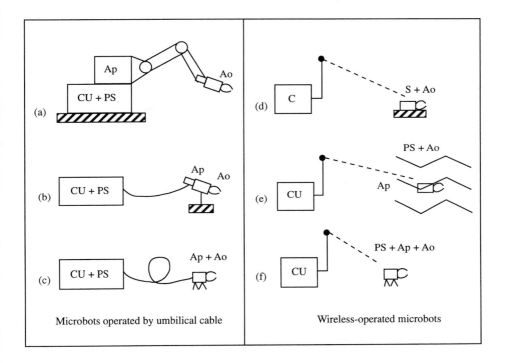

Figure 17.1 *Classification of different microrobots (modified from [9]). CU indicates the control unit, PS the power source, Ap the actuators for positioning, Ao the actuators for operation*

The microrobots illustrated on the right side of Figure 17.1 have the characteristic of being controlled remotely, by radio or other types of signals. The microrobots type (e) and type (f) are the most difficult to fabricate and to control. Particularly important for both is the miniaturisation of the energy source and the ability to generate forces large enough to navigate or walk, and to manipulate objects. Actually, these two types of machines are the configuration closest to the ideal model of microrobot, especially in terms of degree of autonomy.

Functions	Features
1. Maintenance of equipment	Operation inside slender pipes
2. Biological and medical use	Operation inside an organ (closing, inspection, operation)
3. Geological/natural resource sand investigation	Operation in the space between particles
4. Microshepherd	Control of bacteria clusters
5. Microplotter	Writing on paper
6. Investigation in soft and/or dangerous locations	Cliff, buildings, body, tissues
7. Watching, journalistic coverage	Imperceptible/invisible
8. Ocean/space investigation	Inexpensive
9. Working in wall and ceiling	Free from gravitation
10. Flying in the air	Extermination of pest, pollen mediation
11. High performance/intelligent surface	Load distribution, correction of shaping errors
12. Magic carpet (soft)	Vehicle, conveyor
13. Touch T.V.	For the blind
14. Micromanipulator for biology	Work in microterritory
15. Micromanipulator for industries	Operating hand

Table 17.1 *Possible applications of fixed and mobile microrobots (modified from [9])*

It is clear from the considerations presented above that designing and fabricating a real microrobot is a very difficult task, which not only requires substantial progresses in many areas of microfabrication technology and robotics research, but also the ability to integrate these factors into robotic 'microsystems'. This renders the design of microrobots a task even more challenging than the already difficult task of designing 'macro' robots, since in the case of micromachines even basic technologies and components (micromechanical structures, microactuators, controllers, energy sources, microtele-operation, (some) microsensors) are not available on the market, but are research topics on their own. The following research issues can be mentioned to demonstrate current progress in some of the above areas.

The problem of fabricating truly 3-D mechanical structures, which are necessary for robotic machines capable of operating in general circumstances, has been addressed in a very elegant manner at the University of Tokyo [23], and at the University of Berkeley [24]. A real breakthrough in their approach is the idea of obtaining complex 3-D structures by 'folding' simpler planar structures, fabricated by using conventional surface micromachining techniques, just like in the traditional Japanese 'origami' technique. Different approaches, also aimed at obtaining deep micromachined structures are being pursued by other groups, using techniques like Liga [25], or even a combination of classical precision machining, surface micromachining and microelectrodischarge [26].

Controlling microrobots is also a very difficult problem, especially if teleoperation is not used. The control problem for a microrobot cannot be considered without referring to the specific operating environment. Let us consider, for example, the case of the 'real' microrobot as defined previously, that is with an autonomous workspace of a few micrometres: within this workspace the microrobot should exert forces and produce movements. To extend the workspace to the size required to carry out different useful tasks, a microrobot, much more than an ordinary macro robot, must possess mobility. Therefore, the main problem for the microrobot is the ability to navigate in different media.

Unfortunately, navigating in the micro world requires of the microrobot the ability to overcome viscous forces, which are dominant at that scale in comparison with kinetic terms. In fact, the Reynolds number becomes extremely small for micropipes: for example, for the case of blood as the fluid medium, the Reynolds number is between 1 and 10^{-5} for diameters ranging between 1 mm and 1 micron, respectively [18]. Currently available technology does not provide actuators and energy sources capable of generating forces sufficient to overcome viscous effects, for a reasonable time, within the size limits of a microrobot. Given this, a microrobot will not be able to navigate autonomously through a fluid (including air). At present, the only possibility for a single microrobot to reach the desired target is to be confined in a pipe and either to be dragged passively by a fluid flowing in the pipe, or to navigate by receiving energy supply from outside. The 'free floating' behaviour is typical of some biological systems, in which the achievement of a goal is based on redundancy and probability. Seed dissemination and egg insemination are examples of processes when a successful 'docking' is not due to precise navigation, but rather to the law of large numbers. The second approach, based on external energy supply has been explored, at least for the case of miniature robots, by Fukuda *et al.* [13], and by Aoshima *et al.* [12].

When silicon micromachines were proposed, one of their elective areas of application was immediately recognised as the operation in the bloodstream to disgregate, like phagocites, cancer cells or lipids deposited on vein or artery walls. Based on the considerations presented above, we believe that these applications are not feasible for an autonomous microrobot as we have defined it, very difficult for single tele-operated robots, and also very difficult (but perhaps possible) for 'swarms' of co-ordinated microunits. The idea of using a large number of simple and cheap robotic units, which co-ordinate their efforts like a swarm of insects is particularly attractive for the control of microrobots and consistent with the technological constraints posed by small size. This approach, known as 'cellular' robotics, emphasizes simple 'instinctive' responses based on sensory feedback, and co-operation between multiple robot 'agents' [27][28][29], perhaps with no centralised control, rather than the quest for 'intelligent' autonomous behaviour traditionally pursued in 'macro' robotics.

An additional key problem for microrobotics is the availability of suitable sources of energy. Unless connected to a remote power supply by an umbilical cable (Figure 17.1, cases a-c), or powered through radiated electromagnetic or thermal energy (Figure

17.1, cases d-f), an autonomous microrobot should be equipped with some sort of miniaturised battery. Some research efforts are being devoted to address this important topic, but suitable batteries or other types of efficient and miniature energy source are not available yet for practical application, especially for micro and nanorobots.

Actuators are also recognised as critical components of microrobots since the ordinary electromagnetic motors exhibit very serious limitations when miniaturised [6, 30]. For example, a major problem for miniature high speed electric motors used as actuators for microrobots is the need for reduction gears; in fact useful torques can be obtained from those motors using only reduction gears, which are extremely difficult to fabricate in the required size. Additional concerns are the coupling problem of micro reduction gears, and their reliability. A solution to this problem is to use direct drive actuation. Since very small electric 'torque' motors are not available and perhaps even impossible to fabricate, alternative solutions must be devised. A number of innovative types of motors have been proposed and are being investigated for applications in the fields of micromachines and microrobots. These actuators rely on the use of different types of transducing materials and of energy conversion principles.

In conclusion, the field of microrobotics could become in the near future an excellent opportunity for multidisciplinary research and a good chance for demonstrating the potential impact of microtechnologies in the wide range of applications that the intrinsic flexibility of the very same concept of 'robot' makes possible to consider.

17.6 References

1.   MITI Large Scale Project on 'Micromachine Technology', started in 1991.2.NETWORK OF EXCELLENCE in Multifunctional Microsystems 'NEXUS', in Esprit II European Community Research Programme, 1991.

3.   EMERGING TECHNOLOGY INITIATION, Microelectromechanical Engineering, National Science Foundation (U.S.A., 1988).

4.   FEYNMAN, R.P. 'There's plenty of Room at the Bottom', Journal of Microelectromechanical Systems, Vol 1, No.1, pp 60-66, March 1992.

5.   JARA-ALMONTE, J., and CLARK, H. 'Emerging Technologies and Microelectromechanical Systems (MEMS): a White Paper Commissioned by the Panel on Emerging Technologies', Microstructures, Sensors, and Actuators, DSC-Vol. 19, pp 1-10, ASME, New York, 1990.

6.   'MICROMACHINES IN JAPAN', Techno Japan, Vol. 24, No. 4, pp 8-26, April 1991.7.'MICRO-MACHINE', Journal of Robotics and Mechatronics, Vol 3, No. 1, pp 1-64, 1991.

8.   DARIO, P. and VALLEGGI, R. 'Microrobotics; Shifting Robotics Technology Towards a Different Scale World', The Robotics Review 2, MIT Press, 1992.

9.   HAYASHI, T. 'Micro Mechanisms', Journal of Robotics and Mechatronics, Vol.3, No.1, pp 2-7, 1991.

10.   IKUTA, K. 'Micromachine - Its Current State and Future', Micromachine System Vol.3, No. 1, pp 60-64, 1991.

11.   FLYNN, A.M., BROOKS, R.A., WELLS III, W.M. and BARRETT, D.S. 'Intelligence for Miniature Robots', Sensors and Actuators, Vol. 20, pp 187-196, 1989.

12.   AOSHIMA, S. and YABUTA, T. 'A Miniature Robot Using Piezo Elements', IEEE International Conference on Robotics and Automation Piezoelectric Smart Systems, Sacramento, California 1991.

13   FUKUDA, T., HOSOKAI, OHYAMA, H., HASHIMOTO, H. and ARAI, F. 'Giant Magnetostrictive Alloy (GMA) Applications to Micro Mobile Robot as a Micro Actuator without Power Supply Cables', Proceedings of IEEE Micro Electro Mechanical Systems, pp 210-215, Nara, Japan, Jan. 30-Feb. 2, 1991.

14. IKUTA, K. 'The Application of Micro/Miniature Mechatronics to Medical Robotics', Proceedings of 1988 IROS, Tokyo, Japan, pp 9-14, November 1988.

15. TRIMMER, W. and JEBENS, R. 'Actuators for Micro Robots', 1989 IEEE International Conference on Robotics and Automation, Scottsdale, AZ, pp 1547-1552, May 14-19,1989.

16. KIM.C.J., PISANO, A.P., MULLER, R.S. and LIM, M.G. 'Design Fabrication and Testing of a Polysilicon Microgripper', Microstructures, Sensors, and Actuators, DSC-Vol, 19, pp 99-109, ASME, New York, 1990.

17. WASHIZU, M. 'Electrostatic Manipulation of Biological Objects in Microfabricated Structures', in 'Integrated Micromotion Systems', F. HARASHIMA (ed), Elsevier, Amsterdam, pp 417-431, 1990.

18. FUJIMAS, A.I. 'Medical Applications of Micromachine Technology', Technical Digest of the Sensor Symposium, pp 135-139, 1989.

19. DREXLER, K.E. 'Nanomachinery: Atomically Precise Gears and Bearings', Proceedings of IEEE Microrobots and Teleoperators Workshop, Hyannis, Massachusetts, November 9-11,1987.

20. TAKASHI, K. 'Sensor Material for the Future: Intelligent Materials', Sensor and Actuators, Vol. 13, pp 3-10, 1988.

21. Proceedings of International Workshop on Intelligent Materials, Tsukuba, Japan, March 5-17, 1989.

22. KAJIMURA, K. 'STM as a Micromachine', Journal of Robotics and Mechatronics, Vol. 13, No. 1, pp 12-17, 1991,

23. SUZUKI.K., SHIMOYAMA, L., MIURA, H. and EZURA, Y. 'Creation of an Insect-based Microrobot with an External Skeleton and Elastic Joints', Proceedings of IEEE Micro Electro Mechanical Systems, pp 190-195, Travemunde, Germany, Feb. 4-7, 1992.

24. FEARING, R.S. Tutorial M4 on High-Precision Sensors/Actuators and Systems, IEEE International Conference on Robotics and Automation, Nice, France, in 1992.

25. GUCKEL, H., SKROBIS, K.J., CHRISTENSON, T.R., KLEIN, J., HAN, S., CHOL. B. and LOVELI, E.G. 'Fabrication of Assembled Micromechanicai Components Via Deep X-Ray Lithography', Proceedings of IEEE Micro Electro Mechanical Systems, pp 74-79, Nara, Japan, 1991.

26. HISANAGA, M., KURAHASHI, T., KODERA, T. and HATTORI, T. 'Fabrication of 4.8 Millimeters Long Microcar' Proceedings of Second International Symposium on Microemotional Machines and Human Sciences, Nagoya, Japan, October 8-9, 1991.

27. BROOKS, R. 'Micro-Brains for Micro-Brawn; Autonomous Microrobots', Proceedings of IEEE Micro Robots and Teleoperators Workshop, Hyannis, Massachusetts, November 9-11, 1987.

28. FUKUDA, T., NAKAGAWA, S., KAWAUCHI, Y. and BUSS, M. 'Self Organizing Robots Based on Cell Structures - CEBOT, Proceedings of IROS, TOKYO, Japan, pp 145-150, November 1988.

29. DARIO, P., GENOVESE, V., RIBECHINI, F. and SANDINI, G. 'Instinctive Cellular Robots', Proceedings of '91 lCAR, pp 551-555, Pisa, Italy, June 19-22, 1991.

30. TAKAMORI, T. 'Recent Trends in the Development of new Actuators', Journal of Robotics and Mechatronics, Vol 3, No.1, pp 18-27, 1991.

Chapter 18

Characteristics of robot behaviour

A. Lush, J. Rowland and M. Wilson

In this chapter we review some of the recent and more significant contributions to behaviour-based robotics. In doing so, we attempt to identify common characteristics of the various implementations and examine the relationship between potentially conflicting approaches. While most workers have concentrated on the mobile robot domain, our own interest is in exploiting the potential benefits of the behaviour-based approach in robotic handling and assembly applications in appropriate industrial sectors.

18.1 Introduction

Following the pioneering work of Brooks [1,2], behaviour-based robotics has been pursued widely and approaches have diversified, while still retaining the 'behavioural' or 'behaviour-based' label. In this chapter, we examine various approaches and attempt to identify the essential characteristics of behaviour-based robotics. Our aim is to provide a sound basis for the design of new behaviour-based systems that exploit the essential and most beneficial aspects of the concept. The main potential advantages, as compared with traditional approaches to the design of robot architectures, of exploiting the behaviour-based approach in this way are the following (although these are not achieved in all examples of behavioural work):

- improved robustness and reliability,
- easier configuration and reconfiguration,
- easier reuse of designs and individual components of designs.

We explore, in particular, some of the conflicting views as to the relative merits of centralised and decentralised approaches to the design and organisation of behaviour-based architectures. Our specific aim is to combine the robustness of the behaviour-based approach with the predictability and flexibility required in industrial handling and assembly tasks.

18.2   Origins of behaviour-based robotics

Work on robotic behaviours arose from the inadequacies of traditional robot and AI programming, in which the robot requires detailed knowledge of, or a model of, its environment in order to reason about its actions. A premise of the behaviour-based approach is that creatures (e.g. insects) with little or no cognitive ability can survive without reference to an abstract model; their survival cannot be attributed to reasoning. At the same time, any system that is to survive in the real world must, by virtue of the unconstrained nature of the real world, be reactive.

The behaviour-based approach to programming robots attempts to synthesise the overall task from reliable task-achieving modules rather than decompose it into separate information processing functions. It moves away from collecting and processing data that form a single representation of the world in which the robot operates, to integrating sensor information at the lowest possible level to allow more direct, responsive, behaviour. This allows real time control within the individual modules, which can therefore more readily respond to internal priorities and to external world constraints [1-3].

Brooks developed a computational model which he termed the subsumption architecture to investigate the design and implementation of control systems for mobile robots with insect level capabilities, in the belief that insects display the reliability and robustness which is required in mobile robots. Insects are well adapted to the environments in which they live and they manage to act and survive despite the ever changing world around them (see [1] for the insect metaphor). The subsumption architecture consists of layers of task achieving behaviours which together provide the competence of the system. Each behaviour must achieve, or contribute to the achievement of some task; compound behaviours, formed by the combined effect of simpler behaviours, can achieve more complicated goals.

Brooks's subsumption approach provides a complete control architecture for mobile robots that can achieve tasks in the real world. Initially, a control system is built to perform the lowest level task, and is extensively tested. This zero level is now never altered and any additional competences to perform new tasks are added by building other layers on top of it. The new layers can examine data from lower layers and can supply data to them by suppressing the normal data flow within the layers. Each layer is composed of a set of processors, each an augmented finite state machine (AFSM), with no shared global memory. Input signals can be suppressed by signals from the upper layers and output signals can be inhibited. See [4] for details of the hardware for such a system. Brooks has demonstrated the subsumption architecture using real robots working in real environments. Flynn [5] looks back on the early robots developed at MIT and examines their development from January 1985; she examines the development and ideas behind five working robots designed to operate autonomously and robustly. A further example [7] is a soda-can collecting robot designed under the subsumption architecture.

Thus far, behavioural robotics produced robot control mechanisms based on a biological analogy with insect level 'intelligence', that accomplish tasks in a set way, much as an insect does. The task is specified by the designer, who selects and implements the individual behaviours that make up the overall task; variation of overall behaviour is restricted by these initial design decisions, with little adaptive capability beyond that explicitly provided by the initial selection of behaviours, and the availability of alternatives to cope with a restricted variation in the environment.

18.2.1 Behaviour-based automated robotic assembly

Work at Edinburgh [7,8,9] on automated assembly applied a behavioural approach to a more complex, constrained, and less obviously biologically related task, in order to achieve more flexible and reliable systems. In this work, task planning and task achieving were separated. An off-line system planned the task in terms of part motions; a run-time behaviour-based system executed these motions.

Edinburgh's SOMASS system [10] provides a complete assembly system for building Soma shapes (chamfered cubes glued together to form irregular shapes) from the planning stage through to assembly of a specified configuration, by an Adept robot. It is divided into two parts: the planning system and the execution system. The planner does no explicit reasoning about the variation or uncertainty present in the real world; it plans the assembly task in terms of part motions which are realised by behavioural modules engineered to cope with a certain amount of variation in the workcell. The planner uses some implicit assembly rules which help to increase the reliability of the system.

The original system used no sensing, relying instead on variation-reducing strategies to achieve a reliable system. Work has been done to introduce sensing into the system. Some of the behavioural modules have been changed to behaviour-based vision modules [11,12]. Other work, described below, has investigated an architecture to link simple sensory behavioural modules to allow decision making and choice of future actions [9,13].

In the extension to the SOMASS system which includes changing the run-time architecture and introducing new, simple, sensing, the behavioural modules are regarded as modular units which combine hardware and software to achieve a competence in performing an action in the real world. Each behavioural module can be composed of any combination of other behavioural modules in a hierarchical structure (see Figure 18.1). Thus, behavioural modules can be designed by implementing basic low level competences and using these as building blocks to construct higher level behavioural modules with correspondingly higher competences. For example, a height-determining behavioural module can be composed of a series of guarded motion behavioural modules.

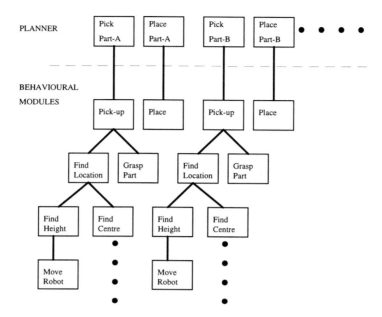

Figure 18.1 *A hierarchical decomposition of planner and behavioural modules*

Emphasis is placed on using information directly from the real world as much as possible, leading to a tight linking of the software, hardware, and the task. As far as is possible, the variation and uncertainty present in the task for which the behavioural module is designed is dealt with inside the module by using variation-reducing strategies devised by the designer and by correctly adapting sensors to the task.

18.2.1.1 Autonomy of behaviours

Since the planner provides only part motions, and does not dictate how they will be performed (unlike more conventional systems where each robot motion is sent down from the planner), the execution system has complete freedom to interpret these part motions according to what behavioural modules are available. This type of system can provide the flexibility to change old strategies with new and improved ones without changing the interface to the planner. The behavioural modules can be designed with minimal complexity at the individual level without limiting the achievable overall complexity of the system. The structure also allows behavioural modules to be slotted into the execution hierarchy to provide alternative competences if these are required.

18.2.1.2 Flexible architecture

Each behavioural module is designed to allow the use of sensors to closely monitor its performance while performing its task. There is no concept of success or failure within a behavioural module, but instead there are a number of exit states which correspond to the control paths available in the behavioural module. Figure 18.2 shows a simple example where a behavioural module BM1 can return an exit state of A or B. A may correspond to there being a part in the gripper fingers, and B may correspond to nothing being present in the gripper.

Figure 18.2 *A behavioural module with two exit states*

Exit states, passed up through the hierarchy, can be used by the calling, higher level, behavioural module to determine whether that low level behavioural module has successfully completed its assigned task. Thus the exit states for a particular behavioural module are determined from local data, from sensor data, and from exit states returned from lower level behavioural modules. They prove useful in monitoring the internal operations of a behavioural module and thus in detecting error conditions, both current and impending, and hence in invoking alternative strategies. Since the modules can be reusable in different situations, the context in which they are called plays an important part in determining whether the correct exit state has been returned for one particular call. In Figure 18.3, in one situation, BM1 may be called and, if there is a part in the gripper (i.e. an exit state A), then BM2 is called. In another situation, it may be that if the gripper is empty (i.e. an exit state B), then BM2 is called. Both are perfectly valid options.

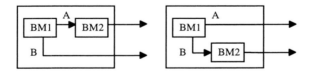

Figure 18.3 *Reuse of behavioural modules in different situations*

The ideal execution path and the alternative execution paths are defined control paths through the hierarchy of the behavioural modules. These paths can be followed by using the exit states of behavioural modules plus local control decisions. The ideal execution path describes the best route, according to the system designer, through the behavioural modules performing the assembly. Even though the behavioural modules are designed and tested to be reliable, it can be demonstrated that they are not always perfect at performing the assembly task, and therefore, alternative behavioural modules, those on an alternative execution path, can be included to restore the system to the ideal execution path without recourse to the planning system. Alternative execution paths can be used when behavioural modules on the ideal execution path fail to correctly perform their task due to the actual workcell conditions being outwith the competence of the modules. Figure 18.4 shows how, with the ideal execution path being defined as through BM1 with exit state A to BM2 with exit state E, if, due to the variation in the real workcell, BM1 produces the exit state B, BM3 can be entered to try to take the route back to the ideal execution path. This will be achieved if the exit state from BM3 is C.

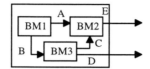

Figure 18.4 *Ideal and alternative execution paths*

The extent to which the system follows the ideal execution path can be taken as a measure of the suitability of the chosen behavioural modules to perform a task. Extensive deviation to alternative execution paths indicates that the behavioural modules on the ideal execution path are not the best suited for the task, whatever the criteria for their original choice; it may also indicate, for example, that a feeder has moved out of position and needs resetting, yet the system may still perform correctly by virtue of the invocation of an alternative execution path. Recovery is possible at different levels of the system, with no outright failure occurring until the top of the hierarchy is reached.

18.3 Adaptation and behaviour

There are two main schools of thought as to how adaptation is best achieved:

•   high level functions such as object recognition, path planning etc. are functionally separate, but are interfaced to the (lower level) behaviours;
•   higher level functionality will emerge from increased behavioural complexity.

These two may conveniently be referred to as *centralised* and *decentralised* approaches respectively. However, the use of the term *centralised* simply reflects the existence of some overall coordination or arbitration mechanism that does not necessarily operate at all times and *may* be invoked only prior to execution, for planning purposes, or in particular circumstances during execution, such as an extreme error condition.

18.3.1 Decentralised adaptation

The centralised approach to behaviour selection is countered by Maes [14] in her work on distributed behaviour selection based on internal motivations, and also by Brooks [15] in his work on hormonal type behaviour selection, inspired by Maes. The premise of Maes's work is that biological systems may respond to similar environmental conditions in different ways. Thus while behaviour is linked to the environment it is, at least partially, internally generated. She quotes an example: a hen presented with an egg may incubate it or eat it, depending on whether the hen is broody or hungry. She argues that the bottom-up distributed approach is more robust and flexible, and follows this strategy when implementing behaviour selection within a behavioural architecture.

Maes considers that a biological system is constructed from a multitude of behaviours, only a few of which (and in some cases only one of which) can be active at a time; all the behaviours then compete for control of the system. In her implementation of a simulated creature, behaviour selection emerges via parallel local interactions among behaviours, and between behaviours and the environment. The whole system exists as a control loop that moves towards a desired state, based on response to the environment and on internal conditions. The behaviours have a specific activation trigger condition and an activation level. The activation levels are adjusted in response to environmental conditions; for example, seeing food increases the activation level of the eating behaviour. Internal motivations such as hunger, fear, thirst can also increase the activation level of behaviours such as 'drink', 'feed', 'run away', that will cause these motivations to be reduced. Behaviours can affect the threshold conditions of other behaviours: a 'go towards food' behaviour may increase the activation level of the 'eat' behaviour. All behaviours desire to be active and those that are inactive will increase the activation levels of their predecessor behaviours so as to increase the likelihood that they will be activated next. Conversely, behaviours that are in contention will decrease each others' activation level.

Brooks and Viola [15] combine the ideas of Maes with the work of Kravitz [16] to produce a behaviour selection system based on a model of hormonal control of behaviour. The combined result is implemented on Genghis, a six-legged walking robot about 35cm long and weighing approximately 1 kg. [3], developed using the incremental levels provided by a subsumption architecture. In this model, 'Releasers' are produced which can inhibit or enhance the action of behaviours much in the way that hormones, produced in response to sensory stimuli, affect biological systems. However, in Brooks's work, the hormones, or releasers, are not directly produced by sensory inputs but by a central summation mechanism that outputs releasers in relation to a set of perceived conditions such as fear or trepidity, which correspond to emotions and are invoked by environmental conditions or by the state of specific behaviours.

Brooks suggests [17] that the centralised approach to behavioural selection is falling from favour, and he continues to advocate his distributed bottom-up approach. He describes a three-level decomposition:

Micro: the study of how distinct couplings between the creature and its environment are maintained for some particular task. The primary concerns are: what perception is necessary and what relationships exist between perception, internal state and action.

Macro: the study of how perceptions and behaviours are integrated into a complete individual creature. Primary concerns are: how independent the various perceptions and behaviours can be, how a competent, complete creature can be built in such a way as to accommodate all the required behaviours, and to what extent apparently complex behaviours can emerge from simple behaviours.

Multitude: the study of how individual creatures in a multitude interact as they go about their business. Primary concerns are: the relationships between individuals' behaviours, the amount and type of communication between creatures and the resulting patterns of behaviour, and their impact on the environment.

The last of these relates to the work of Barnes, [18,19] on co-operant agents, which exploits emergent capabilities from a number of simple behaviours, rather than top-level implementation. His emphasis is to follow the insect analogy into co-operant colonies of relatively simple creatures that can interact to produce more complex functionality, in the way that, for example, ants and bees achieve their goals. He uses set levels of priority to give the individuals different motion strategies based on different criteria; these may be self-preservation strategies, strategies based on the environment or upon the species, and there are common universe goals for all agents. This approach uses sensing to allow the agents to react to each other's actions.

In [17,20] Brooks reiterates that it is possible to make complex perceptions without the need for an internal world model. Mataric's work [21] supports this; she describes a completely distributed behavioural system able to make maps of the environment through assigning sensed values and features to elements in a matrix of behaviours, each of which corresponds to a landmark within the environment. Mataric's creature (TOTO) is able to navigate and path plan in a completely distributed way, and to find its way optimally to a stated location in the map. Within the matrix, the active behaviour corresponds to the current location and the next location in the matrix is primed for action as this is likely to be the next to be active.

18.3.2 Centralised, or hybrid, approaches

The work described earlier comes into this category through the provision of different execution paths that are invoked by the behavioural modules themselves, or explicitly by the high level planner. We have, however treated it separately because its target application is robotic assembly rather than mobile robotics, to which our other examples refer.

Kaelbling [22] in describing her own work points out some difficulties with the horizontal decomposition of the subsumption architecture and the tight coupling of sensing with action. The first of these is that special purpose perception mechanisms tend to be weak; raw sensor data are often noisy and open to a variety of interpretations, so that to achieve robust perception it is necessary to exploit redundancy of different sensor systems and integrate the information from many sources. The second is that at higher levels behaviours tend to depend on conditions in the world rather than particular properties of sensor conditions. The architecture that is suggested separates the perception and action processes into two separate modules, in marked contrast to Brooks's original motivation. Within these modules there is a horizontal decomposition and all levels of the action module have access to all levels of the perception module which in turn range from being completely raw sensor data to complex world models. The control of the sensors is done at the action level, on the basis that perception of the environment or the deployment of the sensors depends on the action that is being taken at the time. Later work [23,24] builds on this and demonstrates the resulting capability.

Arkin [25,26] indicates that the key to producing adaptive behaviour is in combining higher level functions with the reactive abilities of the original behavioural work. His approach is implemented in his autonomous robot architecture (AuRA); a high level planner, consisting of pilot, navigator, mission planner and motor schema planner, which can recruit individual motor schemas as required. The motor schemas are primitive behaviours that can cooperate with each other to achieve more complex behaviours, much as in other approaches. However, Arkin avoids layering and subsumption but instead achieves a network of behaviours that is dynamically changed by the high level planner according to the current goals and external environment of the robot. In his more recent work [27] the high level planner uses a model-based qualitative navigation system.

Kweon [28], describes a system based on sonar navigation, whose overall approach is related to that of Arkin. A number of parallel navigation behaviours are implemented, each designed to cope with a specific navigational problem, such as target following, wall following, or dealing with corridor intersections. An overseeing arbitration system selects the most appropriate behaviour for a particular condition based on the internal conditions of the behaviours.

18.3.3 Learning

One of the current problems of the behavioural approach is that each of the behavioural modules must be laboriously designed by hand and, for the behaviour to function acceptably, this means that the designer has to have a detailed knowledge of how that behaviour interacts with the environment. The amount of design effort that this requires means that the process tends to be a long one. Adaptive behaviour goes some of the way to making the design process less demanding, but learning goes one step further. In principle this should permit the use of behaviours with a much broader range of capability and these could, through a process of learning, tend towards an optimum solution. Although the ability to learn would be a desirable characteristic either of individual behaviours or of a behavioural architecture, a discussion of learning issues and potential approaches is beyond the scope of this work.

18.4   Emergent characteristics

The work we have considered, above, reveals a number of distinct characteristics of behaviours, as well as some possibly conflicting indications:

- All cases give rise to a behaviour being characterised as *a competence achieving module* i.e. a robust and reliable mechanism for doing something useful, (although robustness and reliability are not necessarily achievèd in every example of behavioural work). Clearly this is a wide ranging description that includes techniques beyond what would be considered 'behavioural'.

- A behaviour is a *level of abstraction representing a boundary between levels of competence.* This was an early definition given by Brooks that may appear to have been superseded as a result of subsequent work, particularly that of others.   However, if we consider that behavioural modules may be combined so as to provide increasing levels of competence, then such a combination of behaviours constitutes a new level of competence, whose boundary invites the addition of a further competence achieving module. There are various analogies, including the ISO 7-layer model for communications protocols, where each successive layer implements a new level of competence on top of the layer below it, so a given level of capability is defined by the (boundary of) the topmost layer. A different analogy is required for flat networks of behaviours as used by Arkin, for example.

- A behavioural system must react to the real world via sensors; the sensed stimulus, however, must represent a condition in the real world. In other words, *the system must ultimately be grounded in the real world.* The stimulus must come from outside the behavioural module itself although it does not necessarily need to come directly from the real world but may come from the output of another behaviour. In a biological sense, anything that is done that is not a reaction to sensing is a result of reasoning, based on a learned response or a high level model.

- A further characteristic of a behaviour is that *it must help close a loop of control*, although the loop need not be explicitly around one behaviour, and the resultant action need not directly control the cause of the input. The implication of this and the previous point, taken together, is that behavioural modules may be combined both horizontally and vertically, effectively in series and in parallel, where layers may consist of behavioural modules arranged such that the output of one feeds the input of the next, and layers may be stacked such that upper layers may subsume lower layers. The hormonal model, in which outputs of some behaviours help to set conditions that influence the activity of others, can be considered as a development of such a combination of behaviours.

Within this broad framework of potentially robust and reliable reactive modules, strategies for adaptation and learning may be implemented.

18.5 Discussion and conclusions

A typical autonomous creature achieves a task that is designed into it, with no scope for reprogramming. To achieve a different task, a different creature must be designed; it may include many of the same behaviours but they will combine in different ways. Similarly there is no scope for adaptation outside a designed competence level; the *designer* determines the constraints and makes the 'intelligent' decisions about how the robot will respond and act. Nevertheless, behaviours are able to react to real world situations and accomplish simple tasks, often far better than traditional 'intelligent' systems; thus they offer a potentially robust, reliable and reusable solution to many elemental tasks or subtasks.

A supposedly crucial distinction is made in the literature between the centralised and decentralised approaches, with the greatest body of opinion apparently on the decentralised side. However, there are different ways of looking at the issue:

• The so-called decentralised approach requires careful design at the outset and it is difficult to deny that the design and selection of the individual behavioural modules constitutes a process of off-line centralised planning. For example, although Brooks refuted the need for planning, the particular arrangement of the subsumption architecture for a given application could constitute a plan. Thus the selection, or design, of an appropriate configuration of behaviours can be considered to be task planning. The work at Edinburgh used a central planner and thus makes the planning process explicit. Arkin's hierarchical planner dynamically modifies, on the basis of feedback from the behavioural modules, the selection and interconnection of the behavioural modules; offline and dynamic planners effectively coexist in the system. However, where there is a centralised planner, it is nevertheless the behavioural modules, in combination, that provide the task-achieving functionality, albeit subject to the influence of the higher authority in addition to the environmental and other conditions that more immediately influence their actions. Other possibilities provided by a high level central planner include the use of learning strategies that can be decoupled from the real-time constraints of the behavioural modules but can monitor and influence overall task execution over successive cycles of a handling task, for example.

• Some of the adaptive approaches that purport to follow the decentralised philosophy nonetheless conceal a centralised coordination mechanism of some sort, although such mechanisms are very different from the centralised planners discussed above; for example Brooks's 'hormonal' work relies on central summation of stimulus-induced conditions that is then used to influence the behaviour of individual modules in a loosely coordinated way. Such mechanisms may exist even in explicitly centralised systems, in addition to a high level planner, to impart greater flexibility at the lower, behavioural, levels.

Thus this apparent dichotomy of approaches can instead be considered as components in a progression: from autonomous systems that are designed only to attain specific goals, through increasing degrees of competence, to very flexible systems where the accepted benefits of behavioural robotics are retained but whose goals may be specified, and attainment of those goals assisted, by a high level system. At the latter end of this progression lies the basis for further developing behavioural approaches to robotic assembly and handling, with the potential for flexibility and easy reprogramming that are such important targets in manufacturing industry.

18.6 Acknowledgements

Alan Lush holds an SERC (ACME) studentship. The Centre for Intelligent Systems is funded under the Research Quality Initiative of the Higher Education Funding Council for Wales.

18.7 References

1.   BROOKS, R.A.: Achieving artificial intelligence through building robots. AI Memo 899, MIT, Artificial Intelligence Laboratory, 1986.

2.   BROOKS, R.A.: A robust layered control system for a mobile robot. IEEE Journal of Robotics and Automation, 2(1), 14--23, March 1986.

3.   BROOKS, R.A.: A robot that walks: Emergent behaviors from a carefully evolved network. In: Proceedings IEEE Conference on Robotics and Automation, 692-696, 1989.

4.   BROOKS, R.A.: A hardware retargetable distributed layered architecture for mobile robot control. In Proceedings IEEE Conference on Robotics and Automation, 106-110, 1987.

5.   FLYNN, A.M., and BROOKS, R.A.: MIT robots - what's next? In Proceedings IEEE Conference on Robotics and Automation, pp.611-617, 1988.

6.   CONNELL, J.H.: A Colony Architecture for an Artificial Creature. PhD thesis, Massachusetts Institute of Technology, 1989.

7.   SMITHERS, T., and MALCOLM, C.A., A behavioural approach to robot task planning and off-line programming. Journal of Structured Learning, 10, pp.137-156, 1989.

8.   MALCOLM, C.A., SMITHERS, T., and HALLAM, J.: An emerging paradigm in robotic architecture. DAI Research Paper 447, Department of Artificial Intelligence, University of Edinburgh, 1989.

9.   WILSON, M.S.: Achieving Reliability using Behavioural Modules in a Robotic Assembly System, PhD thesis, Department of Artificial Intelligence, University of Edinburgh, 1992.

10. MALCOLM, C.A., Planning and performing the robotic assembly of soma cube constructions. Master's thesis, Department of Artificial Intelligence, University of Edinburgh, 1987.

11. CHONGSTITVATANA, P., and CONKIE, A.D.: Active mobile stereo vision for robotic assembly. In Prof. Luis Basañez, (ed.), Proceedings 23rd International Symposium on Industrial Robots, pp.393-397, Barcelona, October 1992.

12. CHONGSTITVATANA, P.: The Design and Implementation of Vision-Based Behavioural Modules for a Robotic Assembly System. PhD thesis, Department of Artificial Intelligence, University of Edinburgh, 1992.

13. WILSON, M.S.: Behaviour-based robotic assembly systems: Reliability of behavioural modules. In Prof. Luis Basañez, (ed.), Proceedings 23rd International Symposium on Industrial Robots, pp.343-348, Barcelona, October 1992.

14. MAES, P.: A bottom-up mechanism for behavior selection in an artificial creature. In Proceedings International Conference on Simulation and Adaptive Behaviour, pp.238-246, 1990.

15. BROOKS, R.A., and VIOLA, P.A.,: Network based autonomous robot motor control: from hormones to learning. In R.ECKMILLER (ed.), Advanced Neural Computers, pp.341-348. Elsevier, 1990.

16. KRAVITZ, E.A.: .Hormonal control of behavior: Amines and the biasing of behavioral output in lobsters. Science, 241, pp.1775-1781, September 1988.

17. BROOKS, R.A.:. Challenges for a complete creature architecture. In Proceedings 1st. International Conference on Simulation of Adaptive Behaviour, pp.434-443. MIT Press, Cambridge, Mass., 1990.

18. BARNES, D.P., and GRAY, J.O.: Behaviour synthesis for co-operant mobile robot control. In Proceedings IEEE International Conference on Control, pp.1135-1140, March 1991.

19. BARNES, D.P., DOWNES, C.G., and GRAY, J.O.: A parallel distributed control architecture for a hexapodal robot. In Tsavestas, S., Borne, P., Grandinetti, L., (eds), Parallel and Distributed Computing in Engineering Systems, pp.341-346, IFAC, Elsevier, 1992.

20. BROOKS, R.A.: Intelligence without representation. Artificial Intelligence, 47, pp.139-160, 1991.

21. MATARIC, M.J.: Integration of representation into goal-driven behavior-based robots. IEEE Trans. Robotics and Automation, 8 (3), pp.304-312, June 1992.

22. KAELBLING, L.P.: An architecture for intelligent reactive systems. In GEORGEFF, M.P., LANSKY, A.L., (eds), Reasoning About Actions and Plans. Proceedings 1986 AAAI Workshop, pp.395-410, Morgan Kaufman.

23. KAELBLING, L.P.: Specifying complex behaviour for computer agents. In Sycara, K.P., (ed), Innovative Approaches to Planning, Scheduling and Control, pp.433-438. DARPA, Morgan Kaufman, 1990.

24. KAELBLING, L.P.: An adaptable mobile robot. In VARELA, F.J., BOURGINE, P., (eds), Toward a Practice of Autonomous Systems: Proceedings First European Conference on Artificial Life, pp.41-47. MIT Press, Cambridge MA, 1992.

25. ARKIN, R.C.: Motor schema-based mobile robot navigation International J. Robotics Research, 8, (4), pp.92-112, August 1989.

26. ARKIN, R.C.: Navigational path planning for a vision-based mobile robot. Robotica, 7, (1), pp.49-63, January 1989.

27. ARKIN, R.C., and LAWTON, D.T.: Reactive behavioral support for qualitative visual navigation. In Kaynak, O. (ed), Proceedings IEEE International Workshop on Intelligent Motion Control, pp.21-28, Istanbul, Turkey., August 1990.

28. KWEON, I.S., KUNO, Y., WATANABE, N., and ONOGUCHI, K.: Sonar-based behaviors for a behavior-based mobile robot. IECE Trans. Inf. and Syst., E76-D(4), April 1993.

Chapter 19

A behaviour synthesis architecture for co-operant mobile robots

D. P. Barnes

Research into the procedural control of robotic devices has seen a resurgence of interest in recent years. However, work has tended to focus upon solitary mobile robot scenarios as opposed to those situations where multiple mobile robots are required to communicate and co-operate with each other and use their combined functions to achieve a particular task. Researchers at the University of Salford are involved in a long-term research programme into the area of co-operant mobile robots and a novel behaviour synthesis architecture (BSA) has been designed and implemented for the specific purpose of controlling multiple co-operating mobile robots. To illustrate the application of the BSA, the chapter focuses upon a material handling operation, where two mobile robots are given the task of co-operatively relocating an object while having to avoid an obstacle *en route*. Details of the mobile robots' task achieving behaviour are provided firstly via simulation results, based upon our mathematical model of the BSA, and secondly via the results of having implemented the architecture on two real mobile robots. A discussion of our theoretical and experimental findings is presented and the chapter highlights a number of important areas for future research.

19.1 Introduction

Robotics research has for many years concentrated upon solitary robotic devices. Indeed, great effort has been expended in the quest to endow these robots with greater perception and intelligence and thus extend their task achieving capabilities. Anthropomorphism has certainly played a role throughout these endeavours as man has sought to create a machine capable of mimicking human abilities. Nevertheless, if the future of this work can be generalised as a desire to turn science fiction into science fact, then current understanding would indicate that the robots of science fiction are likely to remain as such for a number of years to come.

      In contrast to the solitary robot approach, researchers at the University of Salford have embarked upon a long-term programme into the area of co-operant robotic devices [1,2,3,4]. The Salford approach involves the design and development of multiple robots, that are technically simpler than their solitary 'cousins', yet capable of communicating and co-operating with one another and hence able to use their combined functions to achieve a given complex task.

      A source of research inspiration for the Salford approach has been the behaviour of social insects. Ants and termites while cognitively inferior to humans are nevertheless capable of many complex tasks. Bellicose termites, for example, are able to co-operatively construct cathedral-like dwelling places of such complexity, their achievements would shame any human endeavour to build a comparable structure. As foragers, constructors and protectors of their way of life, social insects are able to act together as if they were a single superorganism [5] and it is this ability and emergent swarm intelligence that is of fascination to the researchers at Salford. Yet how is it made possible? How might we be able to gain a deeper understanding of such behaviour so that we could transfer the seeds of this knowledge to the robotic domain?

We realised early in our research that to have 'colonies' of robots capable of co-operative behaviour would allow hitherto unobtainable robotic applications to be made possible. What follows in this chapter is a description of our efforts in this new robotic area. The foundations and subsequent focus of our research are presented and a major activity, the development of the behaviour synthesis architecture (BSA) is detailed. The application of this new control architecture to the area of multiple autonomous mobile robots is presented initially via a mathematical simulation and then via the implementation of the BSA on two real mobile robots. This work has allowed a number of results to be obtained and a discussion of these is presented and important areas for future research are described.

19.2   Research foundations

At the start of our research into co-operant mobile robots, our industrial collaborators had identified a number of future applications for such devices. However, at that time, the knowledge of how to realise this technology was not available. Within the robotics community, research into multi-robot systems had tended to concentrate on the co-ordination of multiple manipulators [6,7,8]. Typically, a powerful central computer running a model based control architecture had been connected to each manipulator via a high bandwidth communications link. A similar, centrally controlled approach had been applied to multiple mobile robot systems [9]. The disadvantage of this approach, particularly in a multiple mobile robot system, had been the requirement for a reliable high bandwidth communications link, maintenance of the real-time performance of the computer and the dependence of the robots on the central computer and communications system. To alleviate these problems, a number of researchers had begun to adopt alternative approaches, where multiple robotic devices were independent and autonomous. These approaches fell into two broad categories; the traditional hierarchical, planning based control systems and the sensor driven, reactive control systems. A hierarchical planning approach was being researched by Albus [10], where multiple autonomous underwater vehicles using world modelling techniques were being investigated. This project however, was subsequently abandoned due to the problems of maintaining world model consistency between the multiple mobile robots. The reactive control approach was being pursued by Arkin [11] and a similar behaviour-based approach was being investigated by Brooks [12]. These sensor driven control systems were showing great promise, especially for those applications that involved a robot having to operate within a dynamic environment. While such environments were envisaged by our industrial collaborators, the sensor driven control approach however, had only been applied to single non- co-operating robots. Thus it was set against this `backdrop' of available robotic know-how and our collaborators' desire for co-operant robotic technology that our research commenced.

19.3 Research focus

To provide pace and focus to our research, work began by concentrating upon a number of commercially important applications that had been identified by our industrial collaborators. There were a total of nine applications ranging from warehouse material handling using communicating mobile automata and multiple welding machines working simultaneously on a given weld, through to multiple road transporters working in unison to transport very large components, hazardous material handling using sophisticated co-operating mobile automata and multiple submersible robots for co-operative ocean bed surveying. It would have been impossible to focus in detail upon all of the applications, however, an examination of these areas allowed a `common denominator' research application to be defined that contained the salient features of these diverse applications. It was also important that the research demonstrator be manageable in terms of research effort and laboratory space. This research demonstrator was defined as follows. Two mobile robots were to be capable of co-operatively moving an object from one location in the laboratory to another, while avoiding obstacles *en route*. Hence, in terms of an analogous biological task, we had decided to focus upon a co-operant 'foraging' problem.

Figure 19.1 shows the basic object relocation task. We termed an individual mobile robot a *monad* , two co-operating monads, a *dyad* and a large undefined number of co-operating robots a *polyad*. Figure 19.1 shows two non-co-operating monads and the object to be relocated being supported by a pedestal. The object effectively represents a 'pallet' that the two mobile robots have to relocate and the purpose of the pedestal is to support this object prior to relocation.

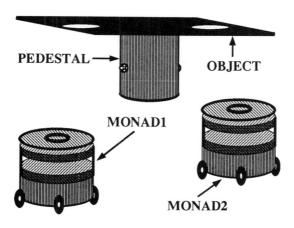

Figure 19.1 *The object relocation task*

Figure 19.2 is a plan view of the task enclosure and this shows the two monads, $M1$ and $M2$, the object to be relocated and several obstacles in the enclosure (these are denoted by a 'brick' in-fill). This figure serves to introduce the many problems that are associated with this object relocation task. These are:

1. the monads must be capable of collision avoidance both with respect to static and dynamic obstacles within their environment, i.e. not only avoiding walls and other similar objects, but also one another;
2. the monads must be able to navigate to a desired location, e.g. to the initial position of the object to be relocated and then to some final goal location where this object has to be deposited;
3. the monads must be able to position themselves, one at either end of the object to be relocated, and then 'acquire' their respective object ends;
4. the monads must be able to co-operatively move the object to a final goal location while negotiating obstacles *en route;*
5. the monads must be able to 'release' themselves from the object and thus leave the object at the final goal location ready for another object to be relocated.

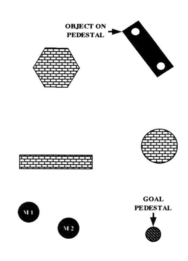

Figure 19.2 *Plan view of task enclosure*

From an examination of these problems, we realised that the activity of the monads could be regarded as a continuum between two basic types of diverse behaviour. At one extreme, the behaviour could be regarded as being essentially *egotistic,* where a monad is concerned purely with self directed behaviour. Obstacle avoidance and energy conservation strategies are examples of such behaviour. At the other extreme their behaviour could be regarded as being essentially *altruistic*, when for example a polyad needs to work together to perform some common task. Multiple robots co-operatively relocating an object is an example of such behaviour. However, these diverse types of behaviour are essentially in conflict! The first would cause a monad to remain stationary and to stay well away from all other objects within its environment, including other robots (hence this is also conflicting behaviour), while the other type of behaviour would cause a monad to team up in close proximity with its fellows to perform an activity of work (again this is conflicting behaviour). Given the many examples of conflicting behaviour that can be found in monadic and polyadic scenarios, the research focused upon the design of a control architecture that could accommodate such diverse and conflicting behaviour types.

19.4 Behaviour synthesis

What was required was a control architecture that could accommodate both egotistic and altruistic behaviour and provide a mechanism for *conflict resolution*. What emerged was the behaviour synthesis architecture, see Figure 19.3 and it constitutes an important addition to the solitary mobile robot control architectures of Arkin [11] and Brooks [12].

For conceptual convenience, four different strategy levels in the architecture were identified.

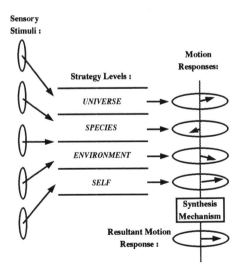

Figure 19.3 *Behaviour synthesis architecture*

● A SELF level contains those strategies concerned with the maximisation and replenishment of internal resources, e.g. remaining stationary to conserve battery power and moving off towards a recharge station only when necessary.

- An ENVIRONMENT level contains those motion strategies associated with activities involving other objects within the robot's environment, e.g. collision avoidance.

- A SPECIES level contains those motion strategies associated with co-operant behaviour, e.g. maintaining a correct position with respect to the demonstrator object while co-operatively relocating this object.

- A UNIVERSE level contains those motion strategies specific to a particular task, e.g. navigating to the initial location of the object to be relocated, then subsequent navigation to the desired goal location.

Sensory stimuli , from our developed monad sensor systems, provide the appropriate internal and external state information needed for the various strategy levels and from each relevant level, appropriate *motion responses* are generated that relate to the desired actuation.

Any strategy level can contain a number of behaviour patterns (*bp*) where

$$bp_t \left\{ \begin{array}{l} r_t = f_r(s_t) \\ u_t = f_u(s_t) \end{array} \right. \tag{19.1}$$

r_t is the particular motion response at time t and this is a function, f_r of a given sensory stimulus, s_t .   Associated to every response is a measure of its *utility* or importance, u_t. This quantity is a function, f_u, of the same sensory stimulus. The use of utility originated from our early research into formalisms for modelling co-operant behaviour. Game theoretic studies [13] showed that single and *n*-player games could be used to model monad versus nature and *monad1 versus monad2* competitive and co-operant scenarios. Strategy selection in these games is dependent upon the information a player may have regarding their opponent's or partner's move and the relative utility (or pay-off) of any counter or co-operative move. As this information is analogous to the sensory stimuli available to a monad and utility is used to great effect in the selection of an appropriate strategy from a set of possible strategies, it was realised that such a concept should be incorporated within our control architecture.

Hence a behaviour pattern defines not only what a monad's motion response should be for a given sensor input, but it also provides a measure as to how the relative importance of this response varies with respect to the same sensor input. The values of r_t and u_t constitute a vector known as a *utilitor* .

Figure 19.4 shows an example of a simple behaviour pattern that might exist at a given strategy level. Consider the situation where the sensory stimulus relates to a monad's forward facing distance to obstacle measuring sensor and the associated motion response relates to the forward translate velocity for that monad. From Figure 19.4 it can be seen that as the monad gets nearer to the object then its forward translate velocity will be reduced to zero. At the same time, the associated utility for this motion response increases. Thus as the monad gets nearer to an object in its path, the more important it becomes for the monad to slow down.

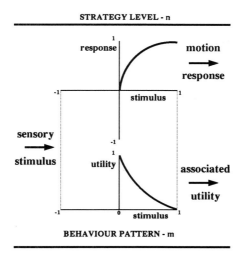

Figure 19.4 *Behaviour pattern example*

At any point in time, multiple conflicting motion responses are typically generated. For example, a monad may be navigating towards a goal location while co-operatively relocating an object when an obstacle unexpectedly appears in its path and at the same time it senses that it must recharge its batteries. In such a situation, what should it do? In the BSA, conflicting motion responses are resolved by a *behaviour synthesis mechanism* to produce a *resultant motion response*. Competing utilitors are resolved by a process of linear superposition which generates a resultant utilitor, UX_t where

$$UX_t = \sum_{n=1}^{m} u_{t,n} \cdot e^{j \cdot r_{t,n}} \qquad (19.2)$$

and m equals the total number of related utilitors generated from the different strategy levels. Given a resultant utilitor, a resultant utility, uX_t, and a resultant motion response, rX_t are obtained from

$$uX_t = \frac{|UX_t|}{m} \qquad (19.3)$$

and

$$rX_t = \arg(UX_t) \qquad (19.4)$$

X corresponds to a particular degree of freedom, e.g. translate or rotate, and the resultant motion response, rX_t, is then executed by the monad. From equation 19.2 it can be seen that generating a resultant utilitor from different strategy levels within the architecture constitutes a process of *additive synthesis*.

Behaviour patterns show some similarity with AI production rules [14] and problems with adding new rules to an existing large rule-base and how to ensure a desired rule-firing sequence, have long been know about. Likewise with our behaviour patterns, we wished incrementally to add new behaviour patterns, when 'programming' the monads for a particular task, without any unwanted interactions occurring at run time and we also wanted to be able to specify sequences of behaviour, such as docking with an object to be relocated, acquiring the object and moving off with this object to a desired goal location etc. To overcome these problems the behaviour synthesis architecture uses a *behaviour script* which serves to package the individual behaviour patterns and hence a variety of *behaviour packages* can typically co-exist. Sensor data are routed to the behaviour script and are used to check against previously defined *pre- and post-behaviour package conditions.* When these conditions are met, then the individual behaviour patterns within a package are activated when appropriate to the task. Details of our behaviour script mechanism are provided in [15] and section 19.5 provides an example of this facility.

The main advantages of our behaviour synthesis architecture are:

1. it is a sensor driven control architecture and hence does not have the problems of world model maintenance;

2. it is able to provide the necessary monad control for dynamic and unstructured environments;

3. it is a very simple architecture that is capable of providing the tactical monad control needed for complex task achievement;

4. it cannot only be used to control an individual monad, but inherent to its design is the ability to generate co-operant behaviour between multiple monads;

5. an end user can easily add or remove behaviour patterns when developing the desired task achieving behaviour;

6. it is capable of resolving multiple conflicting behaviour patterns;

7. an end user can easily modify the utility of the behaviour patterns so as to `hone' the desired task achieving behaviour;

8. a script mechanism is provided so that a desired sequence of monad behaviour can be obtained;

9. behaviour patterns can be added using behaviour packaging and hence a large behaviour-base can be created.

19.5 Modelling co-operant behaviour

To experiment with the behaviour synthesis architecture, we have developed a mathematical model for part of the dyadic object relocation task. This sub-task is depicted in Figure 19.5 which shows a dyad in possession of an object to be moved. The task is to relocate this object from the dyad start position to a location determined by a destination beacon positioned within the enclosure. This sub-task was selected for experimentation as it incorporates the main focus of our research, i.e. co-operant behaviour, and in terms of our defined research demonstrator, only object acquisition and release are omitted. An overview of our mathematical model is presented as a means of both describing how the BSA functions and illustrating what sensory stimuli and behaviour patterns are required for this co-operant task.

In our object relocation sub-task, each monad has been provided with six sensors which form the sensory stimuli for each monad's BSA. These sensors are:

s1   an internal *monad clock sensor* which returns a clock value from the start of the sub-task, to when the maximum velocity of the monad is reached

s2   an *object distance sensor* which returns a distance value from the front of the monad to any object perpendicular to the face of the sensor. The sensor has an effective working range

s3   a *beacon bearing sensor* which returns any angular displacement between the heading of the monad and a destination point produced by an external beacon positioned within the monad's enclosure. For this example the beacon destination corresponds to a point on a wall and it is assumed that the beacon can be sensed at all times, i.e. there is no beacon occlusion due to the obstacle within the enclosure

s4   a *beacon distance sensor* which returns a distance value from the front of the monad to a destination point produced by an external beacon positioned within the monad's enclosure (same beacon as for **s3**). This distance measurement is perpendicular to the face of the sensor

s5   is *a capture head velocity sensor* which returns the velocity of a monad's capture head, this mechanism is described below, relative to the translate velocity of that monad

s6   is a *capture head distance sensor* which returns the distance travelled by a monad's capture head relative to the distance travelled by that monad.

It was recognised early in our research that inter-monad communication is essential for co-operant behaviour to occur. However, we did not want to go down the problematic high bandwidth compliant control route as demonstrated by research at that time into co-operating industrial manipulators. Rather we wished to adopt a minimal approach to the area of inter-monad communication. Our main activity in this area focused upon the design of what came to be called the *monad capture head.*

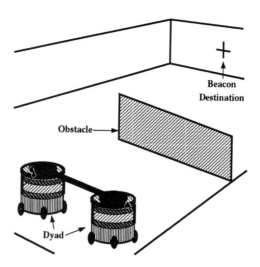

Figure 19.5 *Object relocation sub-task*

This is essentially an instrumented X - Y table, see Figure 19.6, that was designed to be located on top of each monad. With two monads suitably positioned under an object that they are co-operatively relocating, each capture head was designed so that the relative motion of one monad would be 'communicated' to the other monad and *vice versa*. The capture heads were also designed to allow rotational movement of an object with respect to a monad (hence the monads could orientate themselves relative to this object), and they were designed to be self centering, with the aid of a simple spring arrangement. This was to ensure that in the absence of an object they always centered relative to their location on top of each monad.

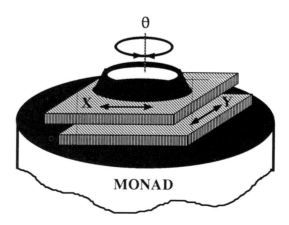

Figure 19.6 *Schematic of monad capture head*

In our mathematical model for the relocation sub-task, the capture heads for each monad, when attached to the object, were represented as a *coupled, driven, damped, spring-mass system.* Figure 19.7 shows the 1D case for our coupled capture head model. Each possesses a given mass ($m1$ and $m2$), restoring springs and appropriate friction. Hence as a monad moves, its capture head mass is displaced by dx_n and this displacement is transferred to the other monad's capture head via the coupling object and *vice versa.* It can be seen from Figure 19.7 that our model contains a *coupling coefficient C.* When $C = 1$, perfect mechanical coupling is modelled between the two monads, when $C = 0$, then the monads are uncoupled and no longer act as a dyad. $Fext_n$ is present in our model so that any additional external force on a capture head could also be incorporated, e.g. collision with an obstacle.

Using our capture head model, sensors $s5$ and $s6$ for each monad provide dx_n/dt and dx_n data respectively. All six sensors are normalised and provide an input into a monad's BSA. Contained within the mathematical model for each monad are the four strategy levels and associated to these levels are a number of behaviour patterns. Each monad in this example is an omni-directional device and therefore has the ability to translate and rotate. Hence the behaviour patterns are appropriately assigned to these two degrees of freedom.

Figures 19.8 and 19.9 show graphically the functions $f_r(s_t)$ and $f_u(s_t)$ that were designed for each behaviour pattern in this dyadic sub-task. The left and middle columns of these figures show the stimulus/response and stimulus/utility functions for each bp, while the right column shows how the motion response generated from each bp varies with respect to time, t, during the object relocation sub-task. Examples of seven behaviour patterns are shown, ($bp1,.....,bp7$), and these have been assigned to an appropriate strategy level within the BSA. Figures 19.8 and 19.9 show the bps for just one monad as both monads possess an identical set. Also shown is the resultant utility and motion response generated by the behaviour synthesis mechanism for the translate (uT_t, rT_t) and rotate (uR_t, rR_t) DOF respectively. Each monad executes its own values for rT_t and rR_t during the object relocation sub-task. A brief description of each bp is as follows

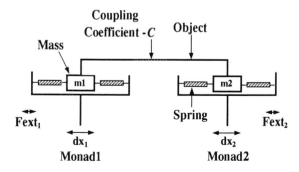

Figure 19.7 *Coupled capture head model (1D)*

bp1 is an example of a **SELF** level behaviour pattern. Its input is from $s1$ and its output contributes to a monad's resultant *translate strategy*. The function of this *bp* is to ensure a steady translate acceleration, at the start of the sub-task, should any 'step function' translate behaviour be generated by another *bp* within the behaviour synthesis process.

bp2 is an example of an **ENVIRONMENT** level behaviour pattern. Its input is from $s2$ and its output contributes to a monad's resultant *translate* strategy. The function of this *bp* is to decelerate the monad when $s2$ detects an object to be within a given range.

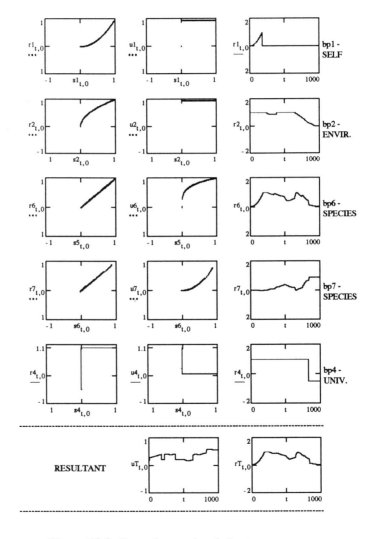

Figure 19.8 *Example translate behaviour patterns*

bp3 is an example of an **ENVIRONMENT** level behaviour pattern. Its input is from *s2* and its output contributes to a monad's resultant *rotate* strategy. The function of this *bp* is to cause the monad to rotate away from an object when detected by *s2* to be within a given range.

bp4 is an example of a **UNIVERSE** level behaviour pattern. Its input is from *s4* and its output contributes to a monad's resultant *translate* strategy. The function of this *bp* is to ensure that the monad translates to the external beacon destination in the shortest possible time, i.e. travels at a maximum given velocity.

bp5 is an example of a **UNIVERSE** level behaviour pattern. Its input is from *s3* and its output contributes to a monad's resultant *rotate* strategy. The function of this *bp* is to orientate the monad so that it is facing towards the external beacon destination point.

bp6 is an example of a **SPECIES** level behaviour pattern. Its input is from *s5* and its output contributes to a monad's resultant *translate* strategy. The function of this *bp* is to maintain a zero capture head velocity. This *bp* and its use of dx_n/dt is analogous to a derivative action in a conventional control system.

bp7 is an example of a **SPECIES** level behaviour pattern. Its input is from *s6* and its output contributes to a monad's resultant *translate* strategy. The function of this *bp* is to ensure that the position of the capture head remains central to its location on top of a monad. This *bp* and its use of dx_n is analogous to a proportional action in a conventional control system.

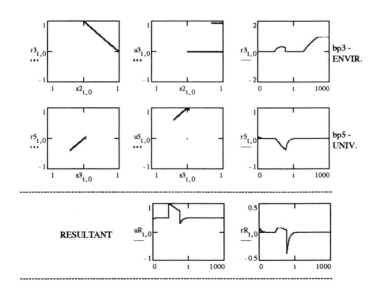

Figure 19.9 *Example rotate behaviour patterns*

From *bp*3 in Figure 19.9, it can be seen that the rotate motion response for $1000 > t > 650$ does not contribute to the resultant rotate response for the same time period. *bp*3 generates this motion response due to the fact that, as the monad is approaching the beacon destination, it is also approaching a wall within its environment (see Figure 19.5). Hence *bp*3 is trying to rotate the monad away from this object and if this were to occur, then the beacon destination would never be reached. A *precondition* associated to *bp*3 is responsible for assigning a utility value of 0 to the behaviour pattern *if the monad is approaching an object* AND *that object is placed at a position which corresponds to the external navigation beacon destination.* With a utility of 0, *bp*3 no longer contributes to the resultant rotate response and hence is an example of how behaviour script preconditions can be used to constrain the monad to conform to an end user defined task.

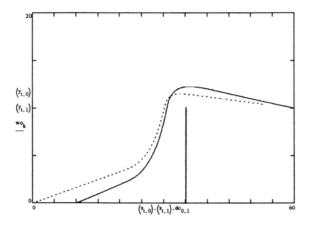

Figure 19.10 *Graph of Cartesian motion for coupled monads*

The graphical output shown in Figures 19.8 and 19.9 rotate was produced using a *Mathcad*^{TM} simulation of the object relocation sub-task. From the generated values of rT_t and rR_t the simulation allowed the Cartesian motion of the coupled monads to be obtained. The values of y_t and x_t for each monad are shown in Figure 19.10 and hence it represents a plan view of the situation depicted in Figure 19.5. It clearly shows how each monad initially rotates to align itself with the beacon, translates towards the beacon, negotiates the obstacle and finally rotates back onto the bearing of the beacon and translates up to the final beacon destination. Most importantly, throughout these activities the monads are acting as a dyad and thus the mathematical model demonstrates the ability of the BSA to generate co-operant behaviour. It can be seen from Figure 19.10 that one monad (whose path is denoted by the dashed line), never reaches the final beacon destination. The reason for this is that for the experiment described in this chapter, the monads were started one in front of the other and perpendicular to the wall containing the beacon destination. Hence one monad arrives at the beacon ahead of the other monad.

Once at the wall a monad will stop, leaving the other monad still attempting to reach the goal location. When this situation occurs, both monads will come to rest as shown in Figure 19.11, because ultimately the competing behaviour patterns of *bp*4 and *bp*7 cancel out each other. Hence *M*1 never arrives at the beacon. By including a capture head rotation sensor (θ in Figure 19.6) and an appropriate behaviour pattern(s), then *M*1 and *M*2 could easily be made to arrive at the beacon in a predefined orientation.

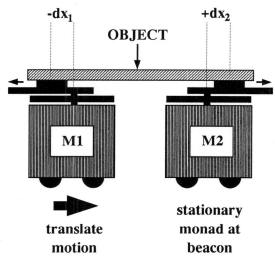

Figure 19.11 *Relative displacement of capture heads for coupled monads*

19.6   Fred and Ginger, our 'dyadic duo'

Not only have we produced a mathematical model of the object relocation sub-task, but we have also fully implemented this work on two real mobile robots [16,17]. These autonomous devices, known as Fred and Ginger, are shown in Figure 19.12 along with a developed pedestal and 'pallet' object.

Figure 19.12 *Fred and Ginger with pedestal and pallet*

It was decided to use two B12 mobile robots as the basis for our research platforms. These robots (obtained from Real World Interface, Inc.) are omni-directional, controllable via an RS232 interface, have a comprehensive set of motion commands, a maximum speed of 2 m/s and are cost effective. However, to work towards the object relocation task required considerable effort in order to enhance the basic B12 platforms. Superstructures have been designed and built to house additional sensors, data acquisition, communications and secondary control hardware.

The superstructures are capable of having several different levels and these currently range from a 'sensor skirt' at level 0 through to a transputer secondary controller at level 2. Level 1 houses data acquisition and sensor preprocessing boards. The secondary controller consists of a transputer mother-board capable of holding up to 16 individual TRAMS. Running through the centre of each superstructure is a common inter-level 'core' which has been designed to allow up to 24 p.c.b.s to be connected at each level. The superstructures can be disconnected from a B12 platform for ease of development and maintenance and similarly, the different superstructure levels have been designed so that they can easily be disconnected from one another and replaced with different levels containing different electronics.

Fred and Ginger both possess the following sensors. A *multi-frequency ultrasonic system* for obstacle proximity detection (multi-frequency was used to overcome multiple robot acoustic interference). A *prototype SiTeK navigation system* which uses a signatured beacon light source and a position sensitive photo-transducer that can be mounted on a mobile robot. The system has been designed to generate x and y values which define the receiver's Cartesian position relative to the light source. An *instrumented, self-centring, X - Y capture head* which uses optical linear encoders to obtain displacement data, and finally, translate and rotate *drive motor position encoders, a battery voltage level sensor and an internal clock sensor.*

The BSA has been implemented on Fred and Ginger and by using the developed sensors and behaviour patterns, we have successfully demonstrated these two mobile robots co-operatively relocating an object while avoiding an obstacle *en route.*

19.7   Discussion of experimental results

During the research we have concentrated upon the area of co-operative object relocation and the design and implementation of the BSA has allowed both empirical and theoretical studies to be undertaken. From experiments with our two mobile robots, a number of important results have emerged. First, the effectiveness of the BSA has been demonstrated, along with the significance of the utility function as a mechanism for radically adjusting the behaviour of the mobile robots. Experiments have shown that it is possible to make the mobiles more altruistic by increasing the relative utility of their species level behaviours. This results in the mobiles moving more slowly and carrying the object more carefully. In the extreme case, a mobile would rather bump into an obstacle than drop the object. Conversely, the mobiles can be made more egotistic or self centred by decreasing the utility of the species level behaviours. This results in the mobiles moving more quickly and carrying the object less carefully. This control versatility is important because the level of co-operation required for different tasks will vary and it is vital that the robots can respond to this. The BSA allows utilities to be adjusted at run time. This means that the behaviour of the mobile robots can be adjusted in the light of external conditions, for example, by increasing the utility of the species level behaviours when carrying a particularly fragile object.

Experimental results have also indicated that it is not necessary to pre-define the relative orientations of the two mobile robots when they are relocating an object. Regardless of their initial orientation, the mobile robots respond to the environment and move into the orientation which is best suited to that situation. This emergent behaviour is beneficial because, in highly unstructured environments, it may not be possible or practical to specify in advance the optimum orientation of the two mobile robots throughout the task execution. If a pre-defined orientation is required for a particular task, then this can easily be implemented by the addition of an appropriate behaviour package at the species level. In addition to sensing the actions of another robot at the species level, it is also necessary to sense other robots at the environment level. In the single robot case, all objects detected in the environment are deemed to be obstacles to be avoided. This assumption no longer holds true in the multiple co-operating robot case. In this situation, it is necessary to distinguish between other robots which are involved in co-operant task achievement and genuine obstacles to be avoided. To overcome this problem we have also developed a behaviour characterisation technique [15]. The principle of this technique is to characterise the sensor data generated by the presence of other mobile robots and to produce a behaviour template which is then used on the incoming sensor data at run time. If the incoming sensor data match, then the sensor data have been generated by another mobile robot. Otherwise the sensor data must have been generated by an obstacle and appropriate avoidance action can then be taken. Experiments have demonstrated the success of the behaviour characterisation technique on our two mobile robots and its effect can be interpreted as a form of *sensor habituation* .

Our theoretical studies have concentrated upon the generation of a mathematical model of the BSA. The work has focused upon the dyadic object relocation task and has resulted in the creation of a system of equations that when evaluated using the *Mathcad*^{TM} software package, a simulation of this co-operant task is produced. Our mathematical model currently consists of 35 coupled equations which are solved by simultaneous iteration. These equations provide a simple navigation and obstacle detection sensor model, define the behaviour patterns under experimentation, produce the required behaviour synthesis and model the capture heads as a coupled, driven, damped spring-mass system [18]. Graphical data are automatically generated and the system of equations can be used to perform extensive studies into the BSA and resultant co-operant task achievement.

Since we commenced our research into co-operant mobile robots, a number of fellow researchers have also been working in this area and a brief comparison is provided. A hierarchical planning approach is being researched by Noreils [19], where two robots can move in convoy and push a box around the laboratory together. This research is of interest because it differs from our sensor driven approach; however, the robots do not relocate the object in the way that we have achieved. In Japan, 'modest co-operation' has been demonstrated between two mobile robots which avoid each other using a simple communications link and traffic rules [20]. The robots however, are scheduled to work, rather than co-operantly working together. A reactive control approach is being pursued by Arkin, who proposes 'co-operation without communication' [21], and has simulated multiple mobile robots relocating objects. Because there is no communication in the system, the 'co-operation' is actually an artifact of the individual actions of the mobile robots and depends critically upon the initial conditions and the environment. In a similar, behaviour-based approach, Brooks [22] proposes to use multiple robots in lunar construction applications. The robots will communicate with each other using an infra-red system and a simple vocabulary. This work however is at an early stage of development.

19.8   Future research

From the results of our work, we have been able to identify a number of important areas for future research. Firstly, the behaviour patterns that we have developed for the mathematical model and Fred and Ginger have effectively been `handcrafted' based upon experimentation. What is required is a deeper understanding of response selection and utility assignment with the aim of creating a formalism for the generation of behaviour patterns. With such a formalism, generation could be improved and we are currently investigating possible automated behaviour generation techniques [23].

Secondly, this chapter began with a statement that the social insects have been a source of inspiration for our behaviour-based control approach. Yet this chapter has only reported our work with **two** mobile robots and these certainly do not constitute a 'colony' of robotic devices. What is required is experimentation with larger populations of mobile robots. Work is ongoing in this area [24], with twelve mini mobile robots ($\pi \times 5^2 \times 10\text{cm}^3$), and the results of this work will be reported in the literature.

Finally, experimentation with our mathematical model and with Fred and Ginger has involved monads that are strongly coupled during their co-operant task, $C = 1$. Further research is required where this coupling can effectively be varied, $1 > C > 0$, as this would greatly extend the application base for this new robot technology. Two such areas include multiple AGV's for transporting containers between cargo ships and rail/road loading points and multiple autonomous mobile robots for industrial cleaning tasks. We are currently investigating these applications [25].

From the results we have obtained, we believe that we have made a valuable contribution to the research area of co-operant mobile robots. We would argue that the results are not just specific to our co-operant object relocation task, rather we feel that they are generic in nature and hence will find a wider use within the robotics research arena.

19.9   Acknowledgements

This work has been supported by the United Kingdom Science and Engineering Research Council under the ACME Directorate grant GR/F71454, together with industrial collaborators, GRAD Ltd., VSEL and Hunting Engineering Ltd. I would like to acknowledge the contribution of my fellow investigators, Prof John O Gray and Dr Brian Bury, and to extend a special thanks to David Eustace, who was the SERC funded Research Assistant on the ACME project. David made significant strategic and tactical contributions to the research and was responsible for the implementation of the behaviour synthesis control architecture and for the subsequent experimentation that was performed with Fred and Ginger.

19.10 References

1. BARNES, D.P., BURY, B., GRAY, J.O.: Co-operant Mobile Robots in 3rd International Symposium on Robotics and Manufacturing : Recent Trends in Research, Education,and Applications, (ISRAM '90), Vol. 3. Burnaby, BC, Canada, 899-904. Edited by M JAMSHIDI and M SAIF. New York : ASME Press, July, 1990.
2. BARNES, D.P., GRAY, J.O.: Behaviour Synthesis for Co-operant Mobile Robot Control, Proceedings IEE Internationl Conference on Control 91, Vol. 2, 1135-1140, March, 1991.
3. BARNES, D.P., EUSTACE, D.: A Behaviour Synthesis Architecture For Mobile Robot Control, Proceedings IEE Colloquium on Autonomous Guided Vehicles, Digest No. 1 991/164, 3/1-3/4, Salford, Nov., 1991.
4. EUSTACE, D., BARNES, D.P., GRAY, J.O.: Co-operant Mobile Robots for Advanced Manufacturing and Material Handling Applications, Proceedings 24th International Symposium on Automotive Technology and Automation, Vol. 1, Florence, 443-450, May, 1991.
5. WILSON, E.O.: The Insect Societies, Belknap Press of Harvard University Press, Cambridge, Mass, 1971.
6. CHIMES, P.R.: Multiple-arm Robot Control Systems, Robotics Age 7, 5-10, Oct., 1985.
7. FREUND, E.: On the Design of Multi-robot Systems, Proceedings of the International Conference on Robotics, IEEE Computer Society Press, Siver Spring, Md, 1984.
8. CALLAN, J.F.: The Simulation and Programming of Multiple-arm Robot Systems, Robotics Engineering 8, 26-29, April, 1986.
9. NEGATA, T., HONDA, K.,TERAMOTO, Y.: Multi-Robot Plan Generation in a Continuous Domain : Planning by Use of Plan Graph and Avoiding Collisions Among Robots, IEEE Journal of Robotics and Automation, Vol. 4, No. 1, Feb., 1988.
10. HERMAN, M. ALBUS, J.: Overview of MAUV : Multiple Autonomous Undersea Vehicles, Unmanned Systems, Vol. 7, No. 1, 36-52, 1989.
11. ARKIN, R.C.: Motor Schema-Based Mobile Robot Navigation, International Journal of Robotics Research, 81, No.4, 92-112, April, 1989.
12. BROOKS, R.A.: A Robust Layered Control System for a Mobile Robot, IEEE Journal of Robotics and Automation, RA-2, 14-23, March, 1986.
13. VON NEUMANN, J., MORGENSTERN, O.: Theory of Games and Economic Behaviour, Princeton University Press, USA, 1953.
14. DAVIS, R., KING, J.: An Overview of Production Systems, in E.W. Elcock and D. Michie (Eds.), Machine Intelligence 8, New York: Wiley Sons, 300-332, 1977.
15. EUSTACE, D., BARNES, D.P., GRAY, J.O.: Co-operant Mobile Robots for Industrial Applications, 19th Annual Conference of the IEEE Industrial Electronics Society, Hawaii, November, 1993.
16. BARNES, D.P., BURY, B., GRAY, J.O., HILL, S.L., EUSTACE, D.: Research Platforms for Investigating Mobile Robot Co-operancy, 5th International Conference on Advanced Robotics, Robots in Unstructured Environments, Vol. 2, Pisa, Italy, 1642-1645. Edited by P Dario. New York IEEE, June, 1991.
17. EUSTACE, D., BARNES, D.P., GRAY, J.O.: Multiple Co-operant Mobile Robots for Unstructured Environments, 6th International Conference on Advanced Robotics. Tokyo, November, 1993.
18. BARNES, D.P.: Dyadic Behaviour Simulation, Internal Research Report RD012, University of Salford, Jan., 1993.
19. NOREILS, F.R.: Toward a Robot Architecture Integrating Co-operation Between Mobile Robots: Application to Indoor Environment, International Journal of Robotics Research, Vol. 12, No. 1, 79-99, 1993.
20. PREMVUTI, S., YUTA, S.: Consideration on the Cooperation of Multiple Autonomous Mobile Robots; Introduction to Modest Cooperation, Proceedings International Conference on Advanced Robotics, Vol. 1, 545-550, June, 1991.

21.   ARKIN, R.C.: Co-operation without Communication: Multiagent Schema-Based Robot Navigation, Journal of Robotic Systems, Vol. 9, No. 3, 351-364, April, 1992.

22.   BROOKS, R.A., et al.: Lunar Based Construction Robots, IEEE International Conference on Intelligent Robots and Systems, Tsuchiura, Japan, 389-392, 1990.

23.   RUSH, J.R., FRASER, A.P., BARNES, D.P.: Evolving Co-operation in Autonomous Robotic Systems, IEE International Conference Control 94, Warwick, March, 1994.

24.   BARNES, D.P.: Polyadic Mobile Robot Behaviour, Royal Society Research Grant, RSRG 11812, 2/3/92 - 1/3/93.

25.   BARNES, D.P., AYLETT, R., EUSTACE, D., GRAY, J.O., LARMOUTH, J.: ITE: Multiple Automata for Complex Task Achievement, EPSERC IT/ACME, GR/J49785, 1/10/93 - 30/9/96.

Chapter 20

Co-operant behaviour in multiple manipulators

G. Dodds

Soft automation in the form of robotics can be applied to low count production runs, variable tasks in assembly and testing and can achieve increased quality and quantity of production. These necessary abilities are due to the in-built repeatability and relative ease of adjustment and feedback. This chapter describes how, even with robotic execution of tasks, there are still considerable costs in fixed jigging and tools due to the limited dexterity of commercial arms or the cost of possible redundancy in the production of special purpose single-arm robots to carry out complex tasks.

The multi-arm systems solution, described here, removes some of the limitations and peripheral costs of further application of robotics, although this is at the cost of greater complexity in intelligence, planning and control. Future developments in the areas outlined in this chapter will be the key to fully flexible automation and high technology production.

Initially a description of the range of multi-arm systems is given. Then work in enabling research which is required for multi-armed systems is outlined. This is followed by details of research at the Queen's University of Belfast into task planning, path planning and collision detection, multi-arm sensing, hardware architectures for practical implementation and maintenance of reliability.

20.1 Introduction

Multi-manipulator systems can increase the versatility and operating limits of robots. This versatility is due to the possibility of replacement of hardware and jigging set up costs by expanded planning algorithms and controllers. Other reasons for their use include increased reach, loading, speed and reconfiguration to other applications. The advantages of multi-arm systems are coupled with problems involved in their co-ordination.

Work is being carried out on limited aspects of this research area in several international and national laboratories; in addition more conferences, sessions within conferences [1] and books are being produced. Successful developments will lead to significant rewards, as multiple smaller robots would be more acceptable in considerably more applications outside and inside normal industrial environments.

At present most planning algorithms involve collision detection followed by avoidance [2] whether for static systems or mobile autonomous vehicles. Control of the multiple robots is then superimposed onto the planning to give a suboptimal task implementation.

In addition to obstacle avoidance co-operative handling of objects requires the imposition of additional constraints on the arms' dynamics [3]. The efficiency of the joint lifting operation is determined by minimisation of conflicting forces in the held object and the distribution of load among the co-operating robots [4]. The multi-arm system as a result requires significantly higher levels of control bandwidth. The increases in complexity required for planning, control, intelligence [5] and sensing can be provided naturally through multi-processing and the fusion of the necessary data from correlated sensors.

20.1.1 Multi-arm research at QUB

At the Queen's University of Belfast, research has been carried out into co-operative robotic theory and applications for over six years. Work has produced demonstrator systems for validating theoretical results [6], dynamic simulation and control through parallelism [7], solutions to the problems of parallel processing, interfacing and sensing techniques and high level task planning [8]. Two multi-arm systems are presently in operation, a light weight commercial SCARA robot system and an in-house SCARA system which has been extensively modelled, see Figures 20.1 and 20.2.

These industrial based multiple manipulators contrast with the other major class of multi-limbed systems, i.e. the multi-legged walking systems, in that the surface contacts are more complex than simple multi-leg point contact but on the plus side the working environment is more structured. The arms have greater degrees of freedom than the three normally used for legs and more constraints than the c.o.g. stability maintenance of the carried platform [9]. In addition industrial systems have a fixed base and do not have the freedom of movement of the supported platform.

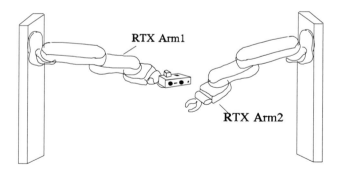

Figure 20.1 *Side view of light industrial arm system*

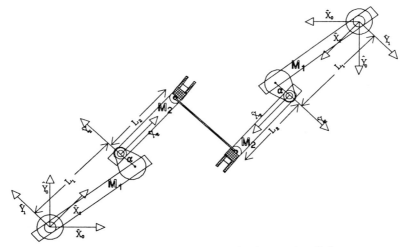

Figure 20.2 *Demonstrator configuration with freely rotating link*

In legged systems there are also only a limited number of contact sensors, with sensor reactive gait control, and the legs are usually programmed to move in a limited number of set sequences. Multi-robot systems have been examined widely in other areas, their usage ranges from autonomous group mission swarms [10] to specially designed force and control testing stations [11], and co-operative force systems [12]. Most of these research facilities have not targeted directly the needs of industry and direct economic application of multi-robot technology.

The QUB light-industrial-multi-arm system consists of two statically mounted RTX robots [13]. These robots are upgraded from the commercial model and have additional, processing capabilities, force sensors and joint position sensors. The original processing has been replaced by efficient interfaces to transputer networks in a manner which could easily be adopted to other commercial robot systems. The additional joint sensors are being used to compensate for the low-cost flexible drive system used in the RTX robots and were developed for use in an associated research project [14]. With the additional sensing, the force and position measurements can be made to 5N and up to 0.02mm. Real-time processing and individual robot measurements of parameters are available to operators in a graphical form, for diagnostic analysis and control evaluation.

20.1.2 Multi-arm research goals

Multi-arm research has highlighted the research needs of multi-arm system technology and its application to commercial tasks. These can be summarised as the following key technologies:

1) Task planning and co-ordinated movement strategies
2) Co-operative force and position control
3) Operating mode switching techniques
4) Operator programming aids, interface during programming
5) Control processing, sensor hardware and data processing
6) Environment, tool and jig design
7) Quality control and reliability.

Demonstrations of a number of these technologies have been produced in isolation and these have concentrated in the main on enabling a small part of the overall system needs and have only functioned properly through use of restrictive environmental constraints. In the control group at QUB, work has attempted to remove unjustifiable constraints and has, for example, examined the important and as yet relatively unexplored problems of fast collision detection strategies [15] and a graphical representation for real-time evaluation of optimal movement strategies [8]. There is thus a need to produce fundamental theory and demonstrations to generalise and make more robust, the controllers and planners for real systems.

In the past co-operative robotic research has examined the use of avoidance and semaphore based monitoring which permits only one arm to enter the jointly accessible space at any time [16]. Increasing the complexity and range of both the available tasks and the movement space were achieved by modelling the manipulator more closely. Co-ordination was achieved with simple collision avoidance, the tasks were simple and primarily directed towards testing the algorithms. Heuristic methods have been used [17] to plan motion in a shared space, and planning aids for multi-robot systems have also been produced for teaching pendant based welding systems [18] and more general systems [19]. Some limited ability co-operative working of robots is already being examined, e.g. special purpose Cartesian manipulators [20] and metal pressing robots.

Optimisation techniques have been used to search for the best paths in a common workspace [21] and other researchers have examined the use of neural networks and learning methods to achieve the same goals. However, all of the previous contributions to the field stop short of actually implementing the proposed systems and engaging them in a practical application.

Transputer and parallel processing have to a limited extent been used for a number of robotics applications [22], although their capabilities and application have been limited. The processing needs of multi-arm theory expand the knowledge of real-time processing and distributed communication needs of transputer and heterogeneous processing networks.

The next section describes the problems of task planning for multi-arm systems and outlines a solution for high dexterity testing tasks. Path planning at a lower level must ensure that the detailed motions of the arms do not collide; this is taken care of through efficient modelling and optimisation as detailed in the next section. In the last section the special needs and architectures for processing and sensing for multi-arm working are investigated.

20.2 Co-operative task planning for object surface processing

This object surface following task [8] uses the benefits of the increased dexterity of multi-arm systems rather than the advantages of increased loading and reach. The task is complex in terms of the required movements, although relatively simple in terms of forcing, sensing and control requirements.

When coating, or moving to test points on the surface of an object, strategies are required to minimise a number of generally diverse costs, such as execution time, amount of joint movement, regripping frequency, overpainting, orientation of contact and contact accuracy. In addition, jig and special fixture manufacturing should be kept to a minimum.

In a versatile handling operation, objects should be gripped by one arm starting from a feeder and manipulated to permit surface testing by the second arm. The gripping point can be changed, if necessary by temporarily locating on a preparation area or re-orientation jig and regripping after the robot configuration has been changed. These abilities impose additional constraints on the arm planning but increase the versatility of the system, provided that the speed of operation is not significantly compromised by the regripping operations.

The application in this case involves the three-dimensional testing or coating of an object, e.g. a multi-layer silicon circuit or a flexible PCB built into a product case. The part must be contacted and tested at various points on its surface. The example task involves complete uniform coverage of the surface of a sphere, which approaches a worst case scenario. There are two parts to the present realisation of the task:

1) path determination in conjunction with inverse kinematic realisation;

2) successive relaxation of constraints with analysis of alternatives in order to satisfy the desired path.

Since the object used in the test is spherical, the initial contact and gripping are carried out easily through joint angles calculated from the inverse kinematics. More complex gripping as required for asymmetrical objects is therefore not necessary. A position in space is chosen where the test piece/applicator and the object/sphere are approximately midway between the robot bases. This is done arbitrarily by holding the object in midspace and searching for the first allowable object contact position of the tool holding arm. Further work will attempt to maximise the usefulness of this first point by ensuring the maximum flexibility for the rest of the tests to be carried out. Although in this case, because of the complete coverage requirement, and resulting uniformity of merit of all positions, this was not attempted.

20.2.1 Tool trajectory planning

After initial contact with the test object the next point for contact must be determined. The direction of movement is based on a vector derived from the distance to be moved on the object surface and the number of points or untested areas in that direction, see Figure 20.3.

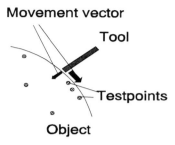

Figure 20.3 *Test points and resultant tool movement*

An inverse relation based on testing points and associated surface travel distance is used to determine the next movement direction. In the algorithm the tangent plane to the surface at the brush contact is found from the vector relationships (see Figure 20.4):

BrshTangent = BrushPos - CentrePos

The direction of each area segment centre from the brush is then found from:

Dirn = TriCent - BrshTangent

This direction is normalised to give a total direction vector of magnitude one, RelDir. The surface distance (Distance) is then found from the arc due to the angle subtended between the brush position and the segment centre position.

    The object surface is effectively flattened and the tool movement direction (2-dimensional in flattened surface) T is found from:

$$T = \sum_{i=1}^{Ntest} RelDir_i / Distance_i \qquad (20.1)$$

where,
$RelDir_i$ - direction in flattened surface to test point i
$Distance_i$ - distance on flattened surface to test point i

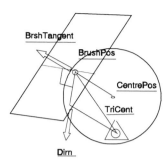

Figure 20.4 *Tool and test point vector description*

The tool movement is constrained to lie on the sphere surface by scaling the movement on the tangential plane to lie at the length of a radius away from the sphere centre.

If the sum of the tangential direction vectors results in a zero direction then a random tangential direction is chosen to enable movement to occur at the next step. In addition when a test point is within the range of step movement then this becomes the next candidate contact point and the vector movement described previously is superseded.

This strategy then permits an optimum coverage and handles movement effectively in terms of object oriented goals. The overall solution is sub-optimal as optimal movement is carried out in object space and this is determined independently of the subsequent required optimisation of the arm movement. Movement from the present attainable position is minimised, and contact is coarsely maintained between points as would be required in a covering application.

20.2.2 Object and contact point features and representation

An elemental modelling method was used to represent the object surface. Triangular elements were used to permit location of the centre of a test area and collisions or contact with this area to be determined in a straightforward manner. The elements were chosen to be as close as possible of equal area. In the case of the examined object a geodesic dome was chosen, see Figure 20.5.

Eighty sections were used to represent the object, these provided sufficiently accurate position segmentation due to the use of the compliant tool. This level of detail also suits real-time graphical simulation on a low cost high-power PC.

The distance moved during a time interval before recalculation of the tangential direction is restricted to within the velocity constraints of the robot pair. When moving to a new test point on the object surface. At the end of the appropriate candidate task trajectory, inverse kinematic calculations on the tool holding arm are used to determine if the desired point is reachable.

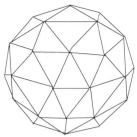

Figure 20.5 *Spherical object representation by 80 segment geodesic dome*

Object size is important in determining strategies and also as to whether a two arm solution is feasible. Large objects require unjustifiable reaching and are costly in terms of the loading and unloading required to engage the object in a reachable orientation, see Figure 20.6. The allowed tolerance for contact angle depends on the application. Other work is required in order to determine the optimum object and robot relative sizing or positioning for a given task sequence.

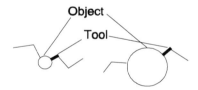

Figure 20.6   *Object size and multi-arm task difficulty*

20.2.3 Contact maintenance strategies

If a test or application point cannot be reached, with orthogonal contact, then alternative strategies are engaged in order to permit automatic, smooth and always achievable contact.

Depending on restrictions based on the application and on the object which is to be tested, the contact angle of the tool can be relaxed within a cone of allowable values, see Figure 20.7. In order to simplify the number of alternatives, angles at the outer limits are chosen, i.e. in this case four values, which are orthogonal on the surface, are chosen.

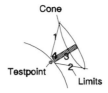

Figure 20.7 Cone of allowable orientations

Unfortunately, as a consequence of the necessity of choosing values which are not normal to the surface, the next movement point is then more likely to fail as the tool may be approaching an unreachable area. But the relaxation of angles does permit the system to carry out an increased number of tests, giving contact within the allowed cone, and reducing the need to resort to more complex and time consuming changes.

Initially only the tool orientation changes to achieve the next test point, but if this single robot action would not be possible, e.g. due to kinematic constraints, then in the next relaxation the object holding robot moves to attain the desired contact. This graduated strategy reduces the short-term complexity of the solution searching strategy, and as a consequence of the reduced number of options, the possibility of a viable solution. In addition, to increase the flexibility and solution probability, if the contact cannot be maintained by a simple rotation of the object, the acceptable contact angles are also relaxed. Thus a total of 10 options are available to be tried before the object must be regripped as described next.

In the worst case and to ensure that all contact points are attainable the object can be set down and picked up in a different orientation, if necessary this can be carried out incrementally and repeatedly, in order that the desired orientation is obtained, see Figure 20.8. During this operation collisions with the tool arm are avoided. At present this simple algorithm maintains the new desired contact in the same direction as the previous point, this limits the availability of further contact. Further theoretical development is required in order to maximise the probability of low-cost contact with further points.

Figure 20.8 *Regripping to reach all points.*

The possibility of repetitive cycling between points is handled by storage of the most recent movement sequence and identifying repetition. When repetition occurs a random direction component is added and this normally breaks the cycle.

20.2.4 Simulation and task execution example

Development of any application requires adequate representation and simulation facilities, during planning and execution. On-screen representations are given in orthogonal projection with a close up of tool contact and an overall view of the robot kinematics, see Figure 20.9. These scenes are produced in real-time and give the co-operative robotic application operator, or programmer, the initial facilities to engage in determining the appropriateness of the approach. Figure 20.10 shows the physical system.

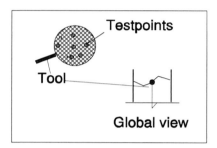

Figure 20.9 *On screen interface for operator*

Figure 20.10 *Physical layout of the multi-arm system*

At present the task execution points are prepared separately and loaded to path planning and collision avoidance algorithms working on a transputer network. These points are stored in a common file and executed sequentially. Since during the planning point reachability is already checked, it is only the inter-point paths which are determined on the transputer network.

The next section describes the development of efficient collision detection strategies for the path planning.

20.3 Collision detection for multiple manipulators

Most path planning schemes allow for avoidance of static objects in the workplace [23]. Time variable obstacles in a multiple manipulator environment make collision free motion planning more complicated, difficult and time consuming. Models of the arms may take the form of the exact link geometry or approximations in the form of uniform objects such as spheres and cylinders.

Sphere and cylinder models, see Figure 20.11 [15], are briefly described. A model using a cylinder and two semi-spheres is made of each link. This model can more efficiently model any length of link and has the advantages of larger volumes of free-space and lower collision detection time.

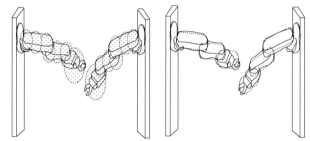

Figure 20.11 *Arm links modelled by spheres*

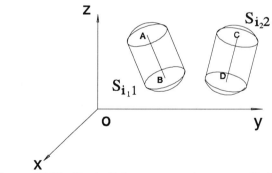

Figure 20.12 *Single link representation as a cylinder*

A link of an arm is represented by the cylinder/semi-sphere combination as shown in Figure 20.12. The arm links have radii r_{i1} and r_{i2} and are represented by the compact sets S_{i1} and S_{i2}. The shortest distance between the cylinders is defined as follows:

$$d_{i_i i_2} = \min \left\{ \sqrt{(x_1-x_2)^2+(y_1-y_2)^2+(z_1-z_2)^2} \; , \right. \tag{20.2}$$
$$\left. (x_1,y_1,z_1)\in S_{i_1 1}\subset R^3, (x_2,y_2,z_2)\in S_{i_2 2}\subset R^3 \right\}$$

The distance between links is checked along the cylinder centre lines

$$d_{i_i i_2} = \min_{s,t\in[0,1]} \left\{ \sqrt{(u_1+v_1 t-w_1 s)^2+(u_2+v_2 t-w_2 s)^2+(u_3+v_3 t-w_3 s)^2]} \right\} \tag{20.3}$$

where,

$$u_1=x_a-x_c, \; v_1=x_b-x_a, \; w_1=x_d-x_c, \tag{20.4}$$
$$u_2=y_a-y_c, \; v_2=y_b-y_a, \; w_2=y_d-y_c,$$
$$u_3=z_a-z_c, \; v_3=z_b-z_a, \; w_3=z_d-z_c$$

$d_{i_i i_2} > r_{i_1} + r_{i_2} + \varepsilon$ implies no collision between $S_{i_1 1}$ and $S_{i_2 2}$
Introduction of Lagrangian multipliers for line inequality constraints, $0<s<1$ and $0<t<1$, gives

$$F \triangleq D_{i_i i_2} - \lambda_1 t + \lambda_2(t-1) - \lambda_3 s + \lambda_4(s-1) \tag{20.5}$$

The minimum distance can then be found using a Kuhn and Tucker theorem [2]

$$\frac{\partial F}{\partial t} = l_1 + m_1 t + n_1 s - \lambda_1 + \lambda_2 = 0 \tag{20.6}$$

$$\frac{\partial F}{\partial s} = l_2 + m_2 t + n_2 s - \lambda_3 + \lambda_4 = 0 \tag{20.7}$$

where

$$l_1 = 2(u_1 v_1 + u_2 v_2 + u_3 v_3) \tag{20.8}$$
$$m_1 = 2(v_1^2 + v_2^2 + v_3^3)$$
$$n_1 = -2(v_1 w_1 + v_2 w_2 + v_3 w_3)$$

$$l_2 = -2(u_1 w_1 + u_2 w_2 + u_3 w_3)$$
$$m_2 = -2(v_1 w_1 + v_2 w_2 + v_3 w_3) = n_1 \qquad (20.9)$$
$$n_2 = 2(w_1^2 + w_2^2 + w_3^3)$$

$$-\lambda_1 t = 0 \qquad (20.10)$$

$$\lambda_2 (t-1) = 0 \qquad (20.11)$$

$$-\lambda_3 s = 0 \qquad (20.12)$$

$$\lambda_4 (s-1) = 0 \qquad (20.13)$$

The search for the stationary terms is best defined in terms of zero and non-zero λ_i. If $\lambda_i = 0$ (i=1,2,3,4) the stationary point may lie on or within the boundary. This case solves for the unconstrained extreme values, i.e.

$$t = \frac{l_2 n_1 - l_1 n_2}{m_1 n_2 - m_2 n_1}, \quad s = \frac{l_1 m_2 - l_2 m_1}{m_1 n_2 - m_2 n_1} \qquad (20.14)$$

$D_{i_1 i_2}$ can be obtained from (20.4). If $D_{i_1 i_2} > (r_{i_1} + r_{i_2} + \varepsilon)^2$, there is no collision between the i_1th limb and the i_2th limb. If $D_{i_1 i_2} < (r_{i_1} + r_{i_2} + \varepsilon)^2$ and $0 < t < 1$, $0 < s < 1$, a collision occurs. If the Lagrange multiplier is not negative then the collision distance is inspected.

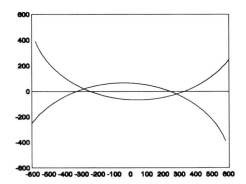

Figure 20.13 *Example planar motion with collision*

An example trajectory is shown in Figure 20.13. This trajectory results in a possible collision. The timing of the collision detection is shown in Figure 20.14; it can be seen the cost is higher during the collision period but the overall timing is still significantly lower than an equivalent sphere representation.

It is possible to use parallel processing to speed up detection by allocating sets of links, or tests of particular sets of Langrangian multipliers, to individual processors. The timing for a 7 sphere model was 1.626s and for the cylindrical model 0.881s.

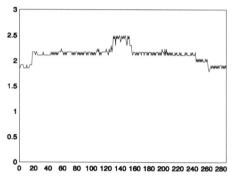

Figure 20.14 *Timing for collision detection at individual sampling periods*

The next section describes the development of optimum path planning strategies.

20.3.1 Optimal path planning

There are two classes of problems: smooth movement and minimum travelling time. The paths must satisfy specified position, velocity and acceleration values at intermediate points and also not exceed joint limits [24].

The path is constrained to pass through the required intermediate points, thus the variability of the inter-knot time can only be used to optimise the trajectory. Given the acceleration and inter-knot time it is then possible to calculate position and velocity, etc. Optimisation with respect to smoothness, i.e. minimise acceleration, can be carried out using Powell Fletcher and Davidon's optimisation algorithm. After calculating the smooth path the maximum values of acceleration, jerk and velocity are then checked. The proportion of the difference from the maximum value is then used to scale the timing of that interval to ensure that the maximum values are attained and not exceeded. The algorithms have been implemented for each arm and they perform optimisation based on all joints simultaneously.

The next section describes the sensing facilities on the present system and enhancements specifically introduced for multi-arm working.

20.4 Sensing requirements for multi-arm systems

Multi-arm systems require high accuracy position sensing and control, otherwise antagonistic co-operative actions could result. The commercial arms used required upgrading in position sensing specifically to deal with flexible transmission and deadband problems, see Figure 20.15.

'On-the-joint' measurements remove the problems of joint inaccuracies at the cost of additional sensing. In addition more expensive high resolution encoders must be used, since the multiplying effect of gearing and belts are now no longer available. Additional problems with respect to maintenance of axial movements of the final drive and lateral forces must also be minimised with additional bearings which possibly increase the frictional loading of the joint. For the RTX the use of these additional bearings has increased position accuracy by decreasing significant joint axial movement.

The basic RTX robot has a resolution typically of 30 counts per degree for its major joints. Uniformity of count resolution on every joint, causes the joints to contribute unequally to the positioning errors. In order to optimise the shoulder joint position measurements, the 'on-the-joint' encoder for the shoulder was given a resolution twice that of the elbow and yaw.

Obvious additional advantages of the new joint encoders included the availability of accurate zero position signals. Prior to the additional encoder use, the initialisation of the resident encoders was carried out by moving to end-stops and counting from this reference position. This approach results in the most significant contribution to absolute error, where this is due to the fact that the zeroing position cannot be attained accurately, causes are the flexibility and lightweight nature of the RTX robot. The major additional problems of the new encoder involve initial calibration or determination of the zero position with respect to arm joint angle.

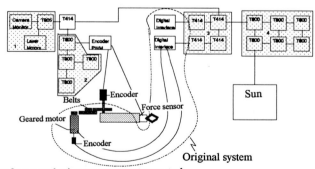

1 Imaging software and mirror stepper motor control
2 Encoder and force sensor interface
3 Low level motor control and motor encoder interface
4 Real-time planning, collision avoidance

Figure 20.15 *Hardware and software for the multi-arm system*

Instead of using this joint sensor as a complete substitute for the motor based encoder the joint based sensors are used in combination with the motor sensor, and this allows additional information on acceleration, torque and higher positioning accuracy to be attained. In order to achieve this a model of the belt dynamics must be produced along with identification of voltage, speed, and current relationships. This would be achieved through step inputs with known arm weight and load parameters.

The response of the system to controllers based on the extra fused information has been explored. The basic approach chosen was a multi-rate variable structure control [14]. This approach permits control of the slow major joint dynamics as well as monitoring and maintenance of the faster belt dynamic system within permissible bounds.

20.4.1 External position sensing

Internal sensing and control must be assisted by external non-contact environment sensing in order to carry out multi-arm tasks, with minimal operator loading.

A camera system on another transputer aids in the location and calibration of the robot configuration, this is in combination with a laser positioning and triangulation system. The compatibility and calibration of this measurement is especially important in order to carry out co-operative tasks, as antagonistic actions and damaging forces can occur if the relative positions of the arms and objects to be co-operatively gripped are not known precisely.

External position measurements are required in order to deal with uncertainty in the relative positions of the two robots and objects held in the jaws. This prevents collisions and impacts between the robot end-effectors and also enables controlled changes, e.g. from single to multi-arm lifting, or reach extension operations to occur, see Figure 20.1.

The position of the end-effectors with respect to the base co-ordinate frame is known accurately due to the internal sensors and these positions can be used as high information content starting points for determination of the size and orientation of held objects. The end-effector centre position is transferred from the internal sensing encoder position sensors and forward kinematics to the imaging and laser position control TRAM. This position is then converted to appropriate line of sight required for the laser. This is found from the following where the variables (x,y,z,ϕ) are defined in Figure 20.16:

$$\text{Horizontal Angle} = \tan^{-1} (x / z)$$
$$\text{Vertical Angle} \ = \tan^{-1} (y / z)$$

Detection of the laser spot in the image is then used to determine if the object is shielding the gripped centre, if this is so, then the spot is stepped over the contiguous object surface and the silhouette (in the laser plane) is determined.   Controlled

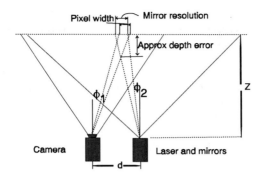

Figure 20.16   *Plan view of laser and camera*

movement of the second robot can now be undertaken in order to move to the edge of the silhouette and into planned contact with the object surface. Touch and force sensing are still required to perform a secure grip. The absolute resolution of the laser ranging system is determined by a number of factors [25].

20.4.2 Detailed multi-arm contact analysis

There are several approaches to measuring multi-arm cooperative control parameters, which each have relative merits in sensing and cost, the appropriateness of these depends on the application to be carried out [26]:

    1)   Global sensing
    2)   Single sensored arm
    3)   Dual arm sensing
    4)   Internal data sensor conditioning

In the global sensing environment there are only certain parameters which can be measured such as relative position from a fixed camera position. This approach is cheap and easily adapted to other robot systems with changes being only required in the robot and environment kinematic description. It is also more robust with fewer requirements for cabling and data transfer.

    The single sensored arm has the advantage of low cost and good accuracy in estimation of most parameters, and in many configurations, for example in Figure 20.17 if A lifts the object, the object can be weighed and the centre of gravity identified. In subsequent co-operative operations the weight and centre of gravity can be compensated for and forces of compression on the object determined. If B had lifted the object first then this information would not be calculable unless available beforehand. Thus with single arm sensing, planning must be used or *a priori* information must be available in order to carry out useful tasks.

    One area where a single sensing arm is viable is in the area of eye-in-hand systems where a single camera can be used to determine contact positioning and also to inspect the held object. This approach again is constrained by planning, since a large object held by the gripper with the camera will effectively hide the second gripper and the environment.

Dual arm sensing can measure most parameters adequately and through conflict can indicate failures. Through the difference of force sensor measurements the relative weight of commonly held objects can be found. More accurate force estimation can be made through use of orthogonal measurements and agreement can be found between conflicting measurements, e.g. due to errors.

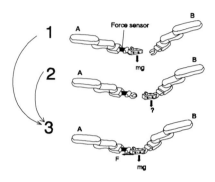

Figure 20.17   *Planning for effective sensor usage*

Use of internal sensors and extra parameters such as belt stretching and motor current can provide adequate measures of force with little hardware cost but significant processing.

In a low-cost geared system, low motor current can indicate that the backlash of a motor was being wound up in the internal gearing and increases in control actuation can speed up this process. The current can also indicate joint torques, and through estimation of velocity and acceleration can indicate the force exerted onto the held object. This would be estimated through use of a motor parameter based model and the forward kinematics.

20.5   Data transfer and processing

When information has been sensed about the environmental and internal condition of the robot it must then be transferred to the appropriate decision processes, which will be at other locations on the network due to:

1) high processing cost of sensing and of condensing data,
2) the natural split of processing, e.g. to each of the robot arms,
3) memory restrictions of individual transputer modules.

The transfer rates can normally be facilitated quite easily on the networks, to almost anywhere, within the required sample rate time. Transmission times are restricted by transputer process time-slicing and effective use of assembly language communication kernels.

The major problems in the data handling area are due to communication of diagnostics data and recorded response data with the host. This causes processing delays even when simple buffering is used, as demonstrated by non-continuous motion of robots during movements if interaction with the host hard disk is required. In order to remove this problem, high memory size trams are placed on the host, and arrays are used for temporary storage and decoupling of the storage cycle.

In addition to the encoders, a force sensor is used to measure the interaction between grippers and contacted objects. This sensor was monitored by part of the same transputer network, see Figure 20.15, and its information was added to the position sensing and belt tension derived torque. This sensed information has been used initially in order to permit overload detection and reduction of possible damage in dangerous configurations.

The distributed nature of the sensing and processing reduces the availability and reliability of the system. In order to monitor the system state and reliability the system state is overseen by a neural learning system, which will ultimately indicate the existence of failure and diagnose its source.

20.6 Discussion

Special sensing is required to carry out multi-arm tasks. This sensing depends on the co-operative task to be executed and more effective use of the presently available sensing. External sensing is required to reduce antagonism in co-operative actions and also to increase reliability in uncertain situations. The large processing requirements and localised nature of processing suit the division the of system processing into a distributed network, the transfer of data to controllers and arbitrators is sufficiently fast to maintain real-time control. Multi-arm systems require high speed processing and formulation of algorithms which minimise the new problems of co-operation. Planning for tasks requires an evaluation of new cost functions for determining movement choices.

20.7 Conclusions

It has been found for multiple sensing and actuation that parallel processing is essential. When using this approach care must be taken to ensure cross network data rates are not detrimental to algorithms processing, condensation of data and close positioning of associated data sources and sinks is required. Initial inter-data source calibration is a significant problem in the fusion of the data sources and is a major workload. Identification of system parameters and adaption to change is essential to increase the effective range of derivative parameters available.

The system has provided a useful platform for parallel processing and multi-arm application development. Simulation facilities and the user interface will give an operator the confidence required to permit the introduction of multi-arm systems to a real application. Further work is investigating the enhancement of CAD aspects for inputting the required tasks and the workspace.

Further work is extending the use of parallel processing and integration of the off-line overall task planning to the transputer based minimum time and smooth motion task, with collision avoidance.

20.8 Acknowledgements

The author acknowledges the contribution of Dr B. Cao to this work. The author also wishes to thank the Industrial Research and Technology Unit for their support under grant ST79.

20.9 References

1.   DODDS, G. and IRWIN, G.W. 'Evaluation and comparison of practical multi-arm systems', invited paper International Symposium on Robotics and Manufacturing, 1994.
2.   FREUND, E. and HOYER, H. 'Pathfinding in multirobot systems', Proceedings IEEE International Conference on Robotics and Automation, 1986, pp 103-111.
3.   TARN, T.J. 'Design of dynamic control of two cooperating robot arms', Proceedings IEEE International Conference on Robotics and Automation, 1987, pp 7-13.
4.   ZHENG, Y.F. and LUH, J.Y.S. 'Optimal load distribution of for two industrial robots handling a single object', Proceedings IEEE International Conference on Robotics and Automation, 1988, pp 344-349.
5.   NAGATA, T. 'Multirobot path generation in a continuous domain', IEEE Journal of Robotics and Automation, 1988, pp 2-13.
6.   DODDS, G and IRWIN, G.W. 'Planning a collision free cooperating robot environment', Proceedings of Conference on Robotics, Institute of Mathematics and its Application, 1989, Oxford University Press.
7.   DODDS, G. and GLOVER, N. 'Simulation of a multimanipulator demonstrator', Proceedings of United Kingdom Simulation Conference, 1990, pp 96-101.
8.   DODDS, G. and IRWIN, G.W. 'Multi-arm robotics in a practical application under transputer based control', IEEE International Conference on Robotics and Automation, 1993.
9.   MACFARLANE, T. et al, 'Development of a PC and Transputer based walking robots', IEE Colloquium on Multi-arm Robotics, 1992, Digest 1992/147, London, UK.
10.   UEYAMA, T. and FUKADA, T. 'Configuration of communication structure for distributed intelligent robotic system', IEEE Proceedings Robotics and Automation Conference, 1992, pp 807-812.
11.   CHOI, M.H. et al, 'An application of the force ellipsoid to optimal load distribution for two cooperating robots', IEEE Proceedings Robotics and Automation Conference, 1992, pp 461-466.
12.   TAO, J.M. et al, 'Compliant coordination of two moving industrial robots', IEEE Trans. Robotics and Automation, Vol. 6, 1990, pp 322-330.
13.   DODDS, G. and GLOVER, N. 'Simulation and control of a cooperative robotic system', Proceedings of IEE Conference on Control, 1991, pp 1218-1223.
14.   DODDS, G., CAO, B., and WILSON, G. 'A flexible joint robot in cooperative tasks with sensor fusion', accepted by the IEE Conference on Control, 1994.
15.   CAO, B., DODDS, G., and IRWIN, G.W. 'Fast collision detection and avoidance for multirobot control', IASTED International Conference on Robotics and Manufacturing, 1993.
16.   FREUND, E. 'On design of multi-robot systems', IEEE Proceedings Robotics and Automation Conference, 1984, pp 477-90.

17. CHU, H. and ELMARGHY, H.A. 'Real-time multi-robot planner based on a heuristic approach', IEEE Proceedings of Robotics and Automation Conference, 1992, pp 475-480.

18. KASAGAMI, F. et al, 'Coordinated motion of arc welding robots using a parallel data processor', IEEE Proceedings of Industrial Electronics Conference, 1992, pp 656-663.

19. GUPTIL, R., and AHMAD, S. 'Multiprocessor based trajectory generation for multiple robotic devices, in 'Parallel Computation Systems for Robotics' edited by FIJANY, A and BEJCZY, A, World Scientific Publishing Company, 1992.

20. DUFFY, N.D., and HERD, J.T., 'A societal architecture for robotic applications', IEE Computing and Control Engineering Journal, November 1991, pp 269-274.

21. NAGATA, T. et al, 'Multi-robot plan generation in a continuous domain: planning by use of plan graph and avoiding collisions among robots', IEEE Trans. Robotics and Automation, Vol. 4, 1988, pp 2-13.

22. ZHENG, Y.L. et al, 'A real-time distributed computing system for coordinated motion of two industrial robots', IEEE International Conference on Robotics and Automation, 1987, pp 1236-1239.

23. BROOKS, R.A. 'Solving the find-path problem by good representation of free space', IEEE Trans. Sys. Man Cyber., Vol.13, 1983, pp 190-197.

24. CAO, B., and DODDS, G. 'Time-optimal and smooth joint path generation for robot manipulators', IEE International Conference Control, 1994, to be published March 1994.

25. DURANT-WHYTE, H.F. 'Integration, coordination and control of multi-sensor robot systems', Kluwer Academic Press, 1988.

26. DODDS, G., IRWIN G.W., ZALZALA, A.M.S., and OGASAWARA, T. 'Basic architectures and processing for multi-robot control', IEE Colloquium on 'Applications of Parallel and Distributed Processing in Automation and Control', November 1992, Digest No: 1992/204, pp 8/1-8/6.

Chapter 21

Neural networks in automation procedures

J M Bishop, R J Mitchell and K Warwick

Human labour remains widely used in many production systems, although this is often not ideal due to such factors as inconsistencies, high costs, and physical dangers. However, replacing a specific individual is often expensive, even when the individual's role appears to be simple and relatively mundane. Tasks that appear to be very simple, such as simple visual inspection, can often be the most difficult to automate. Conversely tasks that to the layman appear very complex, such as the interpretation of mass spectography data, can be performed by a computer *'Expert System'* relatively easily. In between these two domains lies a third area; the fuzzy grey terrain of the professional craftsman. It is in these first and third areas, which to date have proved intractable to conventional computer approaches to automation, that neural methods have shown particular promise. This chapter will describe two industrial problems and illustrate how neural computing is being used to help in their automation.

21.1 Introduction

The discussion begins with a short introduction to neural networks in general. This is followed by a section describing the two forms of neural network used in the ensuing case studies: the n-tuple or weightless network and the multi-layer perceptron with back propagation learning.

A system developed for the visual inspection application is described which is based on the n-tuple net framework and which operates at video frame rate, i.e. 50 frames per second. This speed of operation is more than adequate for most production line inspection, in that up to 50 items per second can be checked as they pass a fixed point. The system is hardware based so that a very simple set up and start procedure is needed in order to make it operative. The overall set up is low cost and simple to use.

A second application is then described, that of replacing the trained colourist in a commercial colour recipe prediction environment. An initial description of the problem is given, together with an outline of conventional, non-neural approaches to its automation. This is followed by a short overview showing how neural techniques can be simply incorporated into standard recipe prediction systems and hence easily integrated into current automation procedures.

21.2   What are Neural Networks ?

The last decade has witnessed the re-emergence of neural networks following a period of academic isolation, and there is considerable interest in the area all over the world: for example, in the USA DARPA have committed over $33 million for work in the field; in the UK there have been several government sponsored research programmes, the most recent being a £5.75 million initiative - the Neural Computing Technology Transfer Programme (NCTTP). It is expected that neural network technology will revolutionise many aspects of industrial computing.

The field of neural networks is variously described as connectionism, parallel distributed processing, connection science, or neural computing. These are all synonymous and define a mode of computing which seeks to include the style of computing used in the brain. In their recent book [1] Aleksander and Morton have the following definition:

'Neural computing is the study of networks of adaptable nodes which, through a process of learning from task examples, store experiential knowledge and make it available for use.'

A neural network, like a computer, is a processor of information. Networks consist of simple processing elements connected together. Each processing element is a very simple model of a neuron in the brain: hence the term neural networks. Thus a neural network could be described as mankind's attempt to create an artificial brain. However, present neural networks only try to mimic the way the brain does things in order to harness its ability to infer and induce from incomplete or confusing information.

What makes these networks powerful is their potential for performance improvement over time as they acquire more knowledge about a problem, and their ability to handle fuzzy real world data. That is, a network *'taught'* certain data patterns, is able to recognise both these patterns and those which are similar: the network is able to *generalise*. Also, neural networks are inherently parallel in their operation, and so have the capacity to operate much faster than conventional computers when implemented on suitable hardware.

A conventional computer operates an algorithm. To program a computer a suitable algorithm is required, and this is developed by a human. Thus a conventional computer can only do those things for which humans can find a suitable algorithm. However, a human information processor is capable of developing behaviour through learning. Neural networks are an attempt to produce machines with such capabilities.

As stated earlier a neural network consists of simple processing elements connected together. There are various ways in which these elements can be connected, in single or multiple layers, fully or partially connected, etc., and there are various forms of processing elements. For the vision system, a single layer network is used and the processing elements are implemented using standard memory cells. For the colour problem a multi-layer perceptron network is used with processing units based on a simplified single neuron model.

A neural network is not programmed to complete a given task, rather it adapts and acquires knowledge over time in order to complete the task. One stage of operation, therefore, requires the network to learn (although this may continue throughout the operation of the network).

Once the network has learned to perform a task, it can then be set to undertake that task. For example, the vision system learns the objects it is to recognise and then is able to report if an unknown object belongs in that class; the recipe prediction system learns a set of reflectance values and their associated chemical recipes, and then is able to predict the reflectance curves of further recipes that it has not been taught.

There are various forms of neural network, each of which has its own advantages and disadvantages. The most commonly used are multi-layer perceptrons [2] (of which there are many forms) and Kohonen networks [3]. Other types include Hopfield networks [4], Boltzmann machines [2], Fukushimas Neocognitron [5], Grossberg and Carpenters ART classifier [6] and n-tuple networks [7]. For the vision application, n-tuple networks are most appropriate as they operate rapidly, are simple to teach, and can be easily implemented in readily available hardware [8]. In the recipe prediction system a multi-layer perceptron was most appropriate.

21.2.1 n-Tuple networks

In the UK Igor Aleksander is the pioneer in the field of n-tuple networks, also called weightless or digital networks, although there is now a large group in the Cybernetics Department at the University of Reading investigating various aspects of these networks. The main difference between these and the more common *perceptron* type networks is their different model of a neuron and hence the different learning strategy required.

The n-tuple neuron operates on binary data and the node is implemented by a standard digital memory. As shown in Figure 21.1, the inputs to the neuron are used as address lines into the memory, and the output of the neuron is the data value stored at the specified address in the memory.

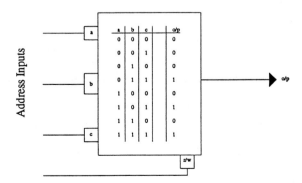

Figure 21.1. *A three input RAM neuron*

In learn mode, the input pattern to be remembered is presented to the address lines of the memory, and a suitable value stored there. In analyse mode, the same procedure for addressing the node is adopted, except the value stored at the address in the memory is output. An advantage of this form of network over the perceptron type is that data values are stored in one presentation of the data (one-shot learning), rather than presenting the training set many times, which is required for perceptron type networks.

n-Tuple networks can be configured and operated in various ways using different topologies with multiple layers and/or with feedback [9,10], and they can store multiple bits of data in each memory location: the so-called probabilistic logic neuron [9]. However, for the vision system described here the simplest configuration with only a single layer of single-bit neurons only is needed.

The following description shows how such a simple configuration of neurons can be used to recognise data patterns, which is the technique used in the industrial inspection application described in this chapter. This also shows how such a network can generalise, that is, it can also recognise patterns similar to, though not identical to, data patterns it has already been taught.

21.2.1.1   *n-Tuple network used as a vision system*
A block diagram of the network is shown in Figure 21.2: each data pattern consists of r bits of information, there are m neurons (or memories) and each neuron has n inputs. Initially,'0' is stored in each location in each neuron.

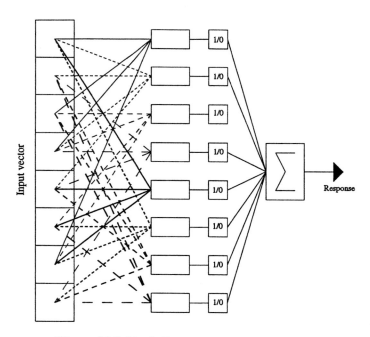

Figure 21.2 *Block diagram of an n-tuple network*

In the learn phase, n bits are sampled from the input pattern (these n bits are called an n-tuple). This tuple is used to address the first neuron, and a '1' is stored at the specified address: effectively the neuron has 'learnt' the first tuple. Another n bits are sampled and these are 'learnt' by the second neuron. This process is repeated until the whole input pattern has been sampled. Note that it is usual to sample bits from the memory in some random manner, as this enables the system to maintain a sensitivity to global features of the image.

In the analyse phase, the same process is used to sample the input pattern so as to generate the tuples. However, instead of writing '1's into the neurons, a count is made to see how many of the neurons had 'learnt' the tuples presented to them: that is, how many neurons output a '1'. If the input pattern had already been taught to the system, the count would be 100%. If the pattern was similar, then the count would be slightly smaller.

In practice, many similar data patterns should be presented to the system, so each neuron would be able to 'learn' different tuples. Therefore the count when a pattern similar to but not identical to any member of the training set is shown to the system might still be 100% because, for example, the first tuple from the image might be the same as one from the third data pattern taught, and the second tuple might be the same as one from the seventh data pattern, etc. In general, the system is said to recognise an object if the count is greater than p%, where p is chosen suitably and is likely to depend on the application. This allows the system to recognise objects despite the noise which will inevitably affect the image.

Therefore such a system is capable of being taught and subsequently recognising data patterns. With a slight extension the system is able to be taught, recognise and discriminate between different classes of data. One set of neurons, as described above, is called a discriminator. If two sets of data are stored in one discriminator, the system will be able to recognise both sets, but not be able to discriminate between them. However, if the system has two discriminators and each data set is taught into its own discriminator, then when a data pattern is presented, a count of the number of recognised tuples is made for each discriminator, and the input pattern is more like the data class stored in the discriminator with the highest count.

21.2.2 The multi-layer perceptron (MLP)

MLPs consist of collections of connected processing elements that are not individually programmable. Each cell usually computes a simple non-linear function, f, on the weighted sum of its input (see Figure 21.3). The output of this function is defined as the activation of the cell. Long-term knowledge is stored in the network in the form of interconnection weights linking these cells.

The MLP used in this case study consisted of an input layer where cell inputs are clamped to external values (a particular dye recipe). A hidden layer, where cell inputs are defined by the weighted activation values from the cells in the previous layer that they are connected to, and an output layer connected to the hidden layer (see Figure 21.4). On presentation of input, the network rapidly settled into a stable state consistent with it.

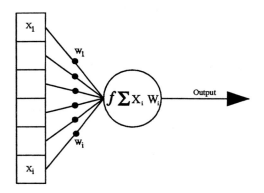

Figure 21.3 *Generalised McCulloch/Pitts neuron*

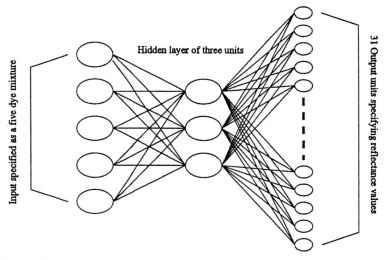

Figure 21.4 *A neural network to compute reflectance values from specified colorants*

In an MLP, learning is the process of defining a set of weights that will produce a desired response to an input pattern. The learning strategy used in this work is known as back propagation of the generalised delta rule [2].

Given enough cells and having learnt the right set of inter-cell weights, a network of this sort can produce input/output mappings of arbitrary complexity and can find them quickly as all of its knowledge operates in parallel. The process of finding the optimum set of inter-cell weights is very computationally intensive, however this initial learning phase is usually only executed once. For the colour work described in Section 21.3, learning took around 48 hours on a fast (386) PC, whereas in use the network took less than a tenth of a second to predict a recipe.

21.2.2.1 Back propagation using the generalised delta rule

Many different learning rules have been developed for neural networks. One of the key developments in neural network research in the past decade has been the development of learning rules that can teach hidden units. One such rule is the generalised delta rule, which propagates an error term backwards through the net from the output units to the input units adjusting inter-cell connection strengths such that its overall prediction error is minimised. A complete introduction to this learning rule [2] is beyond the scope of this chapter, however a brief description will be given.

The generalised delta rule works by performing gradient descent in error/weight space. That is, after each pattern has been presented, the resulting error on that pattern is computed, by comparing the actual output with the desired output, and each weight in the network modified by moving down the error gradient towards its minimum for that input/output pattern pair. Gradient descent involves changing each weight in proportion to the negative of the derivative of the error, defined by this pair.

The generalised delta rule is based on the original delta or Widrow-Hoff learning rule [11]. The weight update procedure for this rule, given an input vector $\mathbf{I}$, an output vector $\mathbf{O}$, a target vector $\mathbf{T}$ and a weight matrix $\mathbf{W}$ is:

$$\Delta w_{ji} = \eta \ (t_j - o_j) \ i_i$$

$$= \eta \ \delta_j \ \mathbf{i_i} \tag{21.1}$$

where Δw_{ji} is the change to be made to the weight linking the jth to the ith unit, given the current input/output vector pair. η is defined as the learning rate constant. δ_j can be considered as an error term describing the difference between the desired output and the actual output. The key development, by the PDP research group at the University of California, was the extension of the above learning rule for single layer networks, to function with multi-layer nets. The basic form of the generalised delta rule is identical to the simple delta rule:

$$\Delta w_{ji} = \eta \ \delta_j \ \mathbf{i_i} \tag{21.2}$$

At the output layer the error signal is:

$$\delta_j = (t_j - o_j) f^1_j \ (net_j) \tag{21.3}$$

and for a hidden layer the error signal is:

$$\delta_j = f^1_j \ (net_j) \Sigma_k \ \delta_k \ w_k \tag{21.4}$$

where $f^1_j (net_j)$ is the derivative of a semi-linear activation function acting on the jth unit, which maps the total input to the unit to an output value. A semi-linear function must be differentiable, continuous, monotonic and non-linear. The activation function used in this study is the logistic activation function:

$$o_j = 1/(1 + e^{-net_j}) \qquad (21.5)$$

where $net_j = (\Sigma_i \, w_{ji} \, o_i + q_j)$. q_j is a bias term and can be learned like any other weight by considering it connected to a unit which is permanently on. The derivative of f is thus:

$$\partial o_j / \partial net_j = o_j \, (1 - o_j) \qquad (21.6)$$

Thus, for an output unit the error signal is:

$$\delta_j = (t_j - o_j) \, o_j \, (1 - o_j) \qquad (21.7)$$

and for a hidden unit the signal is:

$$\delta_j = o_j \, (1 - o_j) \, \Sigma_k \, \delta_k \, w_{kj} \qquad (21.8)$$

21.2.2.2 Learning Rate η

The above learning procedure requires only that the change in weight is proportional to $\partial E / \partial W$, whereas true gradient descent requires infinitesimally small steps to be taken. The constant of proportionality is the learning rate constant, η. The larger this value, the larger the changes in weight at each iteration and the faster the network learns. However if the learning rate is too large, then the network will go unstable and oscillate. It has been shown [2] that one way to increase the learning rate, without leading to oscillation, is to introduce a momentum term, i.e.

$$\Delta w_{ji} \, (n + 1) = \eta \, (\delta_j \, o_i) + \alpha \Delta w_{ji} \, (n)$$

where n indexes the current input/output presentation number, η is the learning rate constant and α is the momentum constant. This defines how much past weight changes affect the current direction of movement, providing a force analogous to momentum in weight space, which acts to filter out high frequency variations in the error surface. For example, if the network arrives at a gradually descending ravine in weight error space, the steepest error gradient may be mainly across, rather than down, the ravine. The use of a momentum term tends to filter out such sideways movement, while compounding movement down the ravine.

21.2.2.3 Using back propagation

In one learning sweep with the generalised delta rule, the following sequence of events occurs:

i)   The network is presented with an input pattern (all the input units of the network are set to the required values).

ii)   This input vector is used to compute the output values by feeding forward through the net and computing activation values for all the other units according to the weight values.

iii)   The output vector for this pattern is compared to the required, or target, pattern and the error term is calculated for every output unit.

iv)   Error terms are recursively propagated backwards through the net to the other units in proportion to the connection strengths between the units.

v)   The weights are then adjusted in such a way as to reduce the error terms, by performing gradient descent in weight error space, in a similar manner to the simple delta rule.

vi)   The process is repeated for all the input/target pattern pairs in the training set.

The process of presenting to the network all the patterns over which it is to be trained is defined as one epoch. Training continues for as many epochs as are necessary to reduce the overall error to an acceptably low value. The theory behind these two types of network has been presented. In the next two sections it is shown how these networks are utilised in the vision and recipe prediction systems.

21.3   A neural vision system for high speed visual inspection

On line visual inspection is an important aspect of modern day high speed quality control for production and manufacturing. If human operatives are to be replaced on production line monitoring however, the inspection system employed must exhibit a number of human-like qualities, such as flexibility and generalisation, as well as retaining many computational advantages, such as high processing speed, reproducibility and accuracy. An important point in considering human replacement though, is total system cost, which should include not only the hardware, software and sensor system, but also commissioning time and problem solving. This section will describe a vision inspection system based around n-tuple neural networks. Important characteristics of the system include its speed of operation, low cost, ability to generalise and, most significantly, the ease with which it can be integrated into real production systems.

21.3.1 A practical vision system

The basic aim in the design of the vision system is that it should be flexible and easy to operate. The n-tuple network method provides the flexibility, the careful system design provides the ease of operation. There are many vision systems on the market, and these often require much setting up or programming, details of which are often specific to a given application. For our vision system, the intention was to allow the user to set up the system using a series of switches and light indicators. These are available on a simple front panel whose layout is shown in Figure 21.5.

Another aim of the system is low cost. Therefore in the configuration described here, there is only one class discriminator: this is sufficient for many applications. However, the system has been designed in such a way that extra discriminators can be added easily, with a minimum of extra hardware.

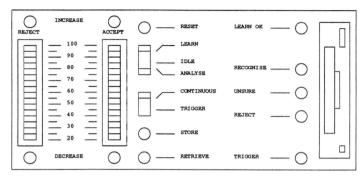

Figure 21.5 *Front panel of vision system*

The basic operation of the system is in one of two modes: learn or analyse. In learn mode the image presented to the system is learnt, whereas in analyse mode the system indicates if the image being presented is sufficiently similar to the image or images it has been taught. Both actions are achieved by simply arranging for the object under test to be shown to the camera.

Each of these modes can be continuous, whereby the system processes an image at each frame time (50 frames per second), or in trigger mode, when the system only processes an image when it receives a suitable signal. These modes are selected using switches. An important point when learning is to teach the system a number of slightly different images, and this is easily accomplished by allowing the object being taught to pass the camera slowly along a conveyor belt; for this continuous mode is useful. However, if an object can come down the production line in various orientations, it is important to train the system by showing it the object in many orientations. This can be achieved in trigger mode, by placing the object in front of the camera, learning it by pressing the trigger button, turning the object slightly, learning the object, etc.

It is important not to saturate the memory, as then the system will recognise almost anything. Therefore the system will automatically stop learning when the discriminator is about 10% full. A light on the front panel indicates when to stop learning.

As regards analysing the object, a decision has to be made as to what is an acceptable image, and what is unacceptable, that is, what percentage of recognised tuples indicates a good object. A novel feature of this vision system is that it provides an extra degree of control: it allows the user to select one level, above which the system accepts the image, and a second level, below which the system rejects the object. Those objects in between are marginal, and perhaps need a human to examine them to determine the appropriate action. By selecting the same level for each, the system just responds pass/fail. The levels are indicated by two LED bars, and these are adjusted up or down by pressing appropriate buttons. The system response is indicated on the front panel with lights for RECOGNISED/UNSURE/REJECT, and there are corresponding signals on the back which can be used to direct the object appropriately.

Although teaching an object to the system is very easy, to save time it is possible to load a previously taught object from disk and, obviously, to store appropriate details about an object. Loading and saving is accomplished by pressing the appropriate buttons. The data files contain both the contents of the discriminator memory and the accept and reject levels.

21.3.2 Implementation

As the system requires hardware for grabbing and processing the input video data, as well as being able to process the front panel switches and access disk drives, it was necessary to include in the system a microprocessor to co-ordinate the action. However, most of the processing is accomplished by the special hardware, so the microprocessor did not need to be very powerful, and its choice was largely due to the available software devlopment facilities. The n-tuple network requires various hardware modules, and the system needs to access the disk and the front panel, and so it was considered sensible to make the device in modules. A convenient method of bringing these modules together is to use a bus system, and a suitable bus is the STE bus, about which the Department has considerable expertise [12]. The main processor board used is DSP's ECPC which provides an IBM-PC compatible computer on one board with all the necessary signals for controlling the disk, etc., and much software is available for the PC.

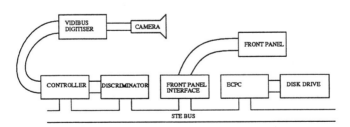

Figure 21.6 *Block diagram of system*

A block diagram of the system is shown in Figure 21.6. The object under study is placed in front of a standard video camera, and a monitor is used to indicate to the user that the camera is aligned correctly: this is only needed in learning mode. The output from the camera is then digitised using a module from the departmental video bus system VIDIBUS [13], and this is then fed to the n-tuple hardware.

There are two parts to the n-tuple network, the controller and the discriminator. The former samples the image, forms the tuples, etc., whereas the discriminator contains the memory in which the tuples are taught and counters which indicate whether the memory is saturated (in learn mode) and the number of tuples which have been recognised (in analyse mode). The discriminator is in a separate module so as to allow a multi-discriminator system to be produced relatively easily.

The ECPC is used to configure the controller, for example to tell it to process the next frame or to specify whether to learn or analyse, and to process the saturation and recognition counters. Such actions are determined by the controls on the front panel which the ECPC processes continually via a suitable interface. The last block is the disk drive, the controlling hardware/software for which is on the ECPC.

The software for the system is quite simple, as most of the processing is performed by special purpose hardware. Essentially the software monitors the front panel, configures the controller, checks the discriminator counters and processes the disk. This is all achieved by a simple program written in Turbo-C.

When developing the system, the ECPC was used as a standard PC running a compiler, etc. When the system was complete, however, EPROMs were blown containing the control program so that the system boots from the EPROM and straight away runs the program. Therefore the system does not require a disk, though it is useful for saving and reloading taught objects.

21.3.3 Performance

The system has been tested on a variety of household products, including coffee jars, packets of cleaning material and hair lotion, and has been shown to operate in the expected manner. One slight problem is that some care is required in setting up the lighting conditions, however recent work has shown this problem can be reduced [14]. The complete system has been compared favourably with other, non-neural systems, for use by Smith Kline Beecham (Toiletries) on their Brylcreem production line [15].

21.4 The application of neural networks to computer recipe prediction

An important aspect of the quality control of manufacturing processes is the maintenance of colour of the product. The use of colour measurement for production and quality control is widespread in the paint, plastic, dyed textile and food industries. An industrial colour control system will typically perform two functions relating to the problems encountered by the manufacturer of a coloured product. First the manufacturer needs to find a means of producing a particular colour. This involves selecting a recipe of appropriate dyes or pigments which, when applied at a specific concentration to the product in a particular way, will render the required colour. This process, called recipe prediction, is traditionally carried out by trained colourists who achieve a colour match via a combination of experience and trial and error. Instrumental recipe prediction was introduced commercially in the 1960s and has become one of the most important industrial applications of colorimetry. The second function of a colour control system is the evaluation of colour difference between a batch of the coloured product and the standard on a pass/fail basis.

Conventional computer mechanisms for colorant formulation (recipe prediction) commonly employ Kubelka-Munk theory to relate reflectance values to colorant concentrations. However there are situations where this approach is not applicable and hence an alternative is needed. One such method is to utilise artificial intelligence techniques to mimic the behaviour of the professional colourist. The following section will describe research carried out at Reading University, originally sponsored by Courtaulds Research, to integrate neural techniques into a computer recipe prediction system.

21.4.1 Introduction

Since the early development of a computer colorant formulation method, computer recipe prediction has become one of the most important industrial applications of colorimetry. The first commercial computer for recipe prediction [16] was an analogue device known as the COMIC (colorant mixture computer) and this was superseded by the first digital computer system, a Ferranti Pegasus computer, in 1961 [17].

Several companies today market computer recipe prediction systems which all use digital computers. All computer recipe prediction systems developed to date are based on an optical model that performs two specific functions:-

i)   The model relates the concentrations of the individual colorants to some measurable property of the colorants in use.

ii)   The model describes how the colorants behave in mixture.

The model that is commonly employed is the Kubelka-Munk theory [18] which relates measured reflectance values to colorant concentration via two terms K and S of the absorption and scattering coefficients respectively of the colorant. In order for the Kubelka-Munk equations to be used as a model for recipe prediction it is necessary to establish the optical behaviour of the individual colorants as they are applied at a range of concentrations. It is then assumed that the Kubelka-Munk coefficients are additive when mixtures of the colorants are used. Thus it is usual for a database to be prepared which includes all of the colorants which are to be used by the system and allows the calculation of K and S for the individual colorants.

The Kubelka-Munk theory is in fact an approximation of an exact theory of radiative transfer. Exact theories are well documented in the literature [18], but have rarely been used in the coloration industry. The Kubelka-Munk approximation is a two-flux version of the many-flux [19, 20] treatment for solving radiative problems.

In order for the Kubelka-Munk approximation to be valid the following restrictions are assumed:

i)   The scattering medium is bounded by parallel planes and extends over a region very large compared to its thickness.

ii)   The boundary conditions which include the illumination, do not depend upon time or the position of the boundary planes.

iii)   The medium is homogeneous for the purposes of calculation.

iv)   The radiation is confined to a narrow wavelength band so that the absorption and scattering coefficients are constant.

v)   The medium does not emit radiation (e.g. fluoresce).

vi)   The medium is isotropic.

There are many applications of the Kubelka-Munk approximation in the coloration industry where these assumptions are known to be false. In particular, the applications to thin layers of colorants (e.g. printing inks [22]) and fluorescent dyestuffs [23, 24] have generally yielded poor results.

The use of an approximation of the exact model has attracted criticism. For example, Van de Hulst [25] when discussing its application to paint layers comments: *'it is a pity that all this work has been based on such a crude approximation...'*

The popularity of the Kubelka-Munk equations is undoubtedly due to their simplicity and ease of use. The equations give insight and can be used to predict recipes with reasonable accuracy in many cases. In addition, the simple principles involved in the theory are easily understood by the non-specialist and rightly form the basis for study for those involved in the coloration industry [26].

The use of the exact theory of radiation transfer is not of practical interest to the coloration industry. The calculations generally require databases and spectrophotometers of a greater complexity than those suitable for Kubelka-Munk calculations.

21.4.2 Computer recipe prediction

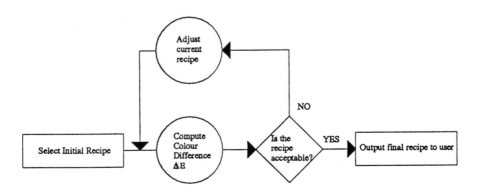

Figure 21.7 *Block diagram of a typical computer colour recipe prediction system*

The operation of a typical recipe prediction system is as follows (see Figure 21.7):

i)   The computer makes an initial *guess* at a suitable recipe for the specified colour, based on historical data and the choice of dyes available.

ii) The reflectance values for the given recipe are computed using Kubelka-Munk theory.

iii) The colour difference (ΔE) between the generated recipe and the desired colour is computed using the CMC 2:1 colour difference equation.

iv) If the ΔE value is within specification the recipe is acceptable and is presented to the user. Otherwise, the recipe is modified (typically using a gradient descent optimisation technique) and the procedure repeated from step (ii).

21.4.3 Neural networks and recipe prediction

As discussed above, one of the problems with conventional recipe prediction is that the application of exact colour theory is not computationally practical and an approximation to it has to be employed. It was expected that a neural network approach to reflectance prediction would offer a novel new approach to this problem, since many problems in artificial intelligence (AI) involve systems where conventional rule based knowledge is not complete or the application of the pure theory is too computer intensive to be used in practical systems.

To investigate whether a neural network can mimic Kubelka-Munk theory and successfully predict reflectance values from specified recipes, a data set was obtained that contains a list of measured reflectance values (at 5nm intervals between 400 and 700nm i.e. 31 wavelengths) with their corresponding recipes (concentrations of colorants used to generate the samples). The total number of colorants in the set was five, and recipes could be combinations of any number of these five dyes. Two of the five colorants were fluorescent, hence some reflectance values are greater than 1.0, and thus the data would present a considerable challenge to Kubelka-Munk theory.

A simple multi-layer perceptron architecture was chosen (see Fig.21.4). The network has five input units, three hidden units in a single layer, and 31 output units. The network is therefore designed to map recipe to reflectance. It is known that this mapping is a many-to-one mapping since there may be more than one way of generating a set of reflectance values from a large set of colorants.

Input and output data were scaled before presentation to the network and 200 samples were used for training the network. The network was trained using the standard back-propagation algorithm. The learning rate (initially 0.3) and momentum (initially 0.5) terms were reduced during training at epochs (200,000; 400,000; 700,000) with the complete training lasting approximately 3,000,000 epochs (an epoch being defined as the presentation of a single sample chosen at random from the 200 sample data set).

21.4.3.1 Performance
After learning, a measure of the effectiveness of the network was made using the test data. The output data range from 0 to 100 and the average output error (averaged over all 31 output units and all 200 training samples) is 3.41. This corresponds favourably with the performance of the best Kubelka-Munk model of the data even though it is known that the training data were not evenly spread through colour space, making successful modelling of the data harder. Analysis of the weights in the trained network showed a trichromatic representation of reflectance data. One of the hidden units was responsible for the blue part of the spectrum, one for the red part, and one for the green part. A chapter by Usui [27] showed that a network could be used to compress reflectance data into three hidden units; two units were insufficient and when four units were used a correlation was found between two of them.

21.4.4 Discussion

The performance of the neural network on the specified reflectance prediction task is very promising - training on a second confidential commercial data set being significantly better than the results quoted above. Earlier attempts at using neural networks in colour recipe prediction concentrated on the inverse problem of predicting dye concentrations from a colour specification [28, 29]. Not only is this a more difficult task (it is a one-to-many mapping), but such a system, even if perfect, would have limited commercial appeal since the user needs to retain control of the colorants to be used. The method outlined above enables, for the first time, a neural solution to be embodied as a black box replacement for the Kubelka-Munk stage in any recipe prediction system, and hence promises to have widespread use within the colour industry.
    The significance of the network's hidden nodes arriving at a trichromatic representation of reflectance data is currently being investigated.

21.5   Conclusion

The examples described in this chapter illustrate two ways in which neural networks are being successfully used in the automation of very complex tasks. Although neural networks cannot be used as a solution to all problems, there is no doubt that they can be a useful tool to aid the system designer in automating otherwise intractable problems.

21.6   References

1.   ALEKSANDER, I. and MORTON, H. An Introduction to Neural Computing. Chapman and Hall 1990.
2.   RUMELHART, D.E. and McCLELLAND, J.L. Parallel Distributed Processing, MIT Press 1986.
3.   KOHONEN, T. Self Organisation and Associative Memory. Springer-Verlag 1984.
4.   HOPFIELD, J.J. Neural networks and physical systems with emergent collective properties. Proc. Nat. Acad. Sci. USA 79, 2554-8.

5.   FUKUSHIMA, K. Neocognitron: a hierarchical neural network capable of visual pattern recognition. Neural Networks, 1, 119-130, 1988.

6.   GROSSBERG, S. and CARPENTER, G.A. The ART of Adaptive Pattern Recognition by a Self Organising Neural Network. IEE Computing Magazine, Mar 1988.

7.   ALEKSANDER, I. and STONHAM, T.J. A Guide to Pattern Recognition using Random Access Memories. IEE Jrn. Cmp. Dig. Tch. Vol 2, No 1, 1979.

8.   ALEKSANDER, I., THOMAS, W. and BOWDEN, P. WISARD, a radical new step forward in image recognition. Sensor review, 120-4, 1984.

9.   ALEKSANDER, I. Neural Computing Architectures. North Oxford Academic Publishers,

10. AITKEN, D., BISHOP, J.M., MITCHELL, R.J. and PEPPER, S. Pattern Separation in Digital Learning Nets. Electronic Letters, 25, No 11, 685-686, 1989.

11. WIDROW, B. and HOFF, M.E.: Adaptive switching circuits. IRE Wescon convention record, New York: IRE, 96-104, 1960.

12. MITCHELL, R.J. Microcomputer systems using the STE bus, Macmillan Education, 1989.

13. FLETCHER, M.J., MITCHELL, R.J. and MINCHINTON, P.R. VIDIBUS- A low cost modular bus system for real time video processing, Electronics and Communications Journal, 1990.

14. MINCHINTON, P.R., BISHOP, J.M. and MITCHELL, R.J. The Minchinton Cell - Analog Input to the NTuple Net Proc. International Neural Networks Conference, INNC-90, p599, Paris, July 1990.

15. GELAKY, R., WARWICK, K. and USHER, M.: The Implementation of a low cost production-line inspection system. Computer-Aided Engineering Journal, 180-184, 1990.

16. DAVIDSON, H.R., HEMMENDINGER, H. and LANDRY, J.L.R., A System of Instrumental Colour Control for the Textile Industry, Journal of the Society of Dyers and Colourists, Vol.79, pp. 577, 1963.

17. ALDERSON, J.V., ATHERTON, E. and DERBYSHIRE, A.N.Modern Physical Techniques in Colour Formulation. Journal of the Society of Dyers and Colourists, Vol. 77, 657, 1961.

18. JUDD, D.B. and WYSZECLO, G., Color in Business, Science and Industry. 3rd ed., Wiley, New   York, 1975, pp 438-461, 1975.

19. CHANDRASEKHAR, S., Radiative Transfer. Clarendon Press, Oxford, 1950.

20. MUDGETT, P.S. and RICHARDS, L.W., Multiple Scattering Calculations for Technology. Applied Optics, Vol.0, pp 1485-1502, 1971.

21. MEHTA, K.T. and SHAH, H.S., Simplified Equations to Calculate MIE-Theory Parameters for use in Many-Flux Calculation for Predicting the Reflectance of PaintFilms. Color Research and Application, Vol.12, pp 147-153, 1987.

22. WESTLAND, S., The Optical Properties of Printing Inks. PhD Thesis, University of Leeds, (UK), 1988.

23. GANZ, E., Problems of Fluorescence in Colorant Formulation. Colour Research and Application, Vol.2, pp 81, 1977.

24. McKAY, D.B., Practical Recipe Prediction Procedures including the use of Fluorescent Dyes. PhD Thesis, University of Bradford (U.K), 1976.

25. VAN DE HULST, H.C.: Multiple Light Scattering; Tables, Formulas and Applications, Academic Press, New York, 1980.

26. NOBBS, J.H.: Review of Progress in Coloration. The Society of Dyers and Colourists, Bradford, 1986.

27. USUI, S., NALAIJCI, S. and NAKANO, M., Reconstruction of Munsell Color Space by a Five-Layer Neural Network. Journal of the Optical Society of America, Series A, Vol 9, no 4, 516-520, 1992.

28. BISHOP, J.M., BUSHNELL, M.J. and WESTLAND, S., The Application of Neural Networks to Computer Recipe Prediction. Color, Vol.16, No.1, pp 3-9, (USA), 1991.

29. BISHOP, J.M., BUSHNELL, M.J. and WESTLAND, S., Computer Recipe Prediction Using Neural Networks. Proceedings Expert Systems 90. (London), 1990.

Chapter 22

Parallel processing, neural networks and genetic algorithms for real-time robot control

A. M. S. Zalzala

This chapter reports on recent research on advanced motion planning and control of articulated and mobile robotic systems. In addition to employing the principles of distributed and parallel processing to produce feasible real-time multi-processor networks, other new theories such as neural networks and genetic algorithms are deployed as possible solutions. All reported algorithms are implemented for either the PUMA 560 arm or the B12 mobile robot. The ultimate aim of the on-going research is to present working architectures for real-time robotic systems by augmenting all developed structures.

22.1 Introduction

Real-time robot control has always presented researchers with great difficulties in terms of both the accuracy of the command actions required and also the efficiency by which the commands are obtained. A very important characteristic of the new generation of robotic systems is the presence of intelligent capabilities which is being rapidly supported by fast computing power and adequate sensory equipment.

A general form of the robot control loop is shown in Figure 22.1, where the required job is first divided by the task planner producing a number of consecutive tasks, followed by the motion planner which gives a time history of positions, velocities and accelerations sufficient and necessary to realise each task. Once the desired motion elements are available, they are used to produce the commands for the individual joint loops via the control module which may or may not include the dynamic model of the system (model reference adaptive controllers vs. simple PIDs). The motion is realised by applying the control commands to the robot system and a feedback module provides the actual motion elements to cater for any uncertainties and/or changes in the system parameters and/or environment set-up. Overall, intelligence may be needed at different parts of the control loop, e.g. in connection with the task planner, dynamic model or sensory feedback. Nonetheless, for complicated multi-joint mechanical chains, the above loop is very hard to accomplish in real-time and provisions must be made for efficient designs.

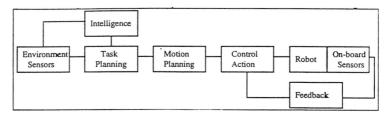

Figure 22.1   *Overview of a robot integrated system: this is from the view point of an upper hierarchical structure and serving to illustrate the main modules;   both the intelligence and feedback blocks can be deployed in other parts of the system as an implementation may require*

This chapter will consider different architectures attempting to realise parts of the robot control loop, and further providing for intelligence in real-time applications. Three approaches are considered. First, medium-grain parallelism of the classical control approaches is investigated, presenting efficient structures on general purpose processing elements, where the transputer is chosen as a vehicle for the realisation of such structures. Second, fine-grain distribution of a module using artificial neural networks is tackled, where the massive parallelism and the learning abilities of the connectionist architectures is of vast interest. In addition, efficient search techniques for fast motion planning of multi-joint manipulators and mobile vehicles are enhanced by using genetic-based algorithms, hence providing a more optimum solution.

In addition to this introduction the chapter is divided into three main sections each reporting on different algorithms using parallel processing, neural networks and genetic algorithms, respectively. In addition, a discussion and conclusions section is included.

22.2   Parallel processing structures on transputer networks

The complexity of the dynamic model of an articulated chain proved to be very computationally demanding for on-line applications. Thus, although simplifications have been proposed in terms of ignoring parts of the model, in particular at low speeds, it is a well established fact that such a practice seriously degrades the performance and tracking accuracy [1]. However, the relatively recent availability of low cost general purpose processors such as the INMOS transputer has presented feasible candidates for the application of the principles of parallel and distributed processing in real-time operation. Nonetheless, due to the lack of any automatic means by which an algorithm can be recast in a parallel form, extreme care must be taken to ensure high utilisation of all processing elements used within a network while keeping the overall system as cost effective as possible. Thus, in applying the concepts of concurrency to design the parallel algorithm, two parameters are of overriding importance, namely the minimisation of the execution time and the maximisation of every processor utilisation, and using a time scheduling procedure is appropriate to maintain the latter requirements [2]. One further issue of interest is communication bottlenecks which must be eliminated to maximise efficiency.

This section reports on several parallel structures for real-time robot control. Although these structures can be applied using any general-purpose processor, the INMOS transputer is chosen as the main processor in all implemented networks due to the availability of a wide range of both software and hardware support. In addition, although recent benchmarks indicated the superiority of other processor architectures, such as the SPARC, in a stand alone mode, the transputer is still the best available for constructing multi-processing networks [3]. A final realisation of the overall integrated system is discussed in Section 22.2.4.

22.2.1 Minimum-time motion planning

The minimum-time control of robot arms has usually been implemented as a two-task procedure, where the appropriate trajectory planning is performed off-line, and then tracking is carried out on-line to achieve the desired motion. However, such an approach is unacceptable whenever the motion is dependent on the on-board sensory equipment operated during the application. In on-line operation, the robot end-effector is required to track a path specified by the on-board sensors during motion. Therefore, the look-ahead point concept has been introduced [4,5] which requires the detection of a via-point to move to, planning the motion segment then tracking the intermediate points produced. While this present segment is traversed, the planner detects a next look-ahead via-point and plans the next segment for the robot to follow once traversing the first segment is completed. The construction of successive motion segments continues in real time provided that the planner is fast enough to provide data to the manipulator, a highly critical requirement for high-speed automated applications. This concept of the on-line alternating between motion planning and trajectory tracking depends on real-time perception and execution thus illustrating an important aspect of intelligent sensory-based systems, where the robot will be able to work in dynamic and unstructured environments.

Nonetheless, for this concept to be implemented, fast computational modules, as well as fast perception abilities are required to maintain continuity of motion, hence the need for multi-processor networks. Algorithms are implemented for the six-joint PUMA 560 arm [6] where minimum-time motion planning in the configuration space is distributed at two distinct levels, as shown in Figure 22.2. In Figure 22.2, a global distribution is achieved by treating each joint of the manipulator separately while a finer local level of parallelism is shown by concurrently constructing and optimising different motion options for each joint. In addition, a control unit maintains the high operational performance of the overall structure. The complete parallel algorithm is implemented on a network of T805 processors as shown in Figure 22.3.

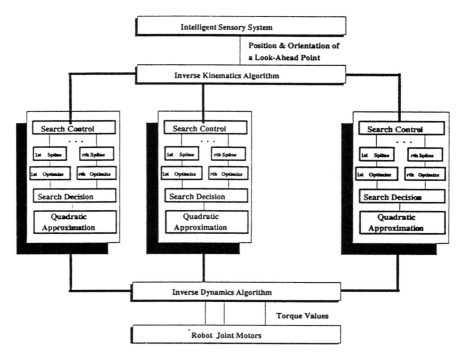

Figure 22.2   *The distributed motion planner: shown within the context of Figure 22.1. A number of parallel modules are designated each for one joint of the robot arm. On each module, a number of concurrent processes are executed to spline, optimise and search for the minimum-time motion option*

22.2.2 Distributed dynamic equations of motion

The motion planner described in the above section provides the time history of positions, velocities and accelerations for the arm to follow. However, the control commands must be computed and downloaded to the joint loops for execution. Although simple single-joint P(I)D controllers are usually used, these are not efficient when high-speed motion is required or when changes in the robot model and/or the environment are encountered, and the dynamic equations of motion must be incorporated in the control module to provide for accurate tracking [7]. Nevertheless, the presence of these highly non-linear equations creates a very computationally expensive problem when combined with the requirements of minimum-time motion, despite several algorithmic simplifications.

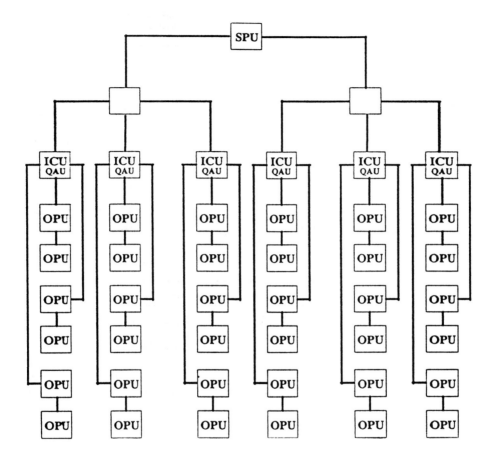

Figure 22.3 *The planner's multi-processor network: six columns of seven processors are set for each joint of the PUMA. The intelligent control unit (ICU) runs different optimisation units (OPUs), gathers all possible options of motion and chooses the minimum-time one, which is in turn communicated to the supervisor unit (SPU). Note the limitation on communication links between the ICUs and their respective OPUs and also between the SPU and the six ICUs. The number of processors assigned to each joint can be modified depending on how fast the planner needs to operate in a particular application*

Hence, the efficient and fast computations of robot arm joint torques is considered, and a solution of the inverse dynamics problem through the design of a distributed architecture for the recursive Lagrangian equations of motion is presented [8]. This form of robot dynamics is a fully recursive formulation having a linear complexity of O(n) with two phases of recursion, backward and forward [9]. Parallelism is exploited at three distinct levels. First, a global level, where a pipelined configuration is imposed to accommodate for the presence of recursion in the equations. Second. a local level, where the procedure at each link is divided into several concurrent jobs. Finally, a sub-local level, yielding parallelism in the operations of a certain job within each link. In addition, a symbolic representation of parts of the relevant equations is developed as a contribution to the reduction of complexity yielding a high utilisation of each processing element. In implementing the algorithm, a two-dimensional torus array of processors is used with the vertical extension employed for the global recursive configuration, while the horizontal extension accommodates for the local distributions, as shown in Figure 22.4.

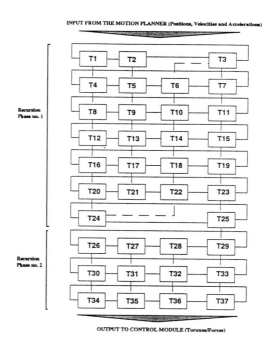

Figure 22.4   *Figure 22.4: Torus multi-processor network for the dynamic equations: the recursion in both phases of the formulation is catered for via a pipeline shown vertically (7 stages in phase 1 and 3 stages in phase 2) while horizontal distribution is exploited within each stage of the pipeline. In all, 37 processors are needed for the six-joint PUMA*

22.2.3 Distributed measurements for the end-effector

As indicated in Figure 22.1, the robot must have some form of sensory equipment to provide a feedback on both the position and orientation of the end-effector. An efficient ultrasonic measurement device have been developed earlier [10] and further modified to present orientation information in addition to 3-D Cartesian position [11]. Different hardware and software design issues have been reported in the literature, while emphasis here is made on the distributed implementation of the inverse kinematics transformations for the PUMA arm, where providing the feedback data must be fast enough to operate with the control module.

The experimental set-up shown in Figure 22.5 employs two sets of eight ultrasound receivers (one at each corner of the volume) along with two sets of five transmitters located at the robot's hand, and the computations are carried out using trigonometric relationships [12]. Two levels of concurrency are identified. First, a procedural level, operating the position detection, orientation detection and inverse kinematics as a pipeline producing the joint values. Then, a measurement level, where measuring the position and orientation of the hand is executed in parallel for all combinations of the receivers.

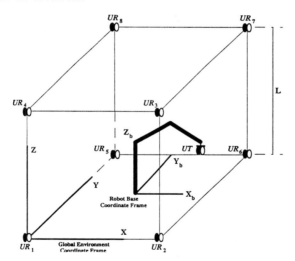

Figure 22.5 *Set-up of the dual ultrasonic systems: Two ultrasonic receivers (UR) are placed in each corner of the volume (denoted as O and O) along with two respective arrays of transmitters (UT) placed at the robot hand. By employing the measurements, both the position and orientation of the robot tip can be calculated. The eight corner receivers' structure also provides for fault tolerance by accommodating for transducer failure and overcoming problems associated with blocking of waves by obstacles in the environment*

The multi-processor system is shown in Figure 22.6 where eight concurrent nodes are employed, one for each corner of the volume, to measure and compute a number of samples, in addition to another four nodes forming a pipeline for computing the joint values. This architecture is tested on a network of ten transputers to be consistent with previous phases of the integrated system, although cost effectiveness may be raised as an issue. However, the aim is to provide for an integrated system, as described in the following section.

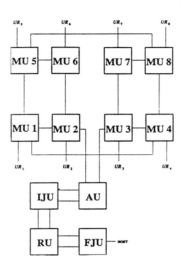

Figure 22.6   *The multi-processor system for ultrasound measurements: eight measurement units (MU) are shown, providing data for an averaging unit (AU), an initial joint unit (IJU), a rotational unit (RU) and a final joint unit (FJU), all operating in a pipeline structure. The final output is a vector of six joint values for the arm*

22.2.4 The integrated system

The three distributed structures reported in Sections 22.2.1, 22.2.2 and 22.2.3 are integrated together to present a massively parallel real-time robotic controller. This is shown in Figure 22.7 with a PD compensator added to facilitate the use of the computed torque values, while current work is progressing towards integrating an alternative adaptive structure [13].

The cost of VLSI structures is dropping continuously, and the above integrated system may not be very expensive considering that it can provide for a far better performance for an automated system which may cost excessively more than the controller. In addition, further modules must be added to enhance the system, in particular a more sophisticated adaptive controller and a high-level sensory architecture.

22.3 The learning control of dynamic systems

The neural network structure used by engineers is generally modelled on the biological nervous system and the brain in particular, where it is hoped to inherit the latter's ability to learn and perform fast computations. Nonetheless, due to the limited knowledge possessed of the brain structure and functions [14], the mathematical model of neurones is more of an *adhoc* approach rather than a well organised one. Unfortunately, neural networks were a fashionable field to follow for many engineering researchers while there was still uncertainty of the ability of the theory of cognition to present a suitable replacement for the existing well-established control theory. Although proving successful in many applications when used as a classifier, the use of an assumed form of neural networks in real-time control of dynamical systems proved a failure. This failure resulted mainly from the fact that an *adhoc* structure of a multi-layered network could not be generalised to accommodate for all operational modes of a non-linear dynamic system, although showing some good results in individual examples.

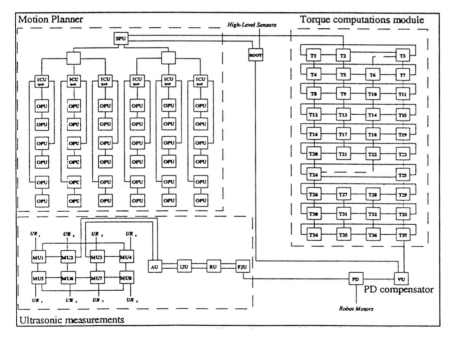

Figure 22.7   *The complete massively-parallel integrated system: including the motion planner, the torque/force computations module, the measurement module and a PD compensator*

A novel approach to neural robot control was developed, where conventional robot control theory was augmented with aspects of the theory of cognition [15]. This approach gives the neural network some initial knowledge of the system model, where the network is expected to behave better once it knows something about its task. Such initial knowledge can be readily available in engineering applications where data of the controlled system are provided for, as is the case in robotics.   In the case of a robot arm, this initial knowledge is presented by including part of the model as the combining function on each of the neurones. The main aim of this approach is to make use of the parallelism for real-time implementations, in addition to the learning abilities to perform adaptive control. However, no model identification is attempted here since the robot model is already defined, while the remaining task is controlling the arm efficiently.

Implementations are accomplished for both the kinematics and dynamics formulations of the full model of the PUMA 560 arm [16], hence illustrating the practicality of the procedure for any robot structure with any number of joints. The complexity of the system can be demonstrated in Figure 22.8 showing the neural structure for the velocity term of the Lagrangian equations of motion. In addition, Figure 22.9 shows the network for the inverse Jacobean. It must be emphasised that the network structures are representatives of the model complexity as both the number of hidden layers and the number of neurones in each are determined by the mathematical model and not assumed *a priori*. Although this implantation is reported here for a robotic system, the same procedure can be applied to solve any problem where real-time dynamic systems are involved.

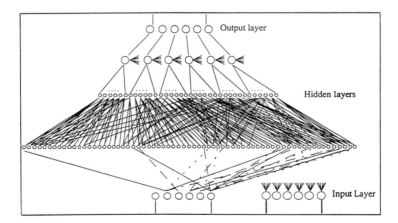

Figure 22.8   *Neural network for the velocity-term of the PUMA's model: employing two hidden layers where the mathematical model is distributed over a total of 103 neurones. The number of layers, number of nodes and the interconnections are imposed by the dynamic formulation while a modified back propagation algorithm is used to accommodate for the changes*

22.4 Genetic-based motion planning

Genetic algorithms are adaptive search techniques mimicking the process of evolution by emulating the concept of the survival of the fittest. Due to different attractive features [17], genetic-based procedures attracted much interest in many fields including robotics. This section is concerned with introducing genetic-based algorithms for the minimum-time trajectory planning of articulated and mobile robotic systems. Although heuristic search techniques are available and can be implemented on parallel structures, as reported in Section 22.2.1, the fact that an infinite number of solutions exist to move from one point to another renders their operation as sub-optimal [18] and even harder to achieve in real time when a large working space is considered.

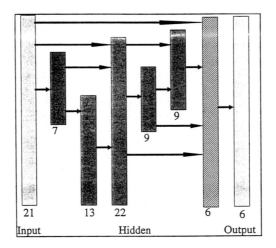

Figure 22.9   *The inverse Jacobean neural structure: employing a total of six hidden layers for which the number of nodes are shown. Note in particular the direct interaction between the third and sixth hidden layers and also the direct input to the sixth hidden layer. The shown structure is, again, imposed by the mathematical model*

In using a genetic based approach, the main aim is to provide an efficient planning algorithm in terms of execution time while respecting the limitations imposed by the design. For an articulated arm, the planning procedure is performed in the configuration space and respects all physical constraints imposed on the manipulator design including the limits on the torque values applied to the motor at each joint of the arm [19]. Consequently, the complete non-linear dynamic robot model is incorporated in the formulation. This feasible algorithm emerged through an evolutionary process while taking account of different concerns related to robot motion planning, and the binary representation of the chromosomes is illustrated in Figure

22.10. Reproduction was controlled to prevent premature convergence, the analogous crossover directed sensible crossover operations, and specially shaped mutation operators promoted new search space. The algorithm is proven to be far more efficient compared to the conventional heuristic search techniques with a reduction in the execution time of 1:20. Although results are presented for a simple two-link manipulator, the algorithm is readily applicable for any number of degrees of freedom.

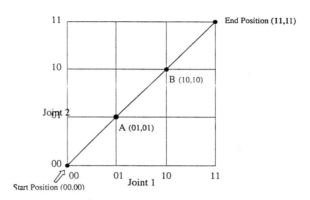

Figure 22.10   *Chromosome representation for the genetic structure: Considering 2-D space, a joint position is represented in binary form as shown, while representations of 00,01,10 and 11 are used to indicate motor saturation in either direction for both joints, respectively. Hence, moving from the starting point to the end point via A and B is represented as(1010\01; 1010\10) with the joints' motor commands saturated at the negative and positive bounds, respectively*

A similar situation, though rather simpler, arises when planning the motion of mobile vehicles. Although many solutions may exist, a condition for obtaining the best (or near best) option may be imposed by the user, where a criterion in terms of the total distance traversed, energy expended or minimum execution time must be achieved. The planning procedure is made more complicated if the robot has to detect and avoid static or dynamic objects in the workcell.

Genetic algorithms are deployed to investigate a more detailed and sophisticated approach to planning in 3-D space involving moving (or disappearing) obstacles, thus accommodating for the concept of intelligent control within a dynamic environment. Prior to the planning process, a global knowledge of the environment is needed and is translated in the form of a terrain map. Firstly, a set of valid random paths are generated as the initial generation. In order to prevent the robot wandering endlessly inside the workcell, a weighted vector of motion is employed during the path construction phase, that is, the direction of motion from the robot's current position towards the goal has more chance to be chosen than other directions. A fitness value is assigned to each path and the one with the best fitness is stored for future usage. Secondly, pairs of paths are chosen randomly for mating, and those with better fitness will have more chance to be drawn out.   The reproduction process repeats until some arbitrary number of generations is reached [20]. To realise the results obtained via the simulation, an actual implementation utilising the B12 mobile robot was accomplished, where maps stored through the on-board sonar array were manipulated by the genetic planner to present a feasible motion. Figure 22.11 illustrates the concept of dynamic environments, while Figure 22.12 shows the genetic planner in operation.

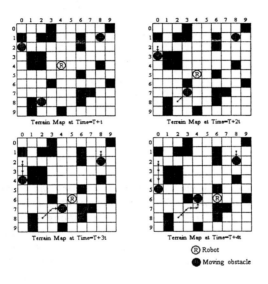

Figure 22.11   *Terrain map at different time intervals: a 10 x 10 map is shown at different time intervals while the planner manoeuvres the robot to avoid a moving obstacle. This procedure may be easy enough to accomplish in simulation, however, a practical system will require updating the map every time interval via data obtained from the sonars where real-time issues must be addressed*

22.5 Discussion and conclusions

This chapter has   reported on different robotic control structure using parallel processing, neural networks and genetic algorithms, with emphasis on real-time operation. The use of distributed processing, whether fine-grain or medium-grain, is deemed as essential to realise an intelligent machine working on-line. The use of new theories such as connectionist and genetics is seen as a feasible development within the framework of the already existing and established control theory. Nonetheless, the researchers' findings lead to the conclusion that although the idea of creating a human brain is both interesting and exciting, a cautious and calculated approach must be followed to ensure a working design. Although an *adhoc* approach may yield some satisfaction in limited examples, it will never be appreciated for practical systems let alone competing with the classical theory.

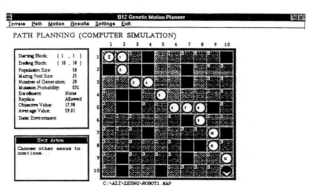

Figure 22.12   *Planning in a static environment: A window-driven graphical interface is designed to monitor the on-line operation of the genetic-driven B12 mobile robot. The size of the terrain is adjustable according to the application and is built via data obtained from the B12's sonars. The system can operate in both manual or auto-detection modes. Analysis of the motion results are stored and can be accessed by the operator via the Results menu*

The design of multi-processing systems utilising general purpose processors is more appreciated where implementation of practical systems are concerned, as these networks are available in VLSI. However, the efficient utilisation of neural structures is still hampered by the lack of a general purpose chip capable of accommodating for the designed algorithms. This is of particular significance when considering any integration between a neural processor and general purpose processors. In this pioneering stage of the research, genetic based approaches to robot motion planning show great potential where the results suggest that their performance may be substantially improved with further work.

22.6 Acknowledgements

The author acknowledges the useful discussions and contributions by colleagues including Alan Morris, Kwong Chan, David Leung, Mike Dickinson, Ian Cornish and Fatollah Shahidi. Parts of this work are funded by Sheffield University Research Fund.

22.7 **References**

1. HOLLERBACH, J.M., Dynamic Scaling of Manipulator Trajectories, Trans. ASME, J. of Dyn. Syst., Meas. and Control , 106, 102.06. 1984.

2. KASAHARA, H. and NARITA, S,. Practical multi-processor scheduling algorithms for efficient parallel processing, IEEE Trans. Computers. Vol.33, No.11, 1984.

3. CORNISH. l.J. and ZALZALA, A.M.S., 'Transputer benchmarks and performance comparisons with other computing machines', In Proceedings Workshop on Parallel Processing in Education. pp 138-143, University of Miskolc, Hungary, March 1993.

4. ZALZALA, A.M.S. and MORRIS, A.S., 'A distributed on-line trajectory generator for intelligent sensory-based manipulators'. Robotica, Vol. 9, No. 2, pp 145-55, 1991.

5. CASTAIN, R.H. and PAUL, R.P., An On-Line Dynamic Trajectory Generator, Int. J. Robotics Research, Vol.3, No.1, pp 68-72, 1984.

6. ZALZALA, A.M.S. and MORRIS, A.S., 'A distributed robot control on a transputer network', IEE Proceedings, Part-E, Vol. 138. No.4, pp 169-76, 1991.

7. SLOTINE, J.J.E. and GUNTER, N., 'Performance in adaptive manipulator control', Int. J. Robotics Research, Vol. 10, No. 2,1991.

8. ZALZALA, A.M.S. and MORRIS. A.S., 'A distributed pipelined architecture of robot dynamics with VLSI implementation', Int. J. Robotics and Automation, Vol. 6, No.3,1991.

9. HOLLERBACH, J.M., A Recursive Lagrangian Formulation of Manipulator Dynamics and a Comparative Study of Dynamics Formulation Complexity, IEEE Trans. Syst., Man, Cyber., SMC-lO, 730-36, 1980.

10. DICKINSON, M., and MORRIS, A.S., Co-ordinate Determination and Performance Analysis for Robot Manipulators' and Guided Vehicles, IEE Proceedings, Part-A, 135, 2, 95-98, 1988.

11. ZALZALA, A.M.S., Ultrasonic location measurement for fast robot control, Research Report no. 380, AC&SE Dept., Sheffield University, 1990.

12. ZALZALA, A.M.S. and MORRIS, A.S., 'Fast ultrasonic location measurement for robot control', In Proceedints Canadian Conference on Industrial Automation, Vol. 1, pp 15.23-15.26, Montreal, 1992.

13. ZIAUDDIN, S.M. and ZALZALA, A.M.S., 'Parallel processing for real-time adaptive control: Theoretical issues and practical implementation', In IEE Colloquium on High performance applications of parallel architectures, 1994

14. CRICK, F.H.C. and ASANUMA, C., Certain Aspects of the Anatomy and Physiology of the Cerebral Cortex, In Parallel Distributed Processing : Explorations in Microstructure of Cognition. Eds.: D. E. RUMELHART and J. L. McCLELLAND, Vol .2, pp 333-4O. MIT Press, 1986.

15. ZALZALA, A.M.S. and MORRIS, A.S., 'A neural networks approach to adaptive robot control', Int. J. Neural Networks. Vol. 2, No. 1, pp 17-35, 1991.

16. ZALZALA, A.M.S., 'A neural network methodology for engineering applications: The learning control of robotic systems', in Neural Computing: Research and Applications, Ed. G. ORCHARD, Institute of Physics Pub., 1993.

17. GOLDBERG, D.E., Genetic algorithms in search, optimisation and machine learning, Addison Wesley Pub., 1989.

18. SAHAR, G. and HOLLERBACH, J.M., Planning of Minimum-time Trajectories for Robot Arms,Int. J. Robotics Research, 5,3,90-100. 1986.

19. CHAN, K.K. and ZALZALA, A.M.S., 'Genetic-based minimum-time trajectory planning of articulated manipulators with torque constraints', In IEE Colloquium on Genetic algorithms for control systems engineering, May 1993.

20. LEUNG, C.H. and ZALZALA, A.M.S., 'A Genetic Solution for the Control of Wheeled Robotic Systems in Dynamic Environments', Proceedintgs IEE International Conference Control'94, 1994.

Index